The Recombinant DNA Debate

The Recombinant DNA Debate

David A. Jackson
Editor

Department of Microbiology
School of Medicine
The University of Michigan
Ann Arbor, Michigan

Stephen P. Stich
Editor

Department of Philosophy
and
The Committee on the
History and Philosophy of Science
University of Maryland
College Park, Maryland

PRENTICE-HALL, INC., Englewood Cliffs, New Jersey 07632

Library of Congress Cataloging in Publication Data

Main entry under title:
The Recombinant DNA debate.

   Bibliography: p.
   Includes index.
   1. Recombinant DNA—Social aspects.   I. Jackson,
David Archer   II. Stich, Stephen P.
[DNLM: 1. DNA, Recombinant.   2. Genetic intervention.
QH438.7 R311]
QH442.R38   574.8'732   78-26385
ISBN 0-13-767442-2

Editorial/production supervision
   by Eleanor Henshaw Hiatt
Interior design by George Whipple
Manufacturing buyer: Ray Keating

Printed in the United States of America

10   9   8   7   6   5   4   3   2

PRENTICE-HALL INTERNATIONAL, INC., *London*
PRENTICE-HALL OF AUSTRALIA PTY. LIMITED, *Sydney*
PRENTICE-HALL OF CANADA, LTD., *Toronto*
PRENTICE-HALL OF INDIA PRIVATE LIMITED, *New Delhi*
PRENTICE-HALL OF JAPAN, INC., *Tokyo*
PRENTICE-HALL OF SOUTHEAST ASIA PTE. LTD., *Singapore*
WHITEHALL BOOKS LIMITED, *Wellington, New Zealand*

For Mit and Jude

Contributing Authors

A. M. Chakrabarty, Ph.D., Environmental Unit of the Physical Chemistry Laboratory at the General Electric Research and Development Center, Schenectady, New York 12301.

Carl Cohen, Ph.D., Professor of Philosophy at the Residential College of The University of Michigan, Ann Arbor, Michigan 48109.

Roy Curtiss III, Ph.D., Professor of Microbiology in the Department of Microbiology, The University of Alabama at Birmingham, Birmingham, Alabama 35294.

Bernard D. Davis, M.D., Adele Lehman Professor of Bacterial Physiology in the Unit for Bacterial Physiology, Harvard Medical School, Boston, Massachusetts 02115.

Roger B. Dworkin, L.L.B., Professor of Law at the Indiana University Law School, Bloomington, Indiana 47401.

Charles R. Eisendrath, M.A., Assistant Professor of Journalism in the School of Journalism, The University of Michigan, Ann Arbor, Michigan 48109.

Rolf Freter, Ph.D., Professor of Microbiology in the Department of Microbiology at The University of Michigan Medical School, Ann Arbor, Michigan 48109.

Max Heirich, Ph.D., Associate Professor of Sociology in the Department of Sociology at The University of Michigan, Ann Arbor, Michigan 48109.

David A. Jackson, Ph.D., Associate Professor of Microbiology in the Department of Microbiology at The University of Michigan Medical School, Ann Arbor, Michigan 48109.

Joshua Lederberg, Ph.D., President of The Rockefeller University, New York, New York 10021.

Donald N. Michael, Ph.D., Professor of Planning and Public Policy in the School of Natural Resources at The University of Michigan, Ann Arbor, Michigan 48109.

Science for the People, a group of scientists concerned with the impact of science and technology on society. Contributors to the chapter appearing in this book are Edward Loechler, Tracy McLellan, Robert Park, David Shore, Scott Thatcher, and Phillip Youderian.

Robert L. Sinsheimer, Ph.D., Chancellor of the University of California at Santa Cruz, Santa Cruz, California 95064.

Stephen P. Stich, Ph.D., Associate Professor of Philosophy in the Department of Philosophy and the Committee on the History and Philosophy of Science at the University of Maryland, College Park, Maryland 20742.

George Wald, Ph.D., Higgins Professor of Biology in the Department of Biology, Harvard University, Cambridge, Massachusetts 02138.

Alvin Zander, Ph.D., Director of the Center for Research on Group Dynamics of the Institute for Social Research, and Associate Vice-President for Research at The University of Michigan, Ann Arbor, Michigan 48109.

Contents

ix

PART II
THE SCIENTISTS DEBATE:

FOR THE OPPOSITION

FOR THE DEFENSE

PART III
PHILOSOPHICAL, LEGAL,
AND SOCIAL ISSUES

AN AFTERWORD

APPENDICES

Preface

The early 1970s saw the emergence of an enormously important and powerful new methodology in the biological sciences. Molecular biologists learned how to remove bits of genetic material (DNA) from various organisms and insert them into bacteria in such a way that the transferred DNA became part of the genetic material of the bacteria. As they grew, the bacterial cells then duplicated and reduplicated this transferred DNA. Thus, researchers could now isolate specific genes and produce unprecedentedly large quantities of them. Recombinant DNA technology, as these techniques have come to be called, provided scientists with a singularly powerful tool for studying the basic mechanisms of genetics in all organisms, and especially in the genetically complex cells of higher organisms, where similarly powerful alternative methods do not exist in general. In addition, the new technology held promise of substantial practical benefits in medicine, agriculture, and industry. However, the new technology also posed many potential dangers. Concern over possible hazards prompted leading scientists in the field to call for a temporary suspension of some classes of experiments using recombinant DNA methodology. This call, in turn, sparked an international debate on whether and how recombinant DNA techniques should be employed.

Naturally enough, the debates flared first in university communities, the places where most recombinant DNA research was being and would be conducted. The first campus to be engulfed in the controversy was The

University of Michigan at Ann Arbor. The history of the Michigan debate, and of the University's struggle to deal with the crucial issues of regulation or suppression of scientific inquiry, are recounted in the essay that serves as our Introduction. A principal event in the Michigan debate was an open Forum aimed at informing the community of the issues involved. It was that Forum that suggested the need for a book such as this. What was needed, and what we have assembled, is a sourcebook laying out the many sides of the debate and the scientific background needed to evaluate the issues intelligently.

The recombinant DNA debate has rapidly escalated from a discussion among scientists holding differing views about scientific questions to a full-blown controversy in which moral, ethical, legal, and sociological issues, as well as scientific questions, have been raised. The debate has come to be a principal focus for many of the most vexing questions concerning the proper social role of modern science and technology. For this reason the book presents the views of philosophers, lawyers, sociologists, political scientists, and others, and provides scientific background so that the reader without technical training can understand the biochemical basis and biological context of recombinant DNA methodology.

The three essays of Part I set out the scientific background in terms accessible to a reader with a high-school background in biology and chemistry. Some of this material, particularly Roy Curtiss' essay, will be slow going for the lay reader. But a careful reading of all three essays will yield a solid understanding of the science that is the focus of the debate. In Part II we have assembled essays by leading scientific opponents and defenders of recombinant DNA research. Part III adds the perspectives of philosophers, social scientists, a lawyer, and a journalist. Carl Cohen's critical look at the logic (or lack of logic) in the arguments that have dominated the debate concludes the book.

Preparing this volume has involved more effort and trauma than either of us had anticipated. There was no insulating our editorial activities from the political storms swirling around recombinant DNA research. Our hope is that this volume will help readers take their own stand on the issues knowledgeably and wisely. If it does, our travails will have been well compensated.

DAVID A. JACKSON
STEPHEN P. STICH

Introduction:
The Debate in a Microcosm

The Discussion of Recombinant DNA at The University of Michigan

Alvin Zander

In May 1976 the Regents of The University of Michigan agreed that research at that school using the techniques of recombinant DNA could "go forward as long as it is submitted to appropriate controls." This decision was the climax of many months' discussion, a discussion so extensive that a Regent remarked, "It has received as much attention as any issue I can recall in the seven and a half years I have been on the Board."

In this chapter we shall examine some of the views presented, consider methods used in moving toward a resolution of conflicting ideas, and identify a few procedural problems that became important during this many-sided argument. Do science and society benefit from a public airing of plans for future research and from attempts by individuals who are not scientists to shape these plans? If so, how and why? What are the relative advantages and disadvantages? The experience at Michigan indicates that such questions do not have straightforward answers.

ASSESSMENT OF RESEARCH IN RECOMBINANT DNA: ORIGINS

Work with techniques in recombinant DNA began at The University of Michigan somewhat sooner than at many other colleges. Several of the younger microbiologists on its faculty had participated in early research into these techniques while they were postdoctoral fellows on the west coast, and they brought their interest and skill with them to Michigan. Soon after their arrival in Ann Arbor, these scientists and a few others founded an informal group in the Medical School, the Ad Hoc Committee for Microbiological Safety, to consider how potential biohazards associated with the new methods might be controlled. When Paul Berg's committee for the National Academy of Sciences published its letter suggesting a voluntary moratorium on some types of recombinant research and recommending the development of national guidelines for controlling this work, this informal local group decided that "The University of Michigan in its entirety, not just the Medical School, should have a cogent policy with respect to this problem," and they asked the Vice President for Research to help them implement this idea. In December 1974 the Vice President accordingly appointed seven faculty persons from three different colleges to a committee on Microbiological Research Hazards, asking them to prepare a policy for the University and to develop guidelines for the conduct of such research at Michigan. Three members of this committee attended the international conference on recombinant DNA held at Asilomar, California, in February 1975.

The group reported in April 1975 that, because of the expertise in recombinant DNA methods already present there, the University had an opportunity to assume a leading position in this country in the use of recombinant DNA technology. They noted that several faculty members were already involved in such research and others soon would be. Five research groups were studying recombination of DNA molecules, and an additional ten to fifteen groups would probably be working with recombinants or tumor viruses in another year or so. The committee recommended that three laboratories be renovated so that they would meet the government's soon-to-be-published requirements for safe conduct of this research and that a new committee be established to monitor these laboratories for compliance with the guidelines. About a year later, as we shall see, the Regents passed resolutions almost identical with the recommendations of this preliminary committee.

The University's Biomedical Research Council, composed of faculty from the health science schools and responsible for overall planning of biomedical research on the campus, endorsed the committee's recommendations, adding that the modification of laboratories would be needed for research not only on recombinant DNA but also on tumor viruses, other pathogenic agents, and the use of many cultured cell systems. "The problem," they stressed, "is larger than recombinant DNA molecule research alone and we should plan accordingly." They foresaw the renovation of three laboratories as insufficient in the long run; eventually, a separate building would be needed for research on a variety of potential injurious agents.

The Council also suggested that the future of research in recombinants be given further and more differentiated study by the campus as a whole. They proposed that three committees be appointed to consider separate aspects in planning for future work in DNA at Michigan. Committee A would be concerned with ways of improving the safety of current facilities and with planning renovations in laboratory space. Committee B would give more extended consideration to the policy aspects of this research. Committee C, to be appointed later, would be the only continuing group and would be responsible for certifying that individual laboratories met the safety requirements for any given type of research. These recommendations were adopted by the Vice President for Research and the relevant Deans.

THE NATURE OF COMMITTEE A

Committee A, composed of eight persons from the biological sciences, was assigned three tasks: (a) to plan the modification of laboratories necessary for moderate-risk containment facilities, (b) to plan and prepare requests for external funding to cover the cost of these modifications, and (c) to monitor federal requirements for research facilities and to apprise the Vice President for Research and the Dean of the Medical School of significant developments in such rules. This committee was also to explore what facilities might be needed for an expanded program of research in this area. It began its work in May 1975.

To accomplish their assigned tasks the members of the committee did many things not originally expected of them. They consulted with scientists in laboratories on other campuses and in government agencies. They had architectural drawings prepared for renovation of three laboratories and asked the National Cancer Institute for the funds needed. They advised managers of campus laboratories on the safety of their facilities. They surveyed the University's biomedical faculty to determine present and future needs for special containment laboratories. They sought advice from the Recombinant DNA Molecule Program Advisory Committee at NIH and provided their own ideas to that group. They presented information about the potential uses and the possible hazards of recombinant DNA procedures to over forty local groups such as churches and luncheon clubs; the Medical Dean's Advisory Council; the Executive Committee of the Medical School; the Faculty of the Division of Biological Sciences in the College of Literature, Science and the Arts; a series of noontime seminars sponsored by the University Values Program. They sponsored a large interdepartmental course on applications of recombinant DNA methodology to research in cell and molecular biology. They helped develop a two-day forum on recombinant DNA and took an active part in it. They provided expert advice to the Regents and other campus groups as needed.

Without planning it that way, this committee became a source of expert knowledge to counteract misinformation or misunderstanding that arose in the community—a truth squad. As time wore on and the need for precise information increased, members of this group at times took the initiative to offer their views and to correct errors advanced by commentators to the public. Examples of these efforts will be seen later.

THE NATURE OF COMMITTEE B

Committee B, appointed by the Vice President for Research, contained eleven members from the varied fields of biochemistry, English, genetics, history, internal medicine, law, philosophy, physics, psychology, religion, and social work. I was its chairman.* No members of the committee worked in the field of microbiology. It was asked to develop and recommend policies or a review process for research in recombinant DNA and related aspects of molecular genetics. It began its work in September 1975.

At the time it was appointed, the issues to be resolved were not clearly identified, here or anywhere else, and so the committee's charge was fairly vague. The group was asked, in effect, to identify the problems in the conduct of this research and to provide its best judgment about whether the work should continue and, if so, under what conditions. As research at The University of Michigan was to be its focus, the committee members set four limits upon the range of their discussion. They decided that they:

Would develop policies and recommendations for The University of Michigan, not for the relevant disciplines as a whole.

Would develop principles that could be used in evaluating specific proposals for research, not for all kinds of research in all the relevant disciplines.

Believed that adherence of researchers to guiding principles could best be monitored by a local committee of faculty peers.

Believed that appropriate principles for guiding work in recombinant DNA and related matters would need to be changed when the field

---

*One member, a historian, was not originally chosen by the Vice President. He was added at the request of the Senate Assembly Committee on University Affairs, who said they wanted "their own man" on this committee because they were suspicious about its autonomy.

advanced and the guidelines changed. It would be necessary, accordingly, to establish a procedure whereby basic principles could be reappraised and changed from time to time.

These limits required the committee to come up with feasible recommendations for guiding local decisions. The members felt that their report would not be adequate if it merely asked and answered broad philosophical questions; they believed they should try to develop practical guidelines that would meet operational requirements.

Overview of Events within Committee B

At the outset of their series of more than two dozen meetings, the members realized they would have to find and sharpen issues as their discussion proceeded. The beginning sessions therefore were devoted to instruction of the members by several microbiologists in the technology of recombinant DNA and in the methods developed for safety in this research. The members were also given reprints of articles on these issues. Several meetings were directed (at the request of the historian) to questioning and establishing the role of the committee. Did it really have any power? Would anyone listen to its recommendations? What if it decided that this research should be forbidden at Michigan: would the work actually be stopped? Each of these questions was tentatively granted an affirmative response, tempered with an inclination to wait-and-see.

The committee published news stories and letters about its purposes and activities and requested comments from the university community. It sent letters to deans and made announcements in faculty governing groups about the opportunity to appear before Committee B. The members broadened their vision by listening to a local scientist who was a member of a national committee preparing guidelines for this research; to a number of scientists from out of town, two of whom opposed the research; to a philosopher of science who was writing a book about this field and had traveled widely in collecting her material; to local scholars who were conducting research using recombinant DNA; and to several speakers at a two-day forum on this topic.

All meetings except the very last ones were open to observers, and each was attended by one or two uninvited students who took extensive notes because they were studying, they said, the nature of decision making in a university. The report was finished on March 22, 1976, meeting a deadline set in a Regents' meeting, as will later be recounted. We turn now to a more detailed account of events in Committee B and the community.

Several months after Committee B had been established, no members of the faculty had asked to appear before it or had submitted statements to it, pro or con. This disregard for the committee's work changed, however, after the Regents' November meeting, when the Vice President for Research asked approval for "the renovation of three research laboratory complexes to become 'Moderate Risk Containment Facilities' to permit continuation and expansion of ongoing faculty work with recombinant DNA techniques or oncogenic viruses." A statement accompanying this request briefly explained the nature of recent developments in research on biochemical genetics, the sources of enthusiasm and uneasiness about this research, the work in tumor virology, and the need for physical changes in three laboratories, two in the medical sciences buildings and one in the natural sciences building. The Regents were informed of the objectives of Committees A, B, and C and were told that approval for these renovations was being sought now so that funds could be requested at once from the National Cancer Institute (in order to meet an approaching deadline). About $306,000 was needed, and the Regents were asked to allow the use of special University resources if the federal grant was not large enough to pay for all the construction. This proposal was approved by the Regents without discussion, and it was described in due course in campus and community newspapers.

An interested faculty member, who subsequently devoted much energy to discouraging this research at Michigan, called a Regent to ask about the plans, and the Regent then requested further information about the research. The following month, therefore, during a public session of the Regents, the Vice President for Research reviewed in greater detail the planned renovation of the laboratories as well as the events that had made it necessary. The chairman of the Department of Microbiology (who was also the chairman of Committee A) gave a brief account of research in recombinant DNA and described the plans of Committee A. I, as Chairman of Committee B, described the purposes and activities of that group. The Regents asked only a few questions, which they addressed to the Chairman of the Department of Microbiology.

Immediately thereafter, in a period set aside for comments from the public, five professors gave brief presentations, each expressing uneasiness about research in recombinant DNA and the plans of the University in this area. Three of the speakers were from the Department of Humanities in the College of Engineering; one was a Professor of Physics and director of a small, undergraduate, residential college on the campus; and another was from the Department of Pathology. Four of the five had been concerned for some time about the impact of science and technology of any kind on the future of society. Recombinant DNA was a familiar type of target to them and, for once, it was not the product of the physical sciences or engineering, where most of these speakers had their peers and the usual objects of their criticisms.

Among the points made by these commenters were the following:

Some kinds of knowledge are dangerous and should not be sought by man (example: methods of creating nuclear fission). Work in DNA recombinants is one of these.

There is no need for quick progress in this field; thus, any planning should be done cautiously and based on experience.

Both benefits and risks must be considered in developing new intellectual fields, but these gains and costs are difficult to evaluate in the present case because users of the procedure may not be emotionally stable.

Those making decisions about the nature and conditions of research into recombinant DNA should obtain the views of citizens in this community who are not employed by the University and who, presumably, do not place as high a value upon research as faculty members and administrators do.

The fact that a particular kind of research can be done does not automatically grant the right to do it.

"In pursuing the DNA research we are actually beginning to tamper with the nature of life itself. In order to tamper with life in a fundamental way, we have to have wisdom and moral responsibility; whereas, in my opinion, we have neither."

The use of *E. coli* in any size, shape, or form in this research entails unacceptable risks.

This University should not accept the guidelines that will be proposed by NIH but should formulate more stringent guidelines of its own.

"Means should be found to adequately inform members of the University and the surrounding community. Such means should include an open forum at which both sides of the issue are represented."

Many listeners were pleased with these brief talks, because they suggested that a wider interest in the issues was developing locally and that Committee B might benefit from the give and take among persons with differing views. A problem-solving process finally appeared to be underway in the community, thanks to a handful of citizens who had taken time to organize their thoughts and to express them publicly. Other observers, however, were not happy with the presentations, since a number of the comments were not correct, were misinformed, were out of date, or conveyed a tone of excessive advocacy. C'est la vie.

Committee B was ready to welcome these intellectual stirrings, because a few weeks earlier it had received a long letter from two of those who had just addressed the Regents. The letter, endorsed by 34 faculty persons from eleven academic departments, stated that the hazards associated with the transfer of genetic information from one organism to another presented a serious danger to human and other forms of life and that this danger made necessary the restricting of research in recombinant DNA. "Academic freedom . . . is not an absolute freedom: it does not include the freedom to do physical or moral harm. . . ."

The letter also said that too few people in town were informed about the issues, that Committee B should consult with scientists and lawyers who were opposed to aspects of this research, that a well-publicized forum should be held so that members of the University could become familiar with the issues, that the committee should initiate a thorough investigation into the legal aspects of the matter, and that, when the NIH guidelines became available, the committee should strengthen them if necessary. These suggestions were, on the whole, accepted as sensible ones.

Two other events helped make the purposes of Committee B better known in this scholarly community. First, a resolution proposed by one of the five persons who had addressed the Regents earlier was passed in the Faculty Senate Assembly. The Assembly, a prime governing unit for the faculty, is composed of 60 representatives elected by professors in all schools and colleges at the University. The Regents' recent approval of the laboratory renovations prior to any extensive public discussion had apparently aroused a reaction close to suspicion, as may be seen in the following statement, passed unanimously by Senate Assembly, December 15, 1975:

> Whereas the proposed research at the University in recombinant DNA has both great risks and great potential benefits, and whereas Committee B was created to recommend policies for the conduct of such research, therefore we urge the Committee to explore aggressively the ethical and legal acceptability of such risks.
>
> Moreover, we urge that the funds voted by the Regents in their November meeting for implementing such research not be expended until there has been reasonable time for Committee B to formulate its report and for that report to be discussed by the University community.

At the Regents meeting in December the Vice President for Research said that this resolution was acceptable to him. This meant that Committee B would have to finish its report by March 1976 in order to leave time for adequate discussion before faculty meetings ceased for the summer recess.

The second event was the decision of the planners of a program called the *University Values Year* to consider the recombinant DNA issue in three of a series of regular weekly meetings open to the university community.

During January several members of the small group that had originally addressed the Regents freely expressed their preferences to me. They requested and received a written response to their long letter of December. They telephoned several times to ask for clarification of dates and agreements. They paid me two visits, summarized these conversations in writing in considerable detail, and requested my approval of these minutes.

During their visits the members of the group urged that a public forum on the recombinant DNA controversy be held, so as to provide Committee B and members of the Ann Arbor community with a wider range of views than had previously been available. Detailed discussions also took place about the timing and ground rules for the forum, about who (if anyone) should replace a member of Committee B going on sabbatical leave, about the proper role of the Committee B report with respect to the Regents' decision, about what should be done if Committee B were not unanimous in their findings, and about the desirability of formally involving "the community" in the decision-making process.

During February the leader of the group that had originally talked to the Regents about DNA sent Committee B a second paper discussing relevant issues. This statement had been read at a meeting of the *University Values* seminar. In 15 pages the author emphasized that the risks involved in the research should not be assessed by scientists alone, because scientists' views were too narrow and were based on a vested interest in the outcome. Thus far, however, the decisions about the national guidelines were being made only by scientists. These decision makers, in addition, had assumed that this research would go on no matter what, had never considered the wisdom of discontinuing it, had been disagreeing among themselves, and were unable to reach a group decision. The recommendations to be made by these national experts would therefore be suspect and the forthcoming guidelines from the NIH would not be acceptable. The author also discussed some of the commonly noted worries about the use of *Escherichia coli* in this research and stated that attempts to develop enfeebled strains of this organism were not succeeding.

Although most of this statement, in deploring the procedures being used in creating the national guidelines, was thereby discussing matters beyond the charge of Committee B, the message was a clear one: the author felt that the committee should study the national guidelines carefully and revise them for local use where this seemed indicated, because the national experts either could not be trusted or did not know enough to make wise decisions, whether on scientific, technical, or moral matters. The scientists put their own needs first, said this author; they ignore the dangers they create for society.

Identifying Principles That Justify
Controls on Recombinant Research

During the depth of winter, Committee B's discussions were dedicated to determining whether this research should be stopped or limited and why. It soon became evident that most of the public comments on combining of genes were attending primarily to the most hazardous experiments, apparently assuming that the worst possible case is the most common occurrence, and ignoring the fact that scientists were closely conforming to the voluntary moratorium on dangerous research. Little popular notice was given to the fact that much scholarly endeavor can make use of recombinant DNA methods without creating any danger at all; for example, in making genetic maps of simple organisms or in studying how the expression of individual genes or groups of genes is regulated. Recombinants make it possible to study, much more effectively, many of the basic questions about the mechanisms of inheritance and the regulation of gene function that had interested biologists for decades. For Committee B, then, the central issue was not whether any and all research in recombinant DNA should be forbidden, but what kinds might be allowed or disallowed and under what conditions?

After a few weeks of preliminary skirmishing on this issue, one of the members, Professor Carl Cohen, a philosopher and a specialist in ethics, volunteered to prepare a paper. In it he attempted, as he put it, to lay out "the alternative principles which, if adopted, might reasonably justify prohibition of further research in a given sphere. Recombinant DNA is of course the sphere of immediate concern; but as a matter of principle nothing will prohibit any work in this sphere that might not also prohibit research in other areas presenting problems of equal gravity." So, the major problem was framed in its most general sense, not only in terms of the committee's most immediate interest.

Cohen advanced two basic premises as givens that need no discussion. (a) That freedom of inquiry is a value of the very deepest importance which, without most compelling reasons to the contrary, should not be abridged. (b) In spite of our rational commitment to free inquiry, there are some research enterprises that an institution might appropriately prohibit within it. The purpose of his paper, he observed, was to characterize the enterprise or settings in which prohibition of research may be defensible.

He described a number of possible alternative principles or prohibitions and went on to examine the validity of each.* After considerable discussion, another member of the committee, the professor of English, summarized several conceptual distinctions made within the group: (1) The risks in the *process* of research, or in the conduct of it, may be judged to be absolutely wrong, unreasonably dangerous, or to offer either a calculated or a forbidden degree of risk. (2) The *product* or outcome of research may be potentially wrong or absolutely wrong. During discussion of these distinctions it was assumed that research leading directly to bad products would be banned by rules that already govern research at this University, but that in all fairness both to the research and society, one cannot forbid research simply because the results of a scientist's efforts might, conceivably, in the dim future, in some vaguely imagined way, be used by someone for ill.

A subsequent statement, prepared by the lawyer, sought to explain why he believed research in recombinant DNA should proceed, why there should be reappraisal of any such decision in order that experience and collected wisdom might be taken into account, and why controls were needed for research with DNA recombinants. Another brief essay, by the historian, urged the committee to recommend that some continuing body (after the demise of Committee B) continue discussion of what he termed "the product of the matter." He recognized that the products of this research were difficult to discuss intelligently because they currently were ideas rather than entities. But, he felt, the potential for harmful output was great enough that this topic warranted further treatment in the future. Primary requirements for safety in research were also put forth in other working papers.

These were the kinds of topics examined in Committee B over several months. In February the committee received a draft of the national guidelines for research in recombinant DNA from the National Institutes of Health. This document prescribed approved procedures for safety in the laboratory and revealed that many of the principles Committee B had been trying to establish about best practices in this research were better stated by the experts. The experts' ideas, moreover, made clear what constraints were required in this work. We shall return later to the effect of the national guidelines on the work of Committee B.

*An expanded version of this paper has been prepared by Carl Cohen as a chapter for this volume, entitled "When May Research Be Stopped?"

THE COMMUNITY FORUM

Before presenting the final recommendations of Committee B, we must note an attempt to educate members of the University and the larger Ann Arbor community in modern microbiology. As already mentioned, observers suggested in several places that an effort be made to inform the townsfolk about the nature and significance of the planned research in recombinant DNA. This proposal was also made to the Senate Assembly, and a resolution was passed there urging that such a forum be held, and that it consider technical, ethical, legal, and other aspects of recombinant DNA research.

A subgroup of the Research Policies Committee with the help of two representatives from the University Values Program and two from Committee A organized the program for the forum. No members of Committee B took part in the program. Funds for the forum were provided by the Office of the President of the University, the Vice President for Research, the Dean of the Medical School, the University Values Program, and the Office of Ethics and Religion. The total direct cost was $2700.

The advent of the forum was widely publicized in the press and in a sober mimeographed letter addressed to the citizens of Ann Arbor by members of the group that originally had spoken to the Regents. The meetings of the forum occurred on the afternoons and evenings of March 3 and 4. More time was devoted to audience participation than to presentations by speakers, five of whom were nationally recognized scientific and legal experts from outside Ann Arbor. Members of the audience were provided slips of paper on which they could write questions or comments. Oral statements by those in the audience, from the floor or stage, were also accepted. About 400 people attended each session.

There was a separate theme for each meeting. The first was devoted to a semitechnical description of the nature of recombinant DNA and its uses in research and application. The second concentrated upon the risks in this research and the procedures for controlling them. The third took up the problems involved in making decisions about risky matters as well as the legal liabilities that might arise. The final period concerned ethical aspects and a summary of the views offered during the two days.

Those in the auditorium received a review of the pros and cons in the combining of genetic materials from separate organisms, and those who stayed at home could read full reports of the meetings in the papers. Dozens of questions were asked and answered at every session. There were emotionally toned pronouncements against such research, and these were to be expected, but few were prepared for the effort by one speaker from another university (who climbed to the stage and took over the microphone when it was not his turn) to introduce political issues into the discussion. His argument was that all research, and therefore research in recombinant DNA, is for the benefit of the rich and deprives the poor. Research in recombinant DNA therefore should be forbidden. Many at Michigan recognized for the first time that disapproval of this work could be based on grounds other than the conduct or outcome of the research itself.

Members of Committee B had a separate meeting with the person who advanced this special approach. They had been urged to meet him by one of the pair of persons who had been meeting with the chairman of Committee B and who had wanted this man to have a part in the forum. At this small meeting the visitor's remarks were somewhat less concerned with politics and more directed to the dangers of DNA. Other speakers from the forum were also invited to be present, and they sensibly answered questions by Committee B about problems of safety in microbiological laboratories.

THE REPORT OF COMMITTEE B

Within a few weeks after the forum the members had settled upon the principles and recommendations they would put into their report. Their final statement described, in nontechnical language, the nature of research in recombinant DNA, the origins of the committee, the principles that might guide this research, and recommendations for monitoring the conduct of the research. These monitoring recommendations were taken, for the most part, from the national guidelines.

Highlights among the principles and recommendations the committee proposed are these.

1. There are two extreme possibilities in assessing DNA research. It might be urged, on the one hand, that any and all such research should be permitted because freedom of inquiry is an absolute freedom that must never be abridged. In contrast, it might be held that such research should be prohibited because it is the kind of thing man should not know. We reject both these views. We believe recombinant DNA research should, in principle, go forward so long as it is submitted to appropriate controls.
2. We hope that officers and faculty of the University will use appropriate influence both within the University and within the scientific community to encourage research that will make it possible to calculate risks in this work with greater accuracy, that will minimize these risks, and that will continue to examine guidelines for the control of this research.
3. We believe that the current guidelines prepared by the National

Institutes of Health are an acceptable basis for assuring the safety of experimentation in molecular genetics and viral oncology. We have come to this conclusion with some uneasiness because a risk remains, though it is small. We have concluded, however, that this risk should not bar experimentation and the possible understanding that may be developed concerning the origins of just such risks. We believe that the potential benefits likely to arise from this work are great.

4. To the NIH guidelines we propose further restrictions for research at this University.

   (a) No high-risk experiments requiring the strictest (P-4) containment levels shall be permitted unless this restriction is removed by the appropriate University decision-making body.

   (b) Specially enfeebled organisms which are not likely to survive outside the laboratory will be used in all experiments when they become available, except where an experiment requires otherwise.

5. A review committee should be established to ensure that laboratory equipment and facilities are appropriate for a given experiment. . . . The review committee should have at least one member who is not an employee of The University of Michigan and one person who is a member of the Senate Assembly Research Policies Committee.

6. We recommend that some body occasionally review the work and principles (including the NIH guidelines) underlying the actions of the review committee. The Senate Assembly Research Policies Committee might be appropriate for this purpose.

7. According to the guidelines prepared by the NIH, certain DNA recombinants should not be studied in laboratories at this University or any other institution because of the possible dangers that might follow if containment fails. Five different kinds of experiments or actions must not be initiated at this time. These involve experimentation with virulent pathogenic organisms, the introduction of bacterial genes into host bacteria so that the latter develops a resistance to antibiotics or an ability to form bacterial toxins which they did not originally possess, experimentation that seems likely to increase the virulence or host-range of organisms that cause disease in plants, and release into the environment of any organism containing a recombinant DNA molecule unless a series of controlled tests leave no reasonable doubt of its safety. The moratorium on these activities prohibits the most dangerous research, allowing only those with less hazard to continue.

8. The possibilities of legal liability arising from this research will be met by adherence to the NIH guidelines and other criteria we have proposed.

After the report had been edited, one participant (the historian) announced that he did not wish to support it and that he intended to write a dissenting statement, which he did that same evening. The committee decided that his remarks would be circulated to all members so that anyone might comment on them in writing if he wished. One colleague (the lawyer) did respond. The dissent and the response were placed in the appendix to the report. The dissenter emphasized that he did "not act from a fear of the laboratory hazards that have been so prominent a part of the debate on this question. . . . Although there remains a measure of hazard I believe it lies well within reasonable limits." Instead, his unwillingness to encourage this research was based on his belief "that the limitations of our social capacities for directing such a capability to fulfill human purposes will bring with it a train of awesome and possibly disastrous consequences. Decisions wiii be made . . . that may well have unintended but irreversible effects." The dissenter, in short, opposed the potential product rather than the process of doing research on DNA. The author of the response to this dissent was not persuaded that the above conclusions were correct but agreed that they were legitimate and warranted continued public perusal of the research being done in this field. As often happens, this brief dissent and response received as much or more attention from commentators than did the full report itself.

CIRCULATION AND EVALUATION
OF THE REPORT

Within a matter of days after the committee's report had been released, an 18-page essay, "Critique of the Report of Committee B," was sent to the Regents of the University. This critique was prepared by some of those who had originally spoken to the Regents in December and who, from this day forward, placed their statements only before the Regents. A number of faculty members indicated their support for the critique by signing their names to an attached sheet.

Before the forum, differences of opinion about research on recombination of DNA provided enrichment for the pool of ideas on which a conclusion could be based. Contrasting opinions were taken to be helpful because reaching a judgment about this research was a campuswide and impartial task. During the forum, however, contrasting beliefs were stated more forcefully than they had been put previously, and reaction of the audience aroused a tendency to engage in competitive behavior between the pros and cons. The critique of the report, moreover, fostered a shift toward polarization, because the writers sought to damage the credibility of Committee B; they became critics rather than commentators. The members of Committees A and B believed that a reply was necessary to the critique because it contained misinformation and misunderstandings. Their response was directed to the Vice President for Research. The writers of the response thus became advocates of the research.

Let us note a few of the critique's comments and the responses offered by participants in Committee B and, after the first few, by Committee A.

Critique: The committee did not have enough interaction with persons who are known to be critical toward recombinant DNA or the NIH guidelines.

Response: There are few telling critics of this work to be found throughout the country. Committee B met with some of them, read others' writings, or heard from them directly: its report has addressed itself to the observations of these persons. All the meetings of Committee B were open. No local critics appeared at the meetings, except at the very last session, when they were not welcome.

Critique: Committee B did not devote enough time to discussing ethical issues.

Response: About 80 percent of its time was spent on such topics. (These discussions were not efficient ones, for reasons noted later.)

Critique: The effect of the renovated laboratories upon the building containing them was not assessed.

Response: Such assessment had been the task of Committee A (not Committee B), who did consult engineers at NIH, had visitors here from NIH, visited labs on other campuses, closely examined the buildings with the help of University and consulting engineers, and the like. (The renovations to be made are then described and explained in much detail.)

Critique: The possibility of new organisms being created is real (though rare) and the danger to the earth's ecosystem may be everlasting.

Response: There is substantial uncertainty as to whether the risks associated with the careful application of recombinant DNA methods to a study of living organisms are any greater than those posed by conventional genetic and microbiological research for over 50 years. It is a specious argument to say that research should not be done because there is uncertainty about the danger, if any, in doing it, for that is true of virtually any action.

Critique: It has been difficult to create a strain of enfeebled *E. coli* and therefore the effectiveness of enfeeblement is in doubt.

Response: The difficulty in constructing a disarmed strain of *E. coli* is no proof that the enfeeblement will be ineffective. An enfeebled strain will soon be in hand for use by researchers worldwide. (Several such have been tested since and certified.)

Critique: The national guidelines are weak because they are the result of deep conflict within the NIH committee that developed them, and thus concern for public safety has been compromised.

Response: The debate was helpful, not harmful. Conflict was not deep; only a few differed from the rest on a few matters. The committee's final approval of the guidelines at NIH was unanimous.

Critique: Methods for monitoring the health of laboratory workers are not feasible.

Response: General methods for monitoring are given in the report of Committee B and in NIH guidelines, and these are based on tested experience of many years. Specific methods will vary with particular experiments being conducted.

Critique: The report of Committee B does not attend to the malicious use of the products from recombinant DNA research in warfare or terrorist activities.

Response: There are more convenient and potent tools than DNA for those who have malicious purposes. Placing crippling restrictions on this research or any other is not a rational response to concern about how it might be used malevolently.

Critique: The benefits of this research are speculative and unknown.

Response: A long list of findings are cited to demonstrate that the benefits are real, of significance, are occurring now, and will continue to accrue.

Discussion of the Report in Faculty Meetings

In order to maximize opportunities for appaisal of the Committee's report, copies were given to a number of governing bodies in the faculty, and it was discussed at a regular meeting of each of these units during April and May. These included the Research Policies Committee of the U.M. Senate Assembly, the Academic Affairs Advisory Council (deans of all colleges), the Research Advisory Council (mostly assistant deans for research), the Senate Assembly, the Executive Committee of the Medical School, the faculty of the Biology Division in the College of Literature, Science and the Arts, and the Executive Officers of the University. The report was also described before the U.M. Senate, where the matter of endorsement was not pertinent. The discussion was most lively in the Assembly, as the critics had carefully prepared for that meeting and several of them spoke from the floor. Even so, the majority of the speeches in the Assembly were testimonials on why each intended to vote to endorse the report, and it was so endorsed by a wide majority. The Assembly, in its closing moments, also passed the following resolution.

> That the Research Policies Committee, in consultation with Committees A and B, formulate the charge of Committee C, including procedures for the selection of personnel, and return the recommendations to the Assembly for its consideration.

(The charge for Committee C was written during the summer and accepted by the Assembly in September 1976. Committee C has been meeting regularly since September 1976.)

RESPONSE OF THE UNIVERSITY REGENTS

Initially, the report of Committee B had been prepared at the request of the Vice President for Research, but by now the topic had become a legitimate concern of the Regents. The critics had requested an opportunity to tell the Regents what they thought of the report, as had the supporters of the research. The Executive Officers therefore decided to have a summary presented to the Regents in April, then to provide time for questions by the Regents, and finally to allow equal opportunity for both critics and supporters to speak to the Regents. A month later, in May, it was planned, the Regents would vote on whether they endorsed the recommendations in the report.

At the April meeting of the Regents I briefly summarized the recommendations of Committee B. The Regents asked no questions. Five critics then spoke. A professor of humanities described the committee's methods as "shoddy," containing "unwarranted assumptions" that led to "hasty conclusions," and added that the report failed to consider the "larger" question concerning the greater value of prevention of disease over cure. A professor of religion argued on behalf of "effective community participation in deciding future university involvement" in such research. A mathematician worried about what would be done for the researcher who needed high-risk facilities for his work; would he be advised to take early retirement? A sociologist once again feared that this research would be used for biological warfare. He also urged more community "input" into the decision process. A historian of science warned that tests to determine risks were not valid, that the NIH guidelines were not rigorous, and that it was uncertain that methods of containment in the laboratory would work.

Next, four endorsers of the report talked, all microbiologists. They noted that research in recombination of DNA would lead to a better understanding of the nature of infections, to improve safeguards against disease, and to further knowledge in all fields of biology. They observed that the risks of disease were remote, because any given bacterium must possess a very large and complex set of characteristics in order to colonize man and cause a disease. It was exceedingly improbable, therefore, that an enfeebled organism could artificially be given all the characteristics it would need to be a cause of disease; indeed, recombination would make it less likely to be a cause.

The Regents now faced several issues. Were they ready to vote the following month on approval of the report and on a substantiation of their prior decision to renovate several laboratories? If not, what procedure should they follow to prepare themselves for this vote? Also, what position should they adopt on the proposal that there be widespread community participation in their decision? In the past they had usually taken the view that the Regents are publicly elected, that they are public representatives, and that it is the expectation of the state's constitution that they will act in the public interest.

The discussion among the Regents on the following day revealed that several of them were not yet prepared to vote. They felt that the contrasting evidence available on the campus was not wholly convincing and that they needed more facts. They would like to hear from experts who had no vested interest in the local decision, experts from outside Ann Arbor. Only one Regent had attended meetings of the earlier two-day forum—they wanted something like that for their own group.

Thus, a mini-forum was arranged for the afternoon of May 12. The speakers from outside Ann Arbor were Paul Berg from Stanford University; Joseph Perpich, Associate Director of Program Planning and Evaluation, National Institutes of Health; and Wallace Rowe, Chief, Laboratory of Viral Diseases, National Institutes of Allergy and Infectious Diseases. Those from the University included a member of Committee B (a geneticist), a professor of Zoology, a microbiologist, and a historian of science. The President of the University served as moderator of the meeting.

The procedure allowed each Regent to ask whatever questions he wished. A set of suggested questions most commonly asked was prepared for the Regents by the chairmen of Committees A and B. Most of these queries were taken up at one time or another during the day, and thus the list serves as a convenient summary of the topics addressed. It was a thoughtful and instructive session that obviously helped and settled the Regents.

Some Common Questions about Research in Recombinant DNA

Purpose of the Work

1. What are scientists trying to accomplish with recombinant DNA research? Has any of this been accomplished?
2. What are the consequences if these laboratories are not renovated?

Potential Risks

3. Does this research create new forms of life?
4. Will this research make it easier for terrorist groups and unscrupulous governments to threaten harm?
5. What are the chances that a new and terrible disease will appear as the result of this research?
6. Will this work lead to genetic manipulation of man?

Implementation of Controls

7. Are the proposed federal (NIH) guidelines for this research strict enough?
8. How will the federal government monitor and regulate this research?
9. How can one be sure that scientists will obey the federal rules?
10. How will our local laboratories be checked?
11. Will this work be secret?
12. How expensive is it to run a laboratory with the required safety features?
13. Wouldn't it be preferable for all of this work to be done at a few remote and secure locations?

Role of the University

14. Is a university an appropriate place for work of this kind to go on?
15. What is being done at other universities?
16. What kind of research in recombinant DNA and tumor virology is now being done at the U.M? What will be done in the next year or so?

17. How much will this work cost U.M. in the next few years?
18. Would the U.M. delay a decision on this issue?

On the day of the mini-forum the critics provided the Regents with 60 or more additional pages of material containing ten new essays. Five of these articles concerned the degree of risk involved in current plans for research, and one of the five was a "Position Paper in Recombinant DNA Research" prepared by the critics. This 17-page document stressed the nature of the risks and the weakness of containment procedures in laboratories, arguing that it was unwise to "proliferate" the number of places in which this work was done. The writers urged that decision makers take their time and that all this research be concentrated in one national center. A letter they had requested from Dr. Robert Sinsheimer at California Institute of Technology accompanied the position paper and supported the proposal for a single national lab. The remaining four papers discussed the long-range implications of research in this area. A final brief statement, "A Response to the Replies of Committee A and Committee B to our 'Critique of Committee B'," had been prepared by the writers of the above position paper. They declared that some local citizens believed the opponents of this research were a small group supported by no one else in the community, but that this belief was wrong: numerous other persons agreed with the critics. The rebuttal, it was said, incorrectly asserted that the critics were against research in recombinant DNA. Instead, the important question was not whether this research should go forward but rather, "How much research should go forward, where, under what conditions?"

The regular May meeting of the Regents was to take place a week later. In justice to all concerned, it was felt best that the Regents vote on DNA at this meeting, rather than later, as many of the faculty would not be on duty during June or the rest of the summer. In anticipation of this session the critics requested that one or two scientists known to have unfavorable views toward research on DNA recombinants be invited to Ann Arbor to address the Regents. Two of these had earlier been invited for the mini-forum, but both had declined. In response to this new request the Executive Officers again invited the same two men, but again they refused. No one could think of other suitable detractors of this work. Thus, the final presentations to the Regents would be made only by Ann Arborites. One hour would be available to the cons and another hour to the pros.

The cons came first, four of them. Only two of the usual critics were present. A mathematician reported data on the number of accidents in laboratories at this university during 1975 as a basis for questioning the efficacy of any efforts to prevent recombined molecules from escaping.

A professor of planning urged that research on recombinant DNA be limited to "national laboratories" (which should, of course, be someplace other than Ann Arbor). He suggested this as the only reasonable alternative to invention of a social process that would give all Ann Arbor residents a direct voice in deciding whether such research should be done at Michigan.

The president of the University janitors' union also spoke. He was not opposed to the research but urged care in conducting it. Finally, the assistant director of the Ann Arbor Ecology Center expressed uneasiness about the effect of this research on the environment and asked for "a real opportunity for participation" in this decision. The prior forums and the many faculty meetings were not enough, he said, because some questions remained unanswered. The Regents, it should be noted, had at no time shown the slightest interest in the recommendation that there be communitywide participation in making this decision. One Regent had observed, "If every decision had to be made by a plebiscite, we'd never make any decisions at all."

Six faculty members spoke in support of the question, among them the Dean of the College of Literature, Science and the Arts; the Dean of the Medical School; and the Dean of the School of Graduate Studies. The chairmen of three departments also endorsed the research: Human Genetics, Biological Sciences, and Microbiology. The speakers' remarks included these ideas.

A small but very vocal minority seems intent on interjecting itself between the judgment of the faculty and the Regents (a reference to the vote in the U.M. Senate Assembly in April).

National research laboratories, which exist in some scientific fields, tend to promote centralization and bureaucratization. Universities in the end take a subsidiary role to government in such labs.

Isolation of scholars is not wise, and a national laboratory would cause just that.

Control of research procedures is likely to be more thorough at a university than at a government facility.

Interfaces exist between work on DNA and developments in many disciplines; thus, this research should be done at a university.

A number of points raised by the opponents of this research (six are listed as examples) are based on misinformation, each has been shown to be incorrect, but their incorrectness has never been acknowledged publicly by the opponents.

There have been repeated instances of public misrepresentation (by the critics) of the views of prominent scientists outside of Michigan (five examples are given). This pattern of action confuses interested observers and widens the gap between the two sides.

On the next morning the following resolution was put before the Regents.

Recommendations of Committee B should be approved, in particular:

1. Recombinant DNA research should go forward as long as it is submitted to the controls described below:
   (a) The guidelines prepared by the National Institutes of Health are an acceptable basis for insuring the safety of experimentation in molecular genetics and viral oncology. Revised NIH guidelines shall be reviewed by Committee C as they appear.
   (b) Further restrictions for research at The University of Michigan are recommended as follows:
      (1) No experiments requiring containment levels above P-3 shall be permitted without approval of the Board of Regents.
      (2) EK-2 biological containment will be used when available in all bacterial experiments requiring P-3 physical containment.

The Regents direct that procedures be developed for constant monitoring and safety of the research. In addition, the Regents direct that periodic appraisal of these review mechanisms be carried out by the University and reported at least annually to the Board of Regents.

A motion to table this matter until the June meeting "to afford an opportunity for additional critics of this research to present their case to the Board" was defeated. The resolution then passed by a 6-2-1 vote. The lone Regent who voted against it was not opposed to the research: he simply preferred to wait until June to make the decision, in order to be fair. Thus, the issue was settled.

A few nights earlier a resolution had been brought before the Ann Arbor City Council by one of its members. It requested that the Director of the National Institutes of Health be asked to provide the Council with a copy of a statement describing the impact that DNA guidelines would have on the environment. This resolution was tabled without discussion and has not been brought up since that night. A local political party later sent a resolution to the state offices of the party urging that the platform of the party discourage this research. That move was subsequently defeated.

THE AFTERGLOW

Following the Regents' vote, a number of persons said aloud that the nine months of discussion had been a good thing because the issues had received wide attention and numerous opportunities had been created for people to have their say. One Regent, in explaining his vote, observed that the discussion "sets an example for rational handling of crucial decisions on University matters."

Did it? Was all this activity in fact worthwhile? Did science and society benefit from a public airing of plans for future research and from the efforts of individuals who were not scientists to shape these plans? If so, how and why?

Such questions are hard to answer with confidence, because one man's actions for the good of the world are another man's thoughtless (even hazardous) behavior. Furthermore, and perhaps especially in an academic community, most of us assume that if we talk enough, a consensus will be reached and everyone will go away satisfied. Those who were lavish in their praise of the year's discussion did not know, could not know, the costs that accompanied the benefits, because they had no part in the discussion. Clearly, certain things happened during the year that would not have come to pass unless the committees, reports, papers, and critical comments, had encouraged them. It may be helpful to review some of these events.

1. Committee A developed a set of plans for improving the safety of laboratories in several parts of the campus (and, we should add, it was successful in requesting federal support to pay for these renovations).
2. Committee B provided recommendations for the actions by the Regents and furnished reasons for these actions (with help from the NIH guidelines).
3. The critics had many (formal and informal) opportunities to make their views known and to influence the final decision.
4. The critics took pride in their part and felt that they had had a beneficial influence.
5. Many citizens in the town and University were informed about the nature of research in recombinant DNA. A popular course emphasizing recombinant DNA research had 200 students. All class lectures were filmed on videotape for future use.

6. Faculty members had a variety of opportunities to question, approve, or disapprove the recommendations of Committee B. In only two of the meetings were negative votes cast, one in a council and a handful in the Senate Assembly.
7. The Regents took part in a decision that might have been left, under other conditions, to the administrators of the University. They were pressed to weigh many pros and cons before they reached their decision, as well as to consider the wisdom of doing this research at all.
8. The Regents sought the views of experts from outside Ann Arbor, not only the opinions of campus commentators.
9. The Regents said they were content that they had heard both sides of the question and were not making an ill-informed decision. They made their decision, moreover, knowing that the recommendations had been widely discussed and accepted by the faculty.
10. Dozens of persons put in hundreds of hours of work on papers, reports, meetings, speeches, studying, and the like. Much of this time was stolen from that needed for other duties in teaching, administration, or research.
11. The cost in professional time and energies for many of the scientists involved was substantial.
12. The emotional strain was severe for many persons.
13. The meetings and reports stimulated coverage of the issues by local and distant newspapers, much of it reasonably accurate.

SOME DILEMMAS IN THIS DISCUSSION

Given the interest of society in protecting itself from the actions of scientists and technologists, we shall see more committees appointed in which inexpert persons will decide what experts are allowed to do. Questions and problems arise during such efforts. In the current case many of these were not met as well as they might have been. Perhaps a review of the difficulties will help future groups of nonexperts meet such problems better.

Nonexperts and Nonexpert Decisions

No one expected (perhaps naively) that the members of Committee B would need much technical knowledge of microbiology. They were instead to evaluate the methods and products of this research as they understood them and to determine if these were in accord with important values in the community. They were not to determine what or how great the risks were but whether the risks specified to them were acceptable. The evaluation was to be made in spite of the decision makers' lack of information about the subject—indeed, the discussion was to benefit somehow from the naivete of the committee members.

The committee's task was harder than had been anticipated, because for several reasons the future use of the technology of recombinant DNA was ambiguous.

1. The eventual products of recombinant DNA, and the value of these to science and society, cannot be stated with precision by anyone, either by well-grounded scientists and/or by well-meaning citizens. There are too few normative data in hand.
2. Questions (both sane and foolish) are being raised about the adequacy of safety measures in the laboratory and in the research methods.
3. Expert knowledge about microbiology and the methods of microbiologists is needed to make sound judgments about the effectiveness of these safety measures and about the criticisms being made of them.
4. Adherence to safe procedures depends upon the care taken by personnel working in the laboratories. Does this provide adequate protection against mistakes?

Because of uncertainties such as these, members of Committee B could not be certain what values to invoke or how to assess adherence to the ones they selected. There was no reliable procedure, moreover, for identifying what was true and what was propaganda, given the state of the field and the innocence of Committee B.

The uncertainties arising from our lack of training in microbiology were increased because we had no microbiologists among our members, in order that the committee's work not be influenced by anyone who had a vested interest in its outcome. The desire to avoid unfair influence also kept us from seeking from microbiologists anything beyond limited technical information. Thus, the membership of Committee B turned inward on itself (largely ignoring help available a few blocks away). Probably the committee would have been more efficient and effective if several microbiologists had been present at every meeting to correct misinformation arising within the group or delivered to it by persons who were not members. Indeed, the microbiologists raised more thoughtful questions about their own practices than the critics did. It was not sensible, in retrospect, to have excluded them.

Because of their beginners' status, the members (or this writer, at least) often entertained wholly incorrect ideas and engaged in inept speculation during hours of discussion. Their lack of sophistication became most keenly visible to the members themselves when the first draft of the guidelines for research in DNA arrived from the National Institutes of Health in February 1976. These guidelines made it evident that the members had been trying to solve problems without the kind of knowledge they should have had, and that the solutions offered by the experts at NIH on ways of controlling safety in a laboratory were a good deal better than those under consideration in Committee B. The national guidelines, moreover, provided a precise statement about safety methods, which was necessary if the ethical recommendations of Committee B were to make any sense. It probably would have been wiser, if it had been feasible, to have delayed the work of Committee B until the guidelines were available. This would have saved much wasted effort and ill-informed discussion.

Those in the committee were not the only ones who were misinformed. Some of the critics who made public comments upon technical matters employed incorrect information about microbiology. It was embarrassing for local scientists publicly to state that the critics were wrong, yet they had to make corrections in order to keep the record straight. The result was increased polarization on the issue, because the corrections were not always happily received by those who were in error and because the scientists believed the critics had a responsibility to be factually correct in their statements to the public. As is often the case, the corrections received little public attention, and thus the damage caused by spreading of misinformation was not undone.

Who Should Make the Decision?

Who should be invited to participate in making a decision about practices that could (conceivably) have unfavorable consequences for many? In the current instance, some members of the Department of Microbiology initially wished to have renovations made in their laboratories and simply wanted clearance to request funds from Washington for that purpose. To get permission for their plans, they had to approach three local agencies: the Dean of the Medical School, the Vice President for Research, and the Regents of the University. The Regents initially became involved simply because the cost of renovation was so large that they were required to approve the expenditure, not because any danger was anticipated in the research itself.

Clearances by the three just mentioned provided a wide involvement, but many microbiologists and the Biomedical Research Council at Michigan believed an even wider participation would be wise, especially on matters of policy. Thus, Committees A and B were created, and these groups sought advice from an exceptionally large number of individuals and groups. Every person who wanted to express his views to these committees, or to University officials, had ample opportunity to do so.

The critics, nevertheless, wanted still more people to take part in more discussions on these matters. They asked the Regents to give others a say. They urged the Ann Arbor Citizens' Council, the Ann Arbor City Council, the Ann Arbor Ecology Center, the local political parties, and other agencies to ask for a part in the Regents' decision. (Even a community referendum was informally proposed, but no one was sure what size the community should be in this case.) Although the critics repeatedly demanded wide participation, this idea, as we have noted, won little support within or outside the University—perhaps because participation was proposed in general terms with no specific suggestions on how to do it, beyond the numerous meetings that had already endorsed the report. Perhaps wider participation was seen as a ploy for arousing support for the critics, or for politicization of the issue, and citizens of the community did not want to be used in that way. One of the arguments used for wider participation was that decision makers in the University could not be trusted to consider the interests of nonuniversity citizens, and perhaps many Ann Arborites did not believe this, were not convinced that the matter was as dangerous as described, or believed there was no need to take the decision out of the hands of the scientists and university officials.

The critics, during the long debate, provided an interesting problem for decision makers. Here was a small group of colleagues wanting to be heard by the faculty, the administrators, and the Regents, who were given (and took) numerous opportunities to present their case. Almost alone, this small set made the matter into an issue. Eventually, they had little to say that was new, yet they were still asking for more discussion because (they said) they had not been fully heard. How much time is enough for such a group? What criterion should be used to determine when the discussion has gone on so long that the time for a decision is at hand? Answering this question is especially difficult if some of the decision makers (such as the Regents) wish to avoid as long as possible taking a stand. Clearly, the commitment to due process within an academic community can be used to prevent the taking of final action even when it is evident that a large majority of the decision makers have decided what to do and that additional discussion will change few minds in either direction.

Uncertainty Muddles Problem Solving

Those opposed to research in recombinant DNA have frequently commented upon its hazards, even though none of these dangers has yet been observed as a result of this work. Because the outcomes of the technique cannot be reliably predicted, however, we cannot rule out all bad events. In such an uncertain situation fear can easily be aroused.

Locally, fears were not unfairly emphasized, but they were employed to some degree in order to stimulate interest, especially by composers of newspaper headlines. The uncertainties in the conduct of recombinant investigations were the primary reasons that constraints were placed upon the work with DNA. The uncertainties also made it easy for critics to attack the credibility of experts and the methods proposed for controlling this research. The uneasiness of many persons had to be given attention that might better have been devoted to rational discussion.

Controversy Fosters Polarization

As already mentioned, the comments by our critics, prior to the forum, were illuminating and useful, even though interested persons held quite varied views from "in favor" at one extreme to "against" at the other extreme. After the forum, the report of Committee B, the critique of that report, and the responses thereto, the participants in the give and take were more obviously bunched at opposing extremes; their views were polarized.

Six months after the Regents had finally approved of research in recombinant DNA at Michigan, one of the critics stated publicly, during a review of the processes involved in this decision, that the critics had deliberately tried to create a state of polarization between themselves and any supporters of recombinant research, on the assumption that this would make the members of Committee B "think harder." Without such polarization, the critic added, the committee's discussion would have been "too bland" and not sufficiently careful. The critics knew that they were "irritating to certain members of the administration," but they assumed that the administrators would not retaliate in any way that would damage the openness of the discussion. The critics were right about the views of the administrators on both counts.

The critics therefore searched for any debating point, of any size, that they might use in generating doubt in the minds of the Regents and others about the credibility of the local scientists, the members of Committee B, or the group writing the guidelines at the National Institutes of

Health. No one of these points was heavy enough to overcome the recommendations of Committee B, or to inspire the Regents to vote against this research, and the total accumulation of them was not weighty enough to have much impact either.

The procedure followed by the critics is similar to that used in the adversary process, a method of reaching a decision when two contrasting views each have their supporters. Each side presents its case as forcefully as it can before some arbiter or jury that remains alert but says nothing. In a different approach, which we might call the review model, control is in the hands of the arbiter, who calls witnesses and interrogates them while the adversaries remain passive until asked for their comments. Committee B used the review approach but was pushed by the critics toward the adversary procedure. The committee did not yield to that pressure, because the members of Committee A had become the advocates for this research when the critics had placed them on the defensive. Members of Committee B believed it was not proper for them to become advocates for either side. They were to be arbiters, evaluators, and impartial advisers for the Vice President for Research, not propagandists.

Did polarization achieve the results the critics hoped it would? The attacks on Committee B probably did make the members think harder and had some effect on the quality of the report, but these effects were minor. The critics were not a source of useful ideas. The critics' comments made the report of Committee B somewhat longer because it had to reply to some of the criticisms.

The strongest effect of polarization was that the supporters of opposing views developed a desire to have their "side" win, rather than a desire to reach the best conclusion based on a careful weighing of all evidence. The contentiousness that accompanies polarization increased hostility between groups, reducing the likelihood of compromise. All in all, polarization reduced the accuracy and completeness of fact finding while simultaneously enhancing conflict. Polarization can sometimes lead to better ideas by counteracting decision-maker bias. In the present discussion this did not occur. Committee B might have reduced the polarization if it had held its own mini-forum similar to the one the Regents arranged for their education.

Abstract Notions Are Hard to Handle in a Group

Committee B devoted a good part of its time to discussing the criteria for prohibiting research of any kind, not only that in recombinant DNA. These discussions required the use of abstract concepts and hypothetical cases. During any discussion, and especially one in which members are seeking agreement rather than enlightenment, the employment of such ideas can be clumsy. In dealing with these abstractions, members of Committee B had trouble understanding one another, the same word was often used in different ways, and talk sometimes obscured ideas rather than clarifying them. Because definitions of such concepts are not always available, verbal interaction while talking about them was not very precise. Clarity of terms was best obtained by inviting a member to put his ideas on paper. This statement could then be edited and rewritten until all accepted it.

There are numerous articles on the ethics of research in recombinant DNA, but most of these pose profound questions that are left unanswered. Answers would have been more useful than questions.

Controversy Is Awkward for an Elected Board

Trustees elected by popular referendum are expected to make sure that their institution serves the interest of the public at large. Such trustees are especially sensitive, as a result, to issues for which there are strong pro and con views. To perform their duty properly, those on a board must know whether the public's interest is pro or con. Where this interest is not obvious, board members cannot be sure which path to follow, as the views of a constituency or the platform of a party are seldom available to guide their decision. These problems are enhanced if board members are in the midst of a new campaign for office, as several were in the current instance.

In public meetings, therefore, board members must listen well to all sides and must assure everyone that they are being fair. When the final decision has in fact been influenced by the beliefs of those who come before the board, the members must visibly demonstrate that they have been so influenced. Such maneuvering takes time and contributes little to the quality of the decision.

The Forum: Heat Versus Light

The two-day forum teetered on the edge of unruliness several times but it was rescued by the steady help of the moderators, the good sense of the audience, and the poise of those participating in the program. As described earlier, one participant attempted to appeal to emotions rather than to reason, but his numerous attempts were not notably successful. These attempts were distracting and embarrassing, however, for those who were at the forum to consider an important problem. For them such polemics were out of place. Shouting matches developed on several occasions, and applauding was rivalrous on matters that did not seem to merit such excitement.

If the purpose of the forum was to educate or influence the members of Committee B, it failed to do that. Little was said that had not already been aired in meetings of the committee. It also seems unlikely that the forum contributed greatly to educating the community in any useful way, because the issues are too complex to be dealt with rationally by people who have had little previous background in the subject.

A forum is particularly capable of generating emotion. An issue important enough to be the focus of a forum (which costs considerable money and effort) is more likely to encourage the taking of sides than the developing of wisdom. Thus the investigation of a serious matter may not be forwarded by a forum unless the waste of intellectual energy is prevented by a procedure that forbids polemics and encourages learning.

The mini-forum for the Regents, unlike the public forum, and to the surprise of those who planned it, was a true success as an educational activity. Perhaps this was because the Regents were faced with a technically based decision and a need for information. This need fostered seeking and learning. Each Regent asked all the questions he wished in a setting that made it possible to obtain complete answers on both sides and to follow up on these responses as long as necessary. The visiting scientists, it is fair to say, were more convincing than the local critics were, and this fact changed what had previously appeared to be a controversy into a simple issue: an exciting new field of research raises some questions, and these inspire plans to be as careful as possible.

AN APPRAISAL

In the light of these evaluative comments we turn to the question: Was all this activity worthwhile? In my judgment, it was. I do not reach that conclusion brimming with confidence, as the ratio of benefits to costs, during the year's discussion, is not obviously in favor of the former. It is difficult to assess and compare these things. Any appraisal that outcomes were beneficial can only be based on guesses about how members of this university community felt about the events and the final resolutions. The costs, however, need not be guessed at; they were factual experiences of the participants and can be described as such.

The resolutions passed by the U. M. Regents in May 1976 were almost the same as the recommendations made by the first local committee to consider these issues in April 1975. Thus, the year's meetings and reports did not provide new ideas about technical or administrative procedures for safety in this work and in a sense were elaborate and expensive redundancies. (The one exception is the plan to have an annual appraisal of Committee C in order to examine how well that group is doing.) All in all, then, one cannot be sure that outcomes outweighed inputs, as we think they did.

Most of the costs for those who had a part in this experience have already been described, in passing, at least. They include the many hours individuals spent in meeting, studying, writing, telephoning, and worrying in order to produce speeches, reports, forums, classes, petitions, and the like. These activities reduced the effort and time they were able to put into their proper professional duties. The work load, and the polarization earlier described, also created emotional strain for those who took part. The lively questions and debates also provided a platform for many strange and ill-informed ideas which would otherwise never have had a public hearing. Where the means and energy were available, these exotic notions had to be countered, and that effort contributed to fatigue and strain.

Because the scientists here and elsewhere initiated this debate by publicly agreeing to develop ground rules for research in recombinant DNA, because these developments were often taken to be evidence that the scientists were doing something wrong (where there is smoke there is fire), and because they were thereupon accused of acting only in their self-interest, many scientists may have concluded that efforts to educate the public about the risks in their work generate suspicion rather than understanding. When another matter of this sort comes along, therefore, it will not be surprising if many scientists choose not to expose themselves to the caustic comments of professional antiscientists and the fears of the misinformed. Instead, they may stick to their laboratories away from the public eye. That, perhaps, would be a negative consequence of this kind of discussion.

In concluding that the year-long discussion was worthwhile, I assumed that the voluntary moratorium among microbiologists as a profession, and the planning of guidelines by the National Institutes of Health, aroused questions and anxiety-provoking publicity throughout the land, and that in Ann Arbor, as in many other towns, people wanted to be assured about this kind of research. The local debate about DNA provided that assurance. The assurance was worthwhile and, given the larger context, necessary.

A number of visible characteristics of the year's discussion tended, we believe, to give citizens this assurance. The University established a procedure that required all decisions about DNA to be made with great care and after thoughtful attention to the issues involved. All views on the matter were invited and given a hearing—which demonstrated the commitment of the University to methods of due process. There were no secret plans or methods; all meetings were open, written minutes were available, and progress was regularly reported in the local press. The microbiologists on the campus also emphasized in every way that they intended to be careful in this research and that they were willing to be watched and monitored by persons who were not microbiologists.

Members of the University, and nearby residents as well, had means available to them for getting information, asking questions, and expressing their fears. They were able, in these ways, to reduce any tensions they might feel about this research. In addition, they were relaxed by the evidence that a group of colleagues recommended, after thorough study, that the research proceed, and that many groups in separate parts of the University supported that recommendation after their memberships had read and discussed the report. The annual review of the work of Committee C will serve perhaps to reassure interested persons hereafter.

Despite the obvious costs, the procedure was worthwhile. Under the circumstances we could not have done otherwise.

PART I

Scientific Background

Principles and Applications
of Recombinant DNA Methodology

David A. Jackson

Recombinant DNA research, in the sense in which the term is used today, began in 1972 with experiments in which a series of enzymes was used to link two DNA molecules together in the test tube. These experiments demonstrated the feasibility of taking pieces of the genetic material of all cells—deoxyribonucleic acid or DNA—from some organism and joining it by biochemical methods to pieces of DNA from virtually any other organism. In certain instances it was thus possible to construct new combinations of genes, genes being segments of DNA molecules, which were useful in studying one or another of a variety of biological questions. Biological scientists, molecular biologists in particular, were quick to see the applicability of these methods to answering some of the most important questions about how living cells, particularly the exceedingly complex cells of higher organisms, organize their genetic information in their chromosomes and regulate the expression of this information so that metabolic processes and intracellular interactions occur in a controlled and orderly manner. They also realized that these new techniques raised a small but finite possibility that new organisms with unexpected and perhaps dangerous characteristics could be constructed in certain types of experiments in which recombinant DNA methodology was utilized. This realization led ultimately to the call by the scientists themselves for a moratorium on certain types of experiments utilizing recombinant DNA methodology so that a careful risk assessment could be made, to the convening of the Asilomar Conference to discuss what safety measures were required for which types of experiments, and finally to the NIH guidelines for the conduct of all research involving recombinant DNA molecules.

The concerns raised about the possible dangers associated with some experiments utilizing recombinant DNA techniques soon led others, less familiar with the methodology itself and with data available from such fields as pathogenic microbiology, infectious disease research, epidemiology, and microbial ecology, to become seriously concerned that recombinant DNA research could and would lead directly to a wide variety of biological catastrophes. The specters raised publically have included Frankensteins walking out of recombinant DNA labs, new species of bacteria of unparalleled virulence and resistance to control, the likelihood of ecological disasters such as bacteria which would consume the world's oil supplies, and widespread use of genetic engineering by tyrannical governments to alter and enslave populations or subgroups within a population. Most of these and other scenarios for disaster are founded upon an imperfect understanding of what recombinant DNA methodology is and what the limitations inherent in it are.

A central source of confusion lies in the implicit assumption that the risk associated with a particular experiment or organism is somehow associated with the fact that recombinant DNA methods are to be used as part of the experiment or to add some genes to the organism. This view represents a fundamental misunderstanding. As we shall see in this chapter, recombinant DNA methodology is a set of techniques that are a particularly convenient and efficient way of making new combinations of genes in the test tube. As such, it has neither inherent risk nor inherent benefit. Any risk or benefit lies in the nature of the genes that are joined in new combinations and in what is done with the new combinations of genes after they have been formed. Thus, it is what is to be done with the methodology *in each particular instance* that must be assessed for potential risk and benefit.

The fact that new and possibly hazardous combinations of genes *can* be formed using these techniques has somehow been transformed in the minds of many into a conclusion that everything done using these techniques is somehow inherently more dangerous than that done with conventional sorts of experimental procedures in the biological sciences. This conclusion does not take into account a number of important biological facts. Many of these are discussed in the subsequent chapters by Davis and Freter. At this point it is worth noting that genetic recombination is a ubiquitous and essential biological process, occurring in all organisms from bacteria to man as a normal part of genetic processes found in nature. Many recombinant DNA procedures involving bacterial genes simply mimic the kinds of genetic rearrangements that are known to occur in nature on a vast scale. However, the biochemical methods used for recombination in the laboratory are much more efficient at producing specific recombinants where a particular gene of interest is joined to a particular vector than are the essentially uncontrollable processes that occur inside cells. Thus, with respect to isolating and amplifying many genes from bacterial sources using recombinant DNA methodology, one is simply constructing by a very efficient and controllable procedure a recombinant that could form by normal genetic mechanisms in nature. In cases of this sort, then, there is no increase in risk but a considerable increase in benefit in using recombinant DNA methods.

In order to decide for oneself what the risks and benefits of recombinant DNA methodology are, it is necessary to know something about the basis of the technique. The major purpose of this chapter will thus be to describe in nontechnical terms the basis of recombinant DNA methodology and to provide some basic facts about the biological context in which the products of this methodology—new combinations of genes—must act.

The fundamental point to understand about recombinant DNA methodology is that it allows one to construct in the laboratory new combinations of genes. Among these combinations are many that at present have no known mechanism for formation in nature. Ducks and oranges do not mate in nature, and so duck genes have no opportunity to recombine by the normal mechanisms for genetic recombination with orange genes. Recombinant DNA methods allow us to join some genes from a duck to some genes from an orange in the laboratory (if we should want to). However, as will be explained in more detail subsequently, it does not follow from this that recombinant DNA methods can be used to construct oranges with wings or ducks that are orange and have seeds, or even something conceivably useful like duck *à l'orange.* What does follow is that this new technology can be used to insert a few genes from a duck or an orange into a bacterium and to grow them there, using the bacterium as a convenient production facility for these genes. This capability might well be useful if one were trying to understand at the molecular and genetic level how ducks (animals) differ from oranges (plants) or what the molecular events are in the developmental processes by which a fertilized duck egg becomes a duckling or how certain viruses are able to induce cancer in ducks.

The general case for the broad utility of recombinant DNA methodology in studying biological questions can be described as follows. Recombinant DNA methods can be used to join two pieces of DNA carrying genes that code for different but complementary functional properties in which one is interested. This new combination of genes has advantageous properties found in neither set alone. Usually the advantage is a technical one, in that the new combination of genes allows us to isolate some of them in quantities and purities previously unattainable. For instance, we may be studying a bacterial gene that codes for a particular enzyme. The fragment of the bacterial DNA molecule containing this gene will in general not be able to replicate itself. In the bacterium, where there is a single copy of the gene per chromosome (and only one chromosome per cell), the gene can be replicated only in concert with the thousands of other genes in the bacterial chromosome during chromosomal DNA replication prior to the division of the cell into two progeny cells. Using recombinant DNA methodology, it is possible to link the specific gene we wish to study to another small piece of DNA, also derived from bacteria and called a plasmid, which does encode the functions required for replication. This new recombinant DNA molecule can now be reintroduced into bacterial cells, where the entire molecule will replicate under the control of the replication functions encoded in part of it. Because of the special properties of plasmid DNA molecules, the gene in which we are interested will now be carried on a molecule that is present in a number of copies per cell instead of just one, and in a form that can be very easily separated from all the rest of the DNA in the cell. Moreover, methods exist by which the new recombinant DNA molecule can be induced to overreplicate, so that the cell will contain many hundreds of copies of it.

Thus, by joining together two different DNA molecules—one carrying a gene we wish to study, the other coding for a set of replication functions, and neither of great utility to us by itself—we can construct a new DNA molecule with highly advantageous properties. By introducing the new DNA molecule into an appropriate bacterium, we can use the bacterium and its progeny as factories to make large quantities of the desired gene and, in many cases, of the enzyme it codes for, easily and inexpensively.

Having seen why one might want to join two different sets of genes, let us consider how it is done. Even though recombinant DNA methodology involves manipulating submicroscopic bits of DNA by a series of enzymes of which few people have ever heard, the whole process is conceptually very simple. The reason is that the joining methods are straightforward consequences of the basic double helical structure of the DNA molecule itself, and it in turn is quite simple.

DNA molecules consist of two long strings of only four different kinds of basic building blocks or *bases*, abbreviated A, T, C, and G. The two strings are wound around each other into the famous DNA double helix discovered by James Watson, Francis Crick, and Maurice Wilkins in 1953. The bases in the two strands are always paired with each other in accordance with two rules: A pairs with T and G pairs with C. A gene is simply a long sequence of base pairs, a typical gene being about 1500 base pairs long. The information content of the gene is encoded in the sequence

$$\begin{matrix} A\text{--}G\text{--}C \\ T\text{--}C\text{--}G \end{matrix} \quad \text{meaning something different from} \quad \begin{matrix} A\text{--}C\text{--}G \\ T\text{--}G\text{--}C \end{matrix} \text{ or } \begin{matrix} G\text{--}A\text{--}C \\ C\text{--}T\text{--}G \end{matrix}$$

If one thinks of a DNA molecule as a piece of string about ⅛ inch in diameter, then a typical gene would be just about 20 inches long. This represents a scaling up from the actual size by a factor of one million. The simplest organisms, such as single-celled bacteria, typically have enough DNA for 3000 to 4000 genes. Looked at again on the scale of the piece of string, the amount of DNA in each bacterial cell is the equivalent of just over one mile of string. (On this same scale, the bacterium itself is a box three feet wide, three feet high, and six feet long.) Cells of complex higher organisms, such as man, have much more DNA per cell than do bacteria, although the size of the typical gene does not change significantly. In fact, each cell in the human body contains about 1000 times as much DNA as a bacterial cell, or enough for 3 to 4 million genes. This amount is equivalent to *1000 miles* on the scale of the string. To get some idea of the genetic complexity of a mammalian genome, imagine walking from New York City to St. Louis watching a piece of string stretched between the two cities and seeing a new gene every two feet of the way.

All this DNA, whether it comes from a bacterium or a buffalo, a man or a mouse, a duck or an orange, is chemically identical in that it consists of two long strings of bases in which A is always paired with T and G is always paired with C. Only in the sequence of base pairs do the DNAs differ. Because of the base-pairing rules, the two DNA strands are said to be complementary to one another. If we know the sequence of bases in one strand, we can immediately write down what the sequence of bases in the other strand must be. Thus, the sequence:

$$\cdots \text{--G--G--T--G--A--A--T--T--C--C--G--A--} \cdots$$

must be paired with the complementary sequence

$$\cdots \text{--C--C--A--C--T--T--A--A--G--G--C--T--} \cdots$$

to give the double helical sequence of base pairs

$$\cdots \text{--G--G--T--G--A--A--T--T--C--C--G--A--} \cdots$$
$$\cdots \text{--C--C--A--C--T--T--A--A--G--G--C--T--} \cdots$$

Moreover, if we take two single strands of bases whose sequences are complementary, such as those shown above, and mix them together in a test tube under appropriate conditions, they will spontaneously pair with one another according to the base-pairing rules to form a double helical DNA molecule. This reaction—the formation of a double helix from single strands of complementary base sequence—is the key to joining genes by recombinant DNA methodology.

Given that two single strands of DNA will form a double helix if their base sequences are complementary, it should be possible to link two different double-stranded DNA molecules by adding single-stranded "tails" to each, the sequence of bases in the tails on one molecule to be complementary to that in the tails on the other. By using an enzyme known as terminal transferase, we can add such tails to DNA molecules. Terminal transferase can add DNA bases one at a time to the ends of a double-stranded DNA molecule to give a molecule with a single-stranded tail at each end. If the enzyme is supplied with only one of the bases, say A, then the tails will be of the defined sequence:

A–A–A–A–A–A–A– · · ·

We can then take another DNA molecule and add the complementary tails to it by giving the enzyme only T's to add. These reactions are illustrated in Figure 1. When the two DNA molecules with the complementary single-stranded tails are mixed together under appropriate conditions, the complementary tails will form a double-stranded DNA

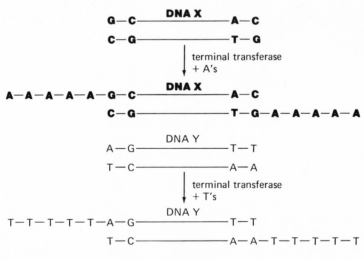

Figure 1

molecule that will hold **DNA X** and DNA Y together. The chemical linkages joining the two DNA molecules can be made precisely the same as in all other DNA molecules if the molecules linked through base pairing of their tails are treated with another enzyme called DNA ligase. DNA ligase causes the formation of a phosphodiester bond at a "nick" in a DNA strand; that is, it joins adjacent bases in the same strand by a covalent bond if they are not already joined by such a bond.

The sequence of reactions diagrammed in Figure 2 illustrates the formation of a recombinant DNA molecule from two different DNA molecules, **DNA X** and DNA Y, by the method just described. The two DNA molecules, having single-stranded tails of complementary sequences, are mixed in a test tube under appropriate conditions. The complementary tails interact through hydrogen bonds between the A and T bases to form two double-stranded A:T regions which link **DNA X** and DNA Y. However, these two DNA molecules are held together only by the relatively weak hydrogen bonds in the A:T regions, since there are still four nicks (shown by arrows) where adjacent bases in the same strand are not joined. DNA ligase is then added to the DNA mixture and forms strong covalent bonds at each nick, just like the bonds that link all the other bases in the DNA strands.

This last molecule is now a recombinant DNA molecule in which all the genes present in both of the parental DNA molecules, **X** and Y, are present in a single molecule. This molecule may look unusual, since it is shown as a circular structure. However, the biologically active form of many DNA molecules, including the chromosomes of bacteria and many viruses, is in fact circular. In particular, the biologically active form of all plasmid DNA molecules is circular, so that, far from being a disadvantage, circularity in a recombinant DNA molecule of this type is a necessity.

For reasons of simplicity, several enzymatic steps have been omitted from the preceding description of how a recombinant DNA molecule is made. The omitted steps are not essential to the conceptual framework of the method; rather, they are reactions used to take care of several technical problems inherent in the method or in the enzymes used.

Although this method is conceptually very simple, it is biochemically rather complex, requiring both a series of highly purified enzymes and a considerable degree of technical expertise in using them. It was thus a major advance in recombinant DNA methodology when an alternative and technically much simpler method for joining two different DNA molecules was developed. The specificity for allowing two DNA molecules to interact with one another still resides in the base-pairing rules for DNA and the interaction between complementary single-stranded tails on DNA molecules, but in the alternative method the tails

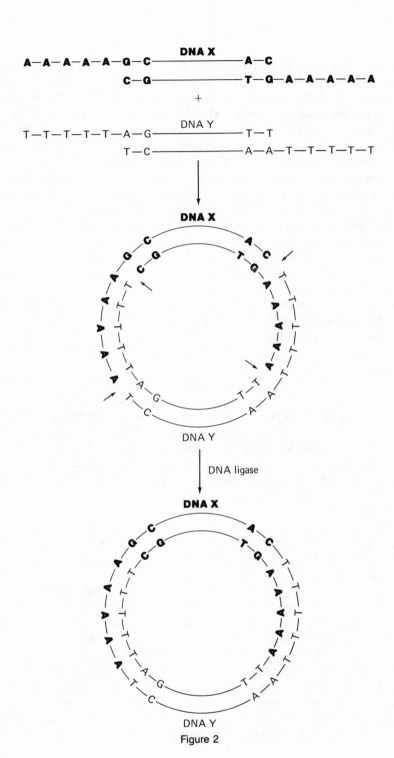

Figure 2

are generated in a way that greatly simplifies the biochemical complexity of the whole tailing and joining process.

The alternative method makes use of a class of enzymes, called restriction enzymes, which cut double-stranded DNA molecules in both strands wherever a specific sequence of base pairs appears in the DNA molecules. Many different bacterial species contain one or more restriction enzymes, and these different restriction enzymes recognize and cut different sequences of base pairs in DNA molecules. Any given restriction enzyme generally recognizes only one specific base-pair sequence, usually four to six base pairs in length, and cuts any DNA molecule containing that sequence wherever that sequence occurs. Over 200 restriction enzymes are now known.

One of the most widely used restriction enzymes comes from the common colon bacterium *Escherichia coli (E. coli)* and is called endonuclease R *Eco* RI. The *Eco* RI enzyme recognizes the base-pair sequence

$$\cdots \text{ –G–A–A–T–T–C– } \cdots$$
$$\cdots \text{ –C–T–T–A–A–G– } \cdots$$

wherever it occurs in a DNA molecule and cuts this sequence between the adjacent A and G in each strand, leaving each end of the cut with the sequence

$$\cdots \text{ –G}$$
$$\cdots \text{ –C–T–T–A–A.}$$

That is, the *Eco* RI enzyme makes a staggered cut in the two strands of the DNA molecule, leaving complementary single-stranded tails four bases long on the DNA molecule segments on both sides of the cut site. The reaction is diagrammed in Figure 3, where the *Eco* RI cleavage sites are indicated by vertical arrows.

These DNA molecules are now precisely analogous to the molecules with complementary single-stranded tails added by terminal transferase. Any two molecules cut by *Eco* RI will be able to interact with one

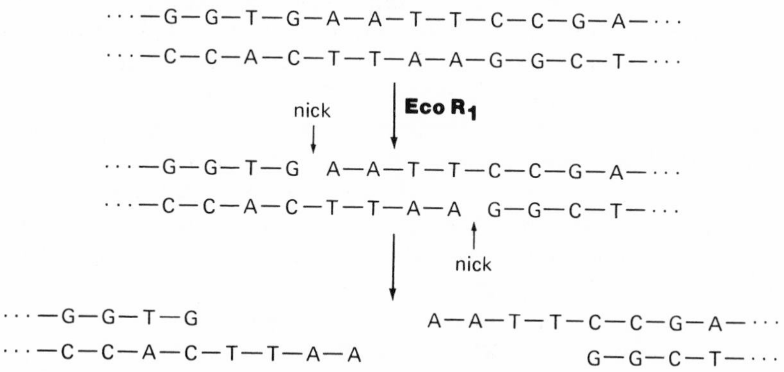

Figure 3

another at the short single-stranded *Eco* RI termini. If DNA ligase is added to the mix, the two molecules will be joined into one, as illustrated in Figure 4. Before incubation with DNA ligase, the two DNA molecules are held together only by the weak hydrogen bonds that hold the two members of a DNA base pair together. These bonds are more stable at low temperature, so the mixing of the two molecules to be joined at *Eco* RI termini and their incubation with ligase are usually done at temperatures below 16° C, whereas *Eco* RI cleavage reactions, where one wishes the molecular segments to separate, are usually performed at 37° C. Once DNA ligase has acted on the DNA molecules to join them together by covalent bonds between adjacent bases in each strand, the recombinant DNA molecule is as stable as any other DNA molecule.

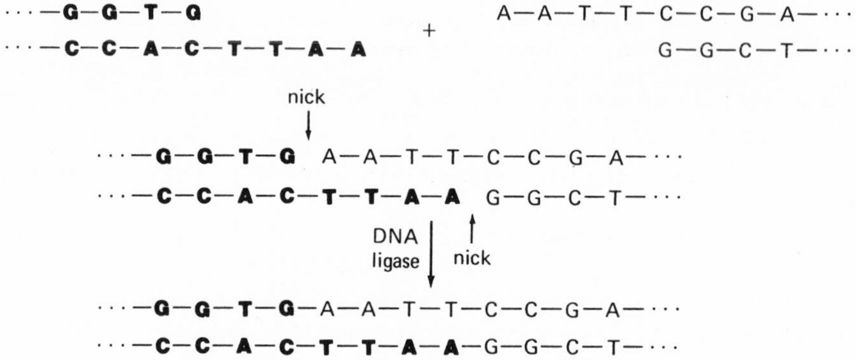

···—G—G—T—G   A—A—T—T—C—C—G—A—···
                                    +
···—C—C—A—C—T—T—A—A   G—G—C—T—···

                              nick
                               ↓
        ···—G—G—T—G A—A—T—T—C—C—G—A—···
        ···—C—C—A—C—T—T—A—A G—G—C—T—···

                     DNA | ↑
                     ligase | nick

        ···—G—G—T—G—A—A—T—T—C—C—G—A—···
        ···—C—C—A—C—T—T—A—A—G—G—C—T—···

Figure 4

Since DNA molecules from all species are chemically identical, they will all be cut by the *Eco* RI restriction enzyme wherever the sequence
G–A–A–T –T–C
C–T–T–A–A–G
appears, which will be on the average every 4100 base pairs. Since the sequences cut are identical and since they possess a plane of symmetry between the central A and T bases, every DNA fragment with an *Eco* RI terminus will be able to interact with every other DNA fragment cut by this enzyme, whether the DNA comes from ducks or oranges or mice or sweet peas. Thus, to join *some portion* of the genes of a duck to *some portion* of the genes of an orange, we need merely extract DNA from duck cells and from orange cells, cut the two DNAs with the *Eco* RI enzyme, mix the two populations of DNA fragments together in the presence of DNA ligase, and we will certainly construct recombinant DNA molecules, joining some duck genes to some orange genes.

In addition to the *Eco* RI enzyme, a variety of other restriction enzymes also make staggered cuts in DNA molecules but recognize a variety of different base-pair sequences. Using one or another of these enzymes, we can construct yet more different combinations of genes.

In the manner just described, it is thus relatively simple to make recombinant DNA molecules, given access to DNA, restriction enzymes, DNA ligase, some equipment, and the knowledge of how to use these enzymes. But are the duck-orange recombinant DNA molecules, made in this simple way, useful to us? Not very, it turns out.

Consider the complexity of what is occurring in the test tube containing the mixture of duck and orange DNA cut with *Eco* RI. First, there are roughly one billion base pairs of DNA in every cell from a duck and in every cell from an orange. (Essentially all cells in a given organism have the same number and sequence of base pairs.) Since *Eco* RI cuts on the average every 4000 base pairs, there will be 250,000 different duck DNA fragments and another 250,000 different orange DNA fragments. Essentially all of these fragments will have *Eco* RI termini at both ends, and thus each of these 500,000 fragments can either (a) circularize, by having the *Eco* RI terminus at one end of the fragment interact with the *Eco* RI terminus at the other end of that same fragment, or (b) interact with essentially equal probability with any one of the other 499,999 fragments in the mixture, or (c) interact with another copy of itself. Furthermore, these interactions are not limited to single ones. A dimer of two different DNA molecules, joined at an *Eco* RI terminus, still has *Eco* RI termini at each end and can participate in all the reactions described above for single fragments. It is thus clear that an extraordinarily complex mixture of recombinant DNA molecules will be obtained from this procedure, in which duck and orange chromosomes are cut up into tiny fragments and then randomly scrambled before some of them are joined together by DNA ligase. Given that chromosomal organization is crucially important for normal expression and regulation of genes, for replication of DNA, and for the viability of organisms, the specter that a DNA molecule for any sort of viable organism will come out of such a mixture is patently absurd.

There are, however, a variety of modifications to this simpleminded approach to make recombinant DNA molecules that *do* lead to useful products. One of the most important modifications is to use much less complex sources of genes to produce the desired recombinant DNA molecules. In this way the complexity of the recombinant mixture can be reduced, thus increasing the relative proportion of any given recombinant DNA molecule. Another modification is to use the method of putting single-stranded tails on the restriction enzyme fragments. If all duck DNA fragments have A tails and all orange DNA fragments have T tails, when these two populations are mixed all the molecules consisting of one or more fragments joined to some other fragment will of necessity be duck-orange recombinant molecules. No duck-duck or orange-orange fragment interactions are possible, nor can recircularization of any single fragment occur, because in these cases the tails are identical rather than complementary.

Perhaps the most useful and powerful modification of simply joining a mixture of genes together is to choose as one of the sets of genes a DNA molecule having the power of self-replication. DNA molecules can be replicated inside cells only when they are physically linked to special sets of genes coding for replication functions. These replication genes are relatively rare. For instance, the chromosomes of bacteria, which contain 3000 to 5000 total genes, contain only a single set of replication genes. However, many bacteria contain small circular DNA molecules called plasmids, which are capable of autonomous self-replication inside cells, each plasmid carrying a set of replication genes. Plasmids are generally from 0.1 to 2 to 3 percent the size of the bacterial chromosome. Many of them contain, in addition to the genes required for replication of their DNA, other genes whose presence confers some characteristic property (phenotype) on cells containing the plasmid. For instance, many plasmids carry one or several genes that confer resistance on the cell to one or several antibiotics that would normally kill it. That is, a plasmid containing a gene that codes for an enzyme that degrades penicillin will confer resistance to being killed by penicillin on any bacterial cell harboring that plasmid. Similarly, a plasmid carrying the genes responsible for the biosynthesis of the essential amino acid tryptophan will confer upon a bacterial cell unable to make its own tryptophan the ability to survive in a growth medium lacking tryptophan. Bacterial cells containing such plasmids can be *selected* from large populations of cells that lack the plasmid by growing the population under conditions in which the genes carried by the plasmid are essential for survival (e.g., in a medium containing penicillin or lacking tryptophan). Even if the plasmid-containing cells are very rare—only one cell out of every 10 million or more cells in the population—they can nonetheless be selected and grown up by these methods.

The characteristics of bacterial cell growth make it possible to grow billions of cells, each containing ten to 20 copies of a plasmid, in less than a day at a cost of a few cents. A typical bacterial cell grown under optimal conditions divides every 20 minutes, producing two identical daughter cells, which in turn divide in another 20 minutes. In this way, a single cell in a quart of growth medium can give rise to one trillion cells in a period of 14 or 15 hours. If the single starting cell contained a plasmid molecule, so will each of the progeny cells. In this manner it is possible to amplify a plasmid, including a recombinant plasmid, to essentially any degree desired in a rapid and inexpensive manner.

DNA molecules other than those of plasmids can also act as vectors for isolating and amplifying foreign genes. There is a class of viruses called temperate bacteriophage which grows on bacterial cells. The DNA molecules of these viruses have long been known to be capable of acquiring by recombination genes from the chromosomes of the cells they infect. The bacterial genes in these naturally occurring recombinant DNA molecules can then be transferred to the chromosomes of other bacterial cells, again by recombination. Recently, biochemical techniques similar to those used with plasmids have been developed to use the DNA molecules of these bacterial viruses as vectors for the isolation, amplification, and transfer of foreign genes.

Thus, the ability to couple a population of DNA molecules containing genes that one wishes to isolate to a vector molecule that has its own replication functions, and then to select for the rare cells containing the recombinant DNA molecule carrying the desired genes, makes it possible in one step both to overcome the enormous complexity of the initial population of recombinant DNA molecules and to produce essentially as much of the molecule of interest as is desired. It is, of course, necessary to be able to introduce recombinant DNA molecules that have been joined in the test tube back into cells so that they can reestablish themselves there as autonomously replicating plasmids. The conditions for doing this have been worked out, and for certain species of bacteria the procedure is now a routine one, albeit one that works only under highly artificial conditions in the laboratory.

Another useful technique for isolating a specific gene involves quite a different approach. The primary product of the vast majority of genes is a ribonucleic acid (RNA) molecule called a messenger RNA (mRNA). mRNA molecules are single-stranded and contain the bases A, G, and C, plus U (uracil), which is substituted for T. An mRNA molecule synthesized from a gene has the same sequence of bases in it as does one of the strands of the DNA of that gene (except that U is substituted for T), and its sequence is thus complementary to that of the other DNA strand. In certain highly differentiated cells of higher organisms—e.g., the cells making hemoglobin or antibodies or muscle protein—the mRNAs for these proteins make up 10 to 30 percent of the total mRNA in the cell, even though the gene from which they are synthesized is still only one part in one million of the total DNA. Because these specific mRNAs are such a large fraction of the total, they can be purified biochemically by conventional techniques, whereas the gene cannot. The ability to isolate mRNAs from specific genes is extremely useful, because there is an enzyme called reverse transcriptase which will synthesize DNA from RNA. Therefore, if one can isolate the mRNA from a particular gene, it is possible to reconstruct the DNA sequence of a large portion of the gene using reverse transcriptase. This DNA can be inserted into a plasmid by standard recombinant DNA methods and then amplified by growing the recombinant DNA in bacteria. This method has already been used to isolate and amplify portions of a variety of genes from higher organisms, in which the complexity of the genome is such that direct isolation of specific genes is not feasible.

It should be clear from the foregoing discussion that recombinant DNA methodology represents a real technical breakthrough in being able to isolate and make large quantities of a variety of genes from a variety of sources. It should be equally clear, however, that at present there are very substantial limitations on both the kinds of genes and the sources from which they can be isolated. It is possible to isolate quite a wide variety of different genes from bacterial sources, because (a) bacterial genomes are only moderately complex and (b) recombinant DNA molecules containing specific bacterial genes can be selected for in bacterial cells by requiring that the function coded for by the gene be essential for the survival of the cell. In contrast, the number of specific, functionally defined genes from higher organisms that can be isolated using presently available methods is quite small, being with a few exceptions limited to those for which an RNA copy of the gene can be isolated by conventional techniques. This is because (a) the genomes of higher organisms are very complex, (b) it has not yet proved possible in general to select for the genes of higher organisms in bacteria because they do not seem to be expressed properly there, and (c) it is not yet possible to select for the genes of higher organisms in the cells of higher organisms because the variety of mutants and the techniques available for selecting for the insertion of particular genes are much less well developed than they are for bacterial cells. So, although it is quite easy to isolate and amplify random bits of DNA from the genomes of higher organisms, it is still very difficult either (a) to identify what genes are carried on one of these randomly selected bits or (b) to select only those recombinant DNA molecules that contain some specific gene.

It should also be clear that recombinant DNA methodology is not a way of constructing fundamentally new forms of life in the test tube. The simplest known microorganisms contain several thousand genes, and the largest recombinant DNA molecules that it is practical to construct and work with transfer a few tens of genes into a new environment. The maximum alteration of even the simplest genome to be expected is of the order of one percent. Moreover, the genomes of organisms are not simply grab-bags of genes, in which any physical arrangement of genes will work so long as all the proper ones are present somewhere, any more than a pile of parts from an automobile engine will power a car. On the contrary, the genomes of even the simplest bacterial cells are highly structured, highly ordered aggregates of genes whose interactive functioning has been subject to the pressures of natural selection for millions of years. The probability that one could make this finely tuned biological mechanism work better by randomly inserting a few genes from an unrelated foreign organism is rather like the probability that one could walk into a large hardware store, randomly select a nut or a bolt or a screw from the thousands of kinds found there, and then improve the performance of a fine Swiss watch by forcing that piece of hardware inside its case.

Most of the current applications of recombinant DNA methodology are in basic research studies involved with efforts to understand the way the genetic material of cells is organized, reproduced, and expressed. Understanding these processes is an essential prerequisite to any fundamental understanding of the normal growth and development of organisms and of the abnormal states we call diseases.

The ability to isolate and amplify a particular gene easily and inexpensively is merely extremely helpful for the scientist studying a bacterial gene, because there are other very time-consuming, cumbersome, and expensive ways in which it can be done. But for someone trying to study a particular gene from a higher organism, let us say a mouse, being able to isolate and amplify a gene is the difference between being able to study it in molecular detail and thinking wistfully about how nice it would be to be able to do so. With extremely rare exceptions, there is simply no practical way except by using recombinant DNA methodology to study the genes of higher organisms at the molecular level. If we wished to isolate a milligram (0.000035 ounce) of a particular gene from a mouse, we would have to start with the total DNA from about 300,000 adult mice. We would somehow have to identify and purify this one milligram of the one gene we wanted from the approximately ten pounds of DNA we would extract from these mice, and we would have to do it in 100 percent yield. This is simply not possible now or in the foreseeable future. On the other hand, if we had our mouse gene linked to an appropriate plasmid vector, we could easily produce a milligram of it, in a form that could be readily purified, by growing a few quarts of bacteria containing the recombinant plasmid overnight at a cost of a dollar or so. The actual purification of the gene would take another two or three person-days of work.

In the short time since recombinant DNA technology was first developed, it has already become an important and widely used methodology in biomedical research. As of February 1977 nearly two hundred different major research projects in the United States involved recombinant DNA techniques. The range of problems being studied is already extensive and is continuing to broaden. The techniques are being used to study the molecular mechanisms of differentiation in organisms ranging from single-celled amoebae to fruit flies to frogs to mice. They are being used to study a variety of viruses that infect bacteria, plants, and animals, including viruses that are capable of inducing tumors in mice and which thus serve as a model system for investigating the viral etiology of cancer. Studies of a wide range of problems involving the molecular genetics and regulation of gene expression in bacteria have been significantly expedited by utilization of recombinant DNA methods. The techniques are being used in studies of how nitrogen fixation and photosynthesis occur in plants. Both these processes, of course, are very important in allowing plants to grow and produce food. Recombinant DNA methods are being used to study a rapidly expanding list of genes from higher animals, including man. The list currently includes the genes for hemoglobin, insulin, interferon (a broad-spectrum antiviral protein produced by cells of higher animals), antibodies, ovalbumin (the major protein in egg whites), ribosomal RNA (ribosomes are the components of the cell that perform protein synthesis), fibroin (the major protein in silk), histones (the primary structural proteins of chromosomes), and a variety of proteins from mitochondria and chloroplasts (the cellular organelles that are responsible, respectively, for energy metabolism and photosynthesis). Another important area of investigation that has greatly benefited from these new methods is studies of the nature of the plasmids that cause bacteria to become resistant to a wide range of clinically useful antibiotics, and of the mechanisms by which these plasmids spread in nature and by which they make the bacterial cells resistant. In other experiments, these same methods are being employed in efforts to increase the yield of antibiotics produced by bacterial fermentations.

It should be clear from the foregoing list that recombinant DNA methodology has extremely wide applicability to research throughout the biomedical sciences. There are two primary reasons for this. The first is self-evident: these techniques allow us to manipulate in a controlled way the genetic material of all organisms, and it is in this genetic material that all the information for the growth, development, organization, and reproduction of the organism is encoded. Any technique that helps us investigate the genetic material will thus be of very wide applicability. The second reason may not be so apparent but is no less important. It is that recombinant DNA methodology is the first reductionist approach to be applied successfully to the analysis of the genomes of higher organisms. As we have seen, these genomes are so enormously complex that it is virtually impossible to study any single part of them in detail. Recombinant DNA methods allow us to break down these complex genomes into manageable-sized bits of genetic information, and then to make enough of the bits we wish to study so that the powerful methods of biochemistry and molecular biology can be used to analyze them.

For the past thirty years, molecular biologists have been studying bacteria and their viruses with the hope that if they could understand the functioning of E. *coli*—which they could study with the techniques available—they would have also learned something about how higher organisms function. These scientists have been spectacularly successful in understanding E. *coli* (and other microorganisms), and they have in fact learned much about the basic biochemistry and molecular organization of higher organisms at the same time. But higher organisms are vastly more complex than E. *coli*, and it has become clear that one cannot understand them simply by understanding everything there is to know about E. *coli*. One must at some point begin to study them with the same kinds of techniques and in the same degree of detail that bacteria can be studied. The advent of recombinant DNA methodology makes this beginning possible and assures that it will come to fruition, rather than remaining a frustrating and unreachable goal.

Recombinant DNA:
Areas of Potential Applications

A. M. Chakrabarty

The advent and development of recombinant DNA has been portrayed as very much a mixed blessing for mankind. While proponents have hailed it as a source of technology that will someday solve many of the problems of environmental pollution, food and energy shortages, and human diseases, including inborn genetic disorders, its opponents have bitterly criticized research in this area because of the possibilities for accidental development and release of highly virulent forms of infective agents that may lead to epidemic diseases of unknown proportions. The wisdom of developing bacterial strains capable of expressing genetic segments of eukaryotes has also been questioned. Each view is supported by major groups of scientists.

The basic problem in reconciling such sharply divergent viewpoints is that in most cases both the benefits and the biohazards that have been ascribed to the development of recombinant DNA technology are highly speculative. Today, we are still far from any demonstrated success at producing antibodies or blood-clotting factors by fermentation, or making plants fix their own nitrogen because of insertion of bacterial nitrogen-fixation genes. Conversely, however, the suggestions of E. coli harboring human cancer or cancer virus genes or (2) developing highly virulent traits because of accidental introduction of uncharacterized DNA segments appear equally far-fetched, speculative, and exaggerated.

This article has two purposes. First, I should like to list and discuss some of the major areas where beneficial applications of recombinant

DNA technology are envisioned, whether in the near or in the more distant future. Second, I should like to note certain of those areas where both the barriers to technological success, and the potential hazards attendant upon such success, appear to be of less formidable proportions. I believe that we may harvest some tangible benefits within a relatively short time, and with little risk, if we develop the new technology in a meaningful way.

Apart from scientific applications toward the greater understanding of the nature and mode of regulation of eukaryotic genes, several broad areas of application of recombinant DNA technology are recognized in industry, agriculture, and medicine. Briefly stated, these are (1) in the manufacture of drugs, chemicals, and fuels—specifically, polypeptide hormones, vaccines, enzymes, and low-cost fermentation products such as solvents, alcohol, methane; (2) in the improvement of crop plants and crop yields, both by the extension of existing cross-breeding technologies and by the incorporation of nitrogen-fixation genes into either the crop plants themselves or their normal microbial symbionts; and (3) in the treatment of genetic disease, by deliberately introducing fragments of functional eukaryotic or prokaryotic genes into the cells of the human patients.

DRUG, CHEMICAL, AND FUEL MANUFACTURING

The simple technique of (1) joining segments of DNA derived from prokaryotes or eukaryotes with bacterial phage or plasmid DNA and (2) their subsequent introduction into bacterial cells may mark the beginning of a revolutionary way of manufacturing biologically functional proteins such as interferons, antibodies, blood-clotting factors, insulin, growth hormones, and a host of other pharmacologically important compounds by fermentation. Since production of these substances is stringently regulated in the animal cells, they are available from animal tissues only in very small amounts. Hence, their fermentative production would make them available in large quantities at modest cost and thus would contribute significantly to the betterment of human health and welfare.

Interferons are glycoproteins, occurring as monomers or polymers of a basic unit with a molecular weight of 12,000 to 20,000. Their formation is induced in various animal cells upon infection by any of a large variety of viruses or certain other materials such as intracellular parasites, endotoxins, or $poly(I) \cdot poly(C)$. The released interferons can attach to other cells, thereby making them resistant to subsequent viral infections. The viral inhibitory effect of interferon is presumably at the level of translation of viral RNA, although some studies indicate an effect on the transcription of the viral genomes. The inhibitory effect of interferons on dividing cells, including some tumor cells, is of great importance in eradicating chronic infection by limiting the infectivity of carriers. In this regard, interferons appear to be more effective for prophylaxis than for combating well-established viral infections. They have been shown to be directly responsible in suppressing the spread of a number of diseases including hepatitis B (8). Essentially the same type of resistance against any particular viral infection can be developed by elevating the level of its specific antibodies in the blood stream. It is apparent that cloning of genes specifying interferons or antibodies in bacteria so as to make these important proteins available in sufficient quantity for general medical use could have a tremendous impact on our ability to control the spread of viral diseases.

Similarly, fermentative production of blood-clotting factors such as antihemophilic factor or von Willebrand factor (presently available only in small quantities by fractionation of human blood) or basic regulatory hormones such as growth hormone (presently available only from pituitary glands) would greatly help in the treatment of patients having reduced levels of these critical proteins.

Another interesting area would be the manufacture of pharmacologically important pain-killing compounds by cloning the genes specifying the natural opiate peptides. Pentapeptides with opiate agonist activity have been isolated from porcine brains (11), and a number of such opiate peptides, 5 to 31 amino acid residues long, as well as β-melanocyte-stimulating hormone, apparently derived by proteolysis of a common precursor (β-lipotropin), have been isolated from pituitary glands (13). Recently another natural opiate, β-endorphin, reportedly 20 to 40 times more effective in pain-killing action than morphine, has been isolated from the same gland. The opiate-receptor complex is believed to be a potent inhibitor of adenylate cyclase and thus can suppress the activation of adenylate cyclase by neurotransmitters and hormones. Prolonged exposure of cells to the opiate peptides often leads to an increase in cellular adenylate cyclase activity to compensate for its inhibition by the opiate peptides. Thus, such cells can maintain normal $3':5'$ cAMP levels and appear tolerant to opiate peptides because the increase in adenylate cyclase activity is approximately equal to the inhibition. The cells eventually become dependent upon the drugs to maintain normal cAMP levels, because withdrawal of the drugs reverses the inhibition, leading to the synthesis of abnormally high levels of cAMP (12). This dual regulation of adenylate cyclase by such narcotics accounts for the phenomena of narcotic dependence and tolerance.

The cloning of genes leading to the fermentative production of such analgesic drugs would not only provide us with large quantities of pain killers, but would also lead to a better understanding of narcotic addiction and mental disorders. By subjecting the genetic segment coding for such opiate peptides within the bacterial cells to mutagenesis, we might also be able to isolate highly active analgesic derivatives that might be less addictive because of their high biological potency. Similar improvements in the biological activities of a number of other therapeutically important proteins could presumably also be achieved by recombinant DNA technology.

Another potential area of application appears to be in the manufacture of vaccines. A number of human diseases such as hepatitis, infectious mononucleosis, Burkitt's lymphoma, measles, poliomyelitis, influenza, and the common cold are caused by viruses. The sequence of genes on the chromosome of many of these viruses can now be ascertained by a combination of physical and genetic methods (18). Undoubtedly, with the perfection of such techniques, the DNA fragments coding for their coat proteins will be identified and isolated. The *in vitro* recombination of such DNA fragments with vector plasmids or phages, and their subsequent introduction into bacterial cells, would enable us to prepare these coat proteins by fermentation in large quantities. The use of antigenic viral coat proteins as vaccines would not only enable us to make vaccines much more cheaply, but it would essentially eliminate the present practice of growing large quantities of viruses by the expensive and difficult process of cell cultures. If the genes specifying viral coat proteins are clustered on a portion of the viral chromosome, and are located on a single fragment after digestion with a restriction endonuclease, and if such a fragment does not possess some essential genes for the assembly or production of intact virus particles, then such techniques would also provide a safe and hazard-free way of making vaccines.

Recombinant DNA technology may also permit both the more efficient production of existing antibiotics and also the development of newer, hybrid antibiotics. Although genes specifying the formation of some antibiotics are known to be chromosomal, evidence is gradually accumulating that suggests that genes coding for others might be plasmid-borne (21). Manipulation of such antibiotic-specifying plasmids in order to increase the copy number within the *Actinomyces* cells would be most helpful in the large-scale manufacture of such antibiotics. The use of such plasmids as vectors in recombinant DNA experiments, using other *Actinomyces* cells as hosts, might also allow the incorporation of chromosomal antibiotic biosynthetic genes onto such plasmids, and this could be one way to introduce the genes responsible for the biosynthesis of several antibiotics into a single cell. It would be interesting to find out if chemical separation and subsequent purification of individual antibiotics, obtained from a single fermentation with such multiplasmid *Actinomyces*, would, in fact, lead to a reduction in the cost of fermentative production of antibiotics. It might also be possible to introduce genes coding for portions of biosynthetic pathways of structurally analogous antibiotics so as to produce new hybrid antibiotics having wider spectra and fewer undesirable side effects than those of the individual parent compounds.

A new direction in recombinant DNA research is the functional expression of synthetic genes. Thus, using phage as a vector, a chemically synthesized gene for tyrosine transfer RNA has been cloned in *E. coli*, where it is fully functional. Similarly, a chemically synthesized duplex DNA fragment containing the sequence of the *lac* operator has been cloned in *E. coli* and shown to be perfectly functional in binding the *lac* repressor specifically (10). The use of chemically synthesized genes is a highly versatile technique, for it allows us to make changes in the sequence of genes at will and to study in detail their effect on the functional aspects of such a gene. Specific changes in the regulatory sequences of such a gene will make it possible to learn the varied mechanisms of gene expression and their control. It is clear that where the sequence of deoxyribonucleotides is known, the combination of chemical synthesis of genes specifying useful protein products and their cloning in bacteria will provide a versatile, novel, and powerful technique for the manufacture of therapeutically important biological compounds.

Turning from pharmaceutical manufacturing to bulk materials processing, recombinant DNA technology is believed to have a particularly intriguing potential for the manufacture of methane, fuel-grade ethanol, or inexpensive chemicals from plant wastes. Methane is normally produced by the anaerobic decomposition of carbohydrates, fats, or proteins, by several groups of microorganisms. One group, called "acid formers," converts these substrates to low-molecular-weight organic acids such as acetate, formate, propionate, etc. Another group, called "methane formers," can convert CO_2 and H_2, and sometimes acetate or formate, to methane. The methane formers are strict anaerobes and are extremely sensitive to small amounts of oxygen. Because of their sensitivity toward oxygen, small changes of pH or metal-ion concentration, and their very narrow substrate specificity, the methane formers are very difficult to cultivate. This constitutes a major handicap in the genetic improvement of methane production. Recombinant DNA technology may allow us to introduce methane-forming genes from the methane formers to other cellulolytic anaerobes and thereby facilitate the direct and rapid conversion of cellulosic waste to methane. Similarly, introduction of cellulose-degrading genes from *Trichoderma viride* (14) to yeasts using vectors such as II DNA (9), or alternatively, introduction of ethanol-forming genes from yeasts to suitable cellulolytic anaerobes, might permit the practical production of ethanol from cellulosic wastes. Fermentation processes for the production of amino acids, acetone, butanol, and acetic acid from low-cost cellulosic wastes might also be developed.

It must be stressed that there are still some very basic unresolved problems in many of these potential industrial applications of novel, genetically improved microorganisms. Although we have been able to introduce eukaryotic DNA segments into bacteria, so far there is little evidence that such genes can be functionally expressed therein. Indeed, excepting one or two cases where some genes from yeasts have been reported to be expressed in *E. coli* (19), eukaryotic genes have generally been shown not to be expressed accurately for the synthesis of functional proteins in bacteria (4). The inability of prokaryotes to express eukaryotic genes may arise from a single type of difficulty, such as a difference in a transcription or translation signal, or may involve a number of steps including posttranslational events. If and when it becomes possible to overcome these barriers, the biohazards to people accidentally infected with *E. coli* that produce insulin or other hormones, or that have cellulolytic activities (3), must be contended with. In this regard it seems to me that there are two basic problems inherent in developing an acceptable recombinant DNA technology. One is the use of *E. coli* as a host, and the other is the wisdom of introducing non-characterized segments of eukaryotic DNA into such a host. Certainly, with ongoing research efforts, bacterial strains incapable of thriving in animal or plant systems (such as psychrophilic or thermophilic strains) could and should be developed as hosts for recombinant DNA research.

Even in absence of such hosts, however, and without introducing plant or animal DNA, but only that of fungi or prokaryotes, much low-risk research involving recombinant DNA can still be pursued, and with reasonable expectations that it actually will lead to novel organisms capable of producing drugs, bulk chemicals, or nonfossil fuels. The previously mentioned opportunities for improvements in methane production, direct conversion of cellulosic wastes to alcohol, and expression of synthetic genes could all be pursued without ever having to introduce a plant, animal, or viral gene into a common bacterium. In addition, many *Actinomyces* and fungi produce solvents, steroids, vitamins, and enzymes that have wide commercial uses. Since actinomycetes and fungi grow rather slowly, and genetic manipulations leading to an enhancement of production of such compounds in these microorganisms are difficult to achieve, recombinant DNA technology for introducing genes specifying such products into bacterial strains having rapid growth rates and simple nutritional requirements would certainly be useful. If such genes were cloned as part of a vector that had a relaxed mode of replication, it should be possible to increase the product yields severalfold. Indeed, it is quite possible that the recombinant DNA technology would, at a future time, enable us to make most of the products presently derived from the broth of slow-growing actinomycetes or fungi by fermentation with selected bacterial species capable of growing rapidly on cheap, simple substrates.

Most of the new, genetically modified bacterial strains designed for these industrial applications would not have any particular survival advantage in the general environment. However, any such strains having selective growth advantages, e.g., those with nitrogen-fixation or photosynthetic genes along with cellulolytic activity, might still present possible ecological problems despite the absence of plant, animal, or pathogenic DNA segments. I have previously alluded to such potential biohazards in connection with bacterial strains designed for oil spill clean-up, enhanced cellulolytic activity, or study of the functional expression of bacterial *nif* genes in plants (1). Definitive tests to resolve questions of long-range potential for biohazards, ecological compatibility, and selective survival would have to be conducted before any such strains were deemed suitable for use in industrial production.

CROP IMPROVEMENT

It has been long recognized that there are just two basic approaches to improving the agricultural productivity of a given crop plant in a given environment. One is to improve the breed of the plant, by whatever genetic modification techniques are available, and the other is to add fertilizer. Today, the promise of recombinant DNA technology offers hope of accelerated progress with both approaches.

Traditionally, the plant breeder has been required to draw his genetic material from plant species that were sufficiently closely related to permit hybridization by conventional crossbreeding. Recombinant DNA technologies, coupled with known techniques for growing entire plants from tissue culture cells, should eventually remove this restriction, making possible the development of new crop plants that would represent, in effect, hybrids of very different plant species, or even of plants and nonplants. At present, however, we still know very little about the problem of achieving expression of foreign DNA in plant cells, so that the ultimate limitations of such hybridization techniques are unknown.

The conventional routes of providing nitrogen fertilizer to plants are either to add petroleum-based synthetic nitrogen compounds or to use plant species having associated or symbiotic microflora that are capable of fixing atmospheric nitrogen. There would appear to be several ways to improve upon such practices. One would be to introduce the nitrogen fixation *(nif)* genes from the nitrogen-fixing bacteria that normally inhabit the roots of legumes to other members of soil microflora that are incapable of fixing nitrogen. The other would be to introduce such *nif* genes directly into plant cells to allow the plants to fix their own nitrogen.

In this connection it must be noted that, in addition to the present lack of suitable recombinant DNA techniques for introducing foreign DNA segments into various soil bacteria or plant cells, there are also two basic biochemical problems. One is that nitrogen fixation is a highly energy-consuming process, so that cells carrying it out to a significant extent should be able to produce large amounts of ATP. The other is that all isolated nitrogenase enzymes, whether obtained from facultative anaerobes such as *Klebsiella pneumoniae* or from strict aerobes such as *Azotobacter vinelandii*, are oxygen-sensitive; they do not function at all in presence of oxygen. The same sensitivity also appears to exist for nitrogenase, a key enzyme in nitrogen fixation, within aerobic cultures other than *Azotobacter*. For example, we have observed that the aerobic soil microorganism *P. putida*, when transformed into a His^+ and presumably Nif^+ state by a drug-resistance plasmid, RP41 (6), which harbors *Klebsiella his* and *nif* genes, shows very little nitrogenase activity. Since many of the predominant members of soil microflora are strictly aerobic, mere introduction of functional *nif* genes may not allow such cells to actually fix nitrogen for the plants. Likewise, plant cells themselves normally have oxygen-rich cellular environments, just like cells of *P. putida*, and may not be able to carry out nitrogen fixation unless some anaerobic microenvironment can be identified. Alternatively, it may be possible to identify mutants of nitrogen-fixing bacteria that would produce an oxygen-tolerant nitrogenase, although so far all attempts to isolate such a mutant have failed. Evidently, any solution to the nitrogen fertilizer problem involving use of recombinant DNA technologies to introduce *nif* genes into additional species of bacteria or plants will require resolution of this oxygen-sensitivity problem.

TREATMENT OF GENETIC DISEASES

Although this chapter has been concerned primarily with potential technological applications of recombinant DNA, brief mention must be made of potential medical application as well, particularly in the treatment of genetic diseases, since some of these applications are closely interrelated.

In recent years there has been tremendous progress in our understanding of the specific molecular mechanisms underlying inherited genetic disorders (15). With advancement in our knowledge of genetic engineering technology, it should become possible to complement many of the mutational sites on human chromosomes by insertion of functional genes, or to supplement the defective protein products resulting from such mutations so as to restore the original enzymatic function, and thereby relieve human suffering. There is always, of course, the danger of inadvertent introduction of new genetic traits, resulting in an alteration of human behavior in unpredictable ways, so that any application of recombinant DNA techniques in human genetic medicine must be preceded by extensive discussion and analysis of the relevant ethical, technical, and moral issues. Only when it is convincingly demonstrated that a particular modification of a human gene poses no danger, short- or long-term, to the patient and may indeed ameliorate suffering because of restoration of particular enzymatic functions, may genetic engineering techniques be successfully applied in human gene therapy.

With that proviso, let us note just a few areas where recombinant DNA technology might find useful applications in the treatment of human genetic disorders (7). One such area appears to include the human storage diseases, such as Hunter's and Hurler's syndromes and metachromatic leukodystrophy, where accumulation of toxic products is caused by a mutation in one of the genes involved in the pathway. We previously noted the potential usefulness of the recombinant DNA technology in the production of pure enzymes of animal origin by bacterial fermentation. Availability of such purified enzymes and development of methods whereby missing enzyme activities within animal cells would be complemented by the external application of purified enzymes would be most valuable in the treatment of such diseases. It has already been demonstrated that semipermeable microcapsules containing catalase can be used to counteract catalase deficiency in acatalasaemic mice without allowing the enzymes to leak out and participate in immunological reactions (5). Similarly, methods have recently been developed that allow the introduction of exogenous pure enzymes, by means of immunoglobulin-coated liposomes, into lyzosomes of genetically deficient cells (20). Needless to say, improved methods will continually be developed that will allow the external application of purified enzymes as therapeutic agents. Likewise, as noted earlier, recombinant DNA technology also holds promise for large-scale manufacturing of such therapeutically important components as blood-clotting factors or insulin, which could be similarly supplied to people suffering from hemophilia or diabetes.

Apart from the external application of pure enzymes, recombinant DNA technology may also enable us to complement the mutant genes with their functional counterparts. A number of disorders involving amino acid and carbohydrate metabolism, such as phenylketonuria and galactosemia, are known to be due to faulty enzymes, specified by mutant genes. Present-day treatment is limited to restricting the supply of the precursor amino acids or carbohydrates (e.g., phenylalanine or galactose) in these patients' diet. It is conceivable that progress in recombinant DNA technology will allow us to insert functional genetic fragments specifying phenylalanine metabolism from healthy individuals to selected fragments of viruses such as adenovirus or SV40. Recent evidence indicates that it is possible to isolate fragments of viral DNA that can transform the mammalian cells but lack the otherwise essential genes for the production of intact viral particles. *In vitro* recombination of functional phenylalanine-metabolizing genes with such a fragment and their subsequent chromosomal integration with various target fetus cells may some day help to permanently cure patients suffering from these genetic diseases.

Finally, it is believed that the recombinant DNA technology holds promise for curing autosomal recessive diseases such as sickle cell anemia. There is at present only symptomatic treatment against such diseases, and the chance for finding a permanent cure in the foreseeable future is indeed dim. The primary mechanism underlying sickle cell anemia is the replacement of a single amino acid among the 287 residues in the half molecule of hemoglobin; this represents a particularly clear example of a genetic mutation in the human chromosome that results in a defective protein. The single amino acid substitution is harmless in heterozygotes carrying both the normal and the mutant gene, but it results in pulmonary emphysema in homozygotes carrying defective genes. The change in structure of the hemoglobin molecules because of the amino acid replacement in homozygotes reduces the solubility of hemoglobin in the red blood cells, so that the venous form of hemoglobin actually crystallizes out (16). Practical methods to prevent this crystallization have so far proven to be inadequate. However, recombinant DNA technology has recently been used to introduce DNA copies of purified rabbit globin mRNA into bacteria (17). Further developments of this technique might in the future enable us to introduce a functional form of the gene coding for the beta chain of hemoglobin inside the bone marrow cells of patients suffering from sickle cell anemia, using, perhaps, a transforming viral DNA fragment as vector. Extensive bone marrow transplantation would then allow synthesis of functional hemoglobin molecules, thereby affording substantial relief from this dreaded disease. Such "genetic engineering" techniques, however, would have to await perfection of other techniques for handling functional expression of foreign genes in bone marrow cells, and possibly transplant rejection. It is nevertheless not unreasonable to expect that judicious use of recombinant DNA technology will indeed enable us someday to achieve permanent cures for a number of human genetic disorders for which no good method of treatment now exists.

REFERENCES

1. Anonymous (1976) *Chem. Week,* May 12, p. 65.
2. Cavalieri, L. F. (1976) *The New York Times Magazine,* August 22, p. 8.
3. Chakrabarty, A. M. (1976) *Indus. Res.,* **18,** 45.
4. Chang, A. C. Y., Lansman, R. A. Clayton, D. A., and Cohen, S. N. (1975) *Cell,* **6,** 231.
5. Chang, T., and Poznansky, M. (1968) *Nature,* **218,** 243.
6. Dixon, R., Cannon, F., and Kondorosi, A. (1976) *Nature,* **260,** 268.
7. Friedmann, T. (1976), *Ann. NY Acad. Sci.,* **265,** 141.
8. Greenberg, H. B., Pollard, R. B., Lutwick, L. I., Gregory, P. B., Robinson, W. S., and Merigan, T. C. (1976), *New Eng. J. Med.,* **295,** 517.
9. Guerineau, M., Grandchamp, C., and Slonimski, P. P. (1976) *Proc. Natl. Acad. Sci. USA,* **73,** 3030.
10. Heyneker, H. L., Shine, J., Goodman, H. M., Boyer, H. W., Rosenberg, J., Dickerson, R. E., Narang, S. A., Itakura, K., Lin, S., and Riggs, A. D. (1976) *Nature,* **263,** 748.
11. Hughes, J., Smith, T. W., Kosterlitz, H. W., Fothergill, L. A., Morgan, B. A., and Morris, H. R. (1975) *Nature,* **258,** 577.
12. Lampert, A., Nirenberg, M., and Klee, W. A. (1976) *Proc. Natl. Acad. Sci. USA,* **73,** 3165.
13. Li, C. H., and Chung, D. (1976) *Proc. Natl. Acad. Sci. USA,* **73,** 1145.
14. Mandels, M. (1975) *Biotechnol. Bioeng. Symp.,* **5,** 81.
15. Mekusick, V. (1975) *Mendelian Inheritance in Man.* Johns Hopkins Press, Baltimore.
16. Perutz, M. F. (1976) *Nature,* **262,** 449.
17. Rabbitts, T. H. (1976) *Nature* **260,** 221.
18. Sambrook, J., Williams, J., Sharp, P. A., and Grodzicker, T. (1976) *J. Mol. Biol.* **97,** 369.
19. Struhl, K., Cameron, J. R., and Davis, R. W. (1976) *Proc. Natl. Acad. Sci. USA,* **73,** 1471.
20. Weissman, G., Bloomgarden, D., Kaplan, R., Cohen, C., Hoffstein, S., Collins, T., Gotlieb, A., and Nagle, D. (1975) *Proc. Natl. Acad. Sci. USA,* **72,** 88.
21. Wright, L. F., and Hopwood, D. A. (1976) *J. Gen. Microbiol.,* **95,** 96.

Acknowledgements

The research in the author's laboratory has been supported in part by a grant from the National Science Foundation (BMS75-10978). I am indebted to Drs. John F. Brown, Jr., and Ronald E. Brooks for critically reading this manuscript and for suggesting many constructive changes.

Biological Containment:
The Construction
of Safer *E. coli* Strains

Roy Curtiss III

Editors' note: The subject of biological containment—the means by which safety factors may be built into the genetic structure of the organisms to be studied—has become one of central importance in the debate about whether some forms of recombinant DNA research can be performed with adequate safety. A description of the construction and characterization of a biologically contained derivative of *Escherichia coli* K-12, the bacterial species used for virtually all experiments in which foreign genes are grown in bacteria, necessarily involves discussion of details of genetics and microbiology not familiar to the general reader. In this chapter Dr. Roy Curtiss III, one of the leaders in designing and producing biologically contained bacterial strains, describes the properties of χ1776, the first bacterial strain to be certified as an EK2 host cell by the National Institutes of Health. (An EK2 cell is one that is at least 100 million times less able to survive in nature outside an artificial laboratory environment than is normal *E. coli* K-12.)

The rationale behind Dr. Curtiss' approach to constructing an EK2 derivative of *E. coli* K-12 embodies a number of fundamental facts of genetics and microbiology. It will be useful to consider briefly some of these facts. First, a central fact of bacterial genetics is that when a cell divides to form two progeny, each of the progeny is genetically identical to the parental cell. Continuation of this process leads to populations of millions or billions of genetically identical cells. Thus, from one single cell having desirable characteristics, as many copies as desired can be obtained. A second fact is that each of the 3000 to 5000 genes in a bacterial cell such as *E. coli* is responsible for some facet of that cell's phenotype (phenotype is the constellation of outwardly observable properties of the cells). If any one of those genes is altered by mutation—a change in the nucleotide sequence of the DNA—then some facet of the phenotype of the cell will change. If, for instance, the gene in *E. coli* that allows the cells to metabolize the sugar lactose is mutated in a cell, that cell and all the progeny cells derived from it will be unable to grow if put into an environment where lactose is the only source of carbon atoms and energy. Therefore, to alter in a heritable fashion a particular characteristic of a cell, one need simply isolate a cell containing a mutated version of the gene responsible for that characteristic.

Mutations occur at low frequency (of the order of one cell in every one million to one billion cells will contain a mutation in any given gene). A wide variety of procedures has been developed for isolating the rare cell in a population containing a mutation in one or another particular gene. Most sorts of mutations can revert to their original state at the same low frequency with which they arose. However, one type of mutational change is irreversible, and that is an actual deletion of a portion of the DNA in a gene. Once the nucleotide sequence specifying all or part of a gene has been lost, there is no mechanism by which the cell can regenerate it. Thus, deletion mutations are very stable, and it is for this reason that they are used whenever possible in the construction of $\chi 1776$.

The process of constructing a mutant cell line such as $\chi 1776$ is thus one of replacing the normal genes in a cell's chromosome that are responsible for properties of the cell one wishes to alter with defective, mutant genes for these properties. These replacements can be effected by transfering the defective genes from the cells in which they arose into the cell that will ultimately become $\chi 1776$ by well-established techniques of bacterial genetics. It then remains only to show that the new cell containing the defective genes does in fact behave as would be expected. In the case of $\chi 1776$, this is done by placing the cells in a series of environments in which they should not grow or should not transfer their DNA to other cells or should degrade their DNA, and establishing that they in fact do not grow or transfer their DNA or do degrade their DNA.

A modified version of this paper was presented at the Tenth Miles International Symposium on the Impact of Recominant DNA Molecules on Science and Society, held in Cambridge, Massachusetts, June, 1976.*

*R. Curtiss III, D. A. Pereira, J. C. Hsu, S. C. Hull, J. E. Clark, L. J. Maturin, Sr., R. Goldschmidt, R. Moody, M. Inoue and L. Alexander. 1977. Biological containment: The subordination of *Escherichia coli* K-12. *In*: Recombinant Molecules: Impact on Science and Society. R. F. Beers, Jr. and E. G. Bassett (eds.), Raven Press, New York. p. 45–56.

INTRODUCTION

The idea of genetically manipulating bacterial host strains and vectors to make them safer for recombinant DNA molecule research was unanimously subscribed to at the International Conference on Recombinant DNA Molecules held at the Asilomar Conference Center in Pacific Grove, California, in February 1975. The birth of this concept of biological containment as a means for augmenting the safety afforded by physical containment facilities and procedures was relatively painless, as was the initial formulation of the means to manipulate hosts and vectors genetically to preclude the survival and/or transmission of cloned recombinant DNA to other hosts and/or vectors in the biosphere. In practice, however, the disarming of *Escherichia coli* K-12 and its plasmid and phage cloning vectors has been considerably more difficult, frustrating and time-consuming than was originally anticipated.

Immediately after the completion of the Asilomar meeting, our laboratory group commenced to construct a number of *E. coli* K-12 strains that would have numerous built-in safety features for recombinant DNA research. Our specific goal in attempting to accomplish this objective was to construct strains possessing mutations that would (i) increase their usefulness for recombinant DNA molecule research, (ii) preclude their colonization of and survival in the intestinal tract, (iii) preclude biosynthesis of the rigid layer of the cell wall in non-laboratory-controlled environments, (iv) lead to degradation of cell and vector DNA in non-laboratory controlled environments, (v) permit replication of specific cloning vectors to be dependent on specific host strains, (vi) preclude or minimize transmission of recombinant DNA to other bacteria that might be encountered in nature should the host strain escape from the confines of the laboratory, and (vii) permit monitoring of the strains in the laboratory and the environment.

Our initial expectation was that the task could be easily accomplished, since a substantial amount was known about each of the areas. To the contrary, we have learned that *E. coli* has enormous resiliency and a great capacity to do the illogical and unexpected. With each passing month, our respect for the sophisticated biological mechanisms that have evolved in *E. coli* has increased. Moreover, we have learned a great deal about which mutations do and which mutations do not contribute either to the safety and/or utility of *E. coli* strains for recombinant DNA molecule research.

Toward the end of January 1976 we completed construction of a strain, designated $\chi 1776$, that seemed to possess a sufficient number of safety features to satisfy our goals and meet the standards of an EK2 host for cloning with nonconjugative plasmid vectors as set forth in the NIH Guidelines for Recombinant DNA Molecule Research (16). We then subjected this strain to an exhaustive series of tests to evaluate its safety and utility for recombinant DNA molecule experiments and to uncover new information that would hopefully permit future construction of a completely fail-safe *E. coli* strain. This report will review what we have learned during our more than year long education by *E. coli*.

RESULTS

Mutations that Increase the Usefulness of Strains for Recombinant DNA Molecule Research

As a standard necessary attribute, all strains being constructed must have *hsdR** or *hsdS* mutations. These mutations abolish the cell's ability to recognize and destroy foreign DNA entering the cell from the outside and thus make it possible to introduce foreign DNA sequences into the strain. In studying the ability of foreign DNA to be taken up and expressed by *E. coli* K-12 in the process of transformation, we discovered that transformation frequencies were increased in strains with mutations in the major cellular endonuclease (*endA*, three- to thirtyfold) and with mutations affecting cell envelope structure and biosynthesis (*dapD* and Δ[*bioH-asd*], about threefold, and *galE* or Δ[*gal-uvrB*], four- to tenfold). Although this information makes it possible to construct a strain that is considerably more transformable (i.e., able to take up and incorporate exogenous DNA molecules) than wild-type K-12 strains, we have found that *rfb* mutations, which cause the deletion of carbohydrates from the lipopolysaccharide (LPS) core of the cell envelope, reduce transformability three- to sevenfold. This is unfortunate since these mutations have several desirable features in that they reduce the ability of the cells to act as recipients for many types of conjugative plasmids, thus leading to decreased potential for genetic exchange with other bacteria. They also confer increased sensitivity to bile salts, detergents, and antibiotics, so that strains containing these mutations can be selectively killed by these agents. In this latter regard, strains with *dapD*, Δ[*bioH-asd*] and *rfb* mutations are very easy to lyse, thus facilitating isolation of chimeric plasmid DNA.

Many of the strains being constructed possess mutations that lead during growth of the culture to the continuous production of minicells that lack chromosomal DNA. Minicells result from abnormal cell divisions occurring near the polar ends of cells. Minicells are particularly useful in studies on the expression of DNA cloned on plasmid vectors

*All gene symbol designations are those used and defined by Bachmann et al. (1).

since minicells produced by plasmid-containing strains possess plasmid DNA that can be transcribed and translated (11,12,18). Consequently, such minicell-producing strains have already been useful in studying the expression of eukaryotic DNA sequences cloned on the pSC101 plasmid (5,15).

Mutations that Preclude Colonization of and Survival of Strains in the Intestinal Tract

Unpublished experiments done ten years ago at Oak Ridge National Laboratory indicated that both *pur* and *thy* mutations, which affect the biosynthesis of DNA precursors, caused *E. coli* strains to survive at reduced frequencies during passage through the intestinal tract of mice. We were also interested in both of these mutations for other reasons, since *thy* mutants should undergo thymineless death, which is accompanied by degradation of DNA. Moreover, *pur* mutations should contribute further to the lack of virulence of *E. coli* K-12 since it is known that *pur* mutations in pathogenic microbial species cause avirulence due to interference with the ability of the bacterial cells to grow intracellularly in their hosts. However, in numerous tests in which approximately 10 billion bacteria were suspended in milk and introduced into the esophagus of rats, we could obtain no evidence that *pur* mutations affected the intestinal titers or the duration of excretion of the *E. coli* K-12 strains prior to their elimination from the intestinal flora. We have thus discontinued including a purine requirement in our strains since it could impede the growth of strains in certain environments, thereby interfering with the phenotypic expression of other mutations that abolish synthesis of the rigid layer of the cell wall. Strains with mutations in the *thyA* gene, however, do fail to survive passage through the rat's intestinal tract at normal frequencies. It is interesting to note that strains carrying the *thyA* mutation and a secondary mutation in the *deoC* gene but not the *deoB* gene usually survive substantially less well and are excreted for a shorter duration than the *thyA* strain. Since *deoC* mutations lack the enzyme deoxyriboaldolase, we can surmise that their lower survival rate in the rat intestine is due to the presence of high concentrations of purine deoxyribonucleosides and/or thymidine, which are toxic to cells that possess such mutations (4). Further tests are underway to elucidate the basis for this behavior.

Certain mutations in *rfb* genes, in conjunction with other mutations altering outer membrane structure (*oms*), confer increased sensitivity to a large number of antibiotics, drugs, ionic detergents, and, of greatest importance, bile salts which occur naturally in the intestinal tract. We have not yet tested the effect of these mutations alone on colonization and survival of *E. coli* K-12 strains in the rat intestinal tract, but we know that strains with *rfb* and *oms* mutations in conjunction with *thyA* and mutations that abolish synthesis of the rigid layer of the cell wall do not survive passage through the rat intestinal tract. So far we have tested over 40 rats, with an inoculation dose of approximately 10 billion *E. coli* cells per rat. It is expected, although unconfirmed as yet, that mutations in other genes such as *lpcA*, *lpcB*, and *rfa* that affect the cell envelope structure and which confer bile salts sensitivity, would similarly lead to inability of strains to survive in or colonize the intestinal tract of animals, provided that bile was produced and delivered to the duodenum.

Mutations that Preclude Biosynthesis of the Rigid Layer of the Cell Wall in Non-Laboratory Controlled Environments

Our initial goal was to construct strains that would require diaminopimelic acid (DAP), the precursor in a biosynthetic pathway unique to prokaryotic organisms leading to the synthesis of lysine. Since DAP is an unusual amino acid, we reasoned that it would not be prevalent in nature. Since it is an essential constituent of the rigid inner layer of the *E. coli* cell wall, strains possessing defects in its synthesis would be unable to form strong cell walls in its absence and would thus form spheroplasts and lyse in the absence of DAP ("DAP-less death").

Among numerous *dap* mutations tested, only the *dapD8* allele isolated and characterized by Bukhari and Taylor (2) was found to be genetically stable (reversion frequency of ca. 10^{-9}), and thus this allele was introduced into some of our strains for initial tests. We soon found, to our dismay, that strains possessing this allele did not always undergo DAP-less death and indeed were capable of growth, not only in liquid media but also when replica plated to solid media lacking DAP. The ability to survive in the absence of DAP was dependent on the presence of certain metal ions (sodium, potassium, magnesium and/or calcium) and was also temperature dependent, with survival being noted at 37° C or lower but not at 42° C. Upon microscopic observation, surviving cells growing in DAP-deficient liquid media appeared as spheroplastlike bodies surrounded by a capsular material. Moreover, colonies forming on agar medium lacking DAP were very mucoid and were also composed of spheroplastlike bodies. It was thus evident that these *E. coli* strains, when placed in adverse circumstances, were able to synthesize colanic acid, a polysaccharide composed of galactose, glucose, glucuronic acid and fucose and whose synthesis is regulated by the *lon* gene (14). Consequently, it was necessary to introduce one of a variety of mutations such as *galE*, *galU*, *man*, *non*, etc. that would abolish the ability of cells to make colanic acid and thus minimize if not preclude survival of spheroplast forms resulting from cessation of synthesis of the rigid cell wall layer.

During these studies we also investigated properties of other mutations that would confer a Dap phenotype, such as deletions of the gene for aspartic acid semialdehyde dehydrogenase. Ultimately, a strain with the *dapD8*, Δ[bioH-asd] and Δ[gal-uvrB] mutations was constructed and found to undergo DAP-less death under a variety of conditions. This strain was also shown to be unable to synthesize colanic acid. Such a constellation of mutations also promoted more rapid rates of death during passage through the intestinal tract of rats. The rates of DAP-less death, however, depend on a number of parameters: cells exhibiting faster rates of protein synthesis yield more rapid rates of death and cells inoculated from log-phase cultures die more rapidly than cells inoculated from stationary-phase cultures. The cell density at the inception of DAP-less growth is also important in that cells starting at low density die more rapidly than cells inoculated at high density. Indeed, when starting with cell densities greater than 5×10^7 cells/ml, the culture never completely dies since survivors apparently can scavenge DAP released from lysed cells. We have also observed that the presence of other microorganisms, whether introduced into the culture deliberately or accidentally, accelerates the rate of DAP-less death. Since DAP-less death depends upon concomitant protein synthesis and since *dap* mutations confer an obligate requirement for lysine and *asd* mutations an obligate requirement for both threonine and methionine in addition to DAP, DAP-less death can be achieved only when *dap asd* strains grow in the presence of threonine, methionine and lysine. For this reason we are currently exploring the behavior of mutations that abolish alanine racemase and D-alanyl-D-alanine ligase activities that are necessary for the synthesis of D-alanyl-D-alanine, another unique component of the rigid layer of the *E. coli* cell wall. Since such mutants would be able to synthesize L-alanine and carry out protein synthesis in media containing a metabolizable energy source but devoid of amino acids, they should offer some safety advantages over use of strains with *dap* and/or *asd* mutations.

It is evident from our studies that strains defective in the synthesis of the rigid layer of the cell wall offer a significant safety advantage if they contain additional mutations abolishing the synthesis of the extracellular capsule that balances the osmotic pressure differential between the medium and the inside of the bacterial cell. It appears, therefore, that this experimental approach to safer strain construction is not only highly efficacious for *E. coli* but conceivably valid for application to a wide range of microbial species for future development of safer host-vector systems.

Mutations that Facilitate Degradation of DNA in Non-laboratory Controlled Environments

We expected that the presence of *thyA* mutations, which lead to death in the absence of thymine (thymineless death), would also lead to degradation of DNA due to the accumulation of unrepaired single-strand breaks in the cellular DNA (13). Although thymineless death does occur in thymine-deficient medium and this is accompanied by degradation of DNA, the remaining DNA does not appear to contain single-strand breaks at an abnormally high frequency. Since we have observed that covalently closed circular pSC101 plasmid DNA molecules in this strain are lost during thymine starvation, it is possible that DNA degradation is so closely associated with the introduction of single-strand breaks in double-stranded DNA that we could not expect to detect a net accumulation of single-strand breaks. As an additional or alternative explanation, it should be noted that the strains in which DNA degradation has been investigated most extensively also possess mutations that abolish both dark repair of DNA and restriction, either of which in the mutant or wild-type state might affect detection of single-strand breaks and/or subsequent DNA degradation during thymineless death. These possibilities are currently under study.

We have also investigated the behavior of strains with a temperature-sensitive (TS) mutation in the gene for DNA polymerase I (*polA214* or *polA12* [TS]) in combination with a temperature sensitive mutation in the gene for a major function required for genetic recombination (*recA200* [TS]) to see if these combinations of mutations in the presence of a *thyA* mutation would lead to an increased rate of DNA degradation in the presence or absence of thymidine. Although the presence of *polA* (TS) and *recA* (TS) in *thyA* strains causes an accelerated rate of DNA degradation during thymineless death at 42° C, the presence of these mutations has little noticeable effect at 37° C. The presence of a *polA* (TS) allele does, however, cause *thyA* strains to commence thymineless death immediately at 37° C with no lag in loss of colony forming ability as is usually noted for *thyA* or *thyA recA* (TS) strains. We have not, however, observed that *polA* (TS) and/or *recA* (TS) alleles alter either the rates of death or the time for clearance of *thyA* strains from the rat intestinal tract. Based on these results, it is evident that if *polA* and *recA* mutations are to be introduced into future safer strains to accelerate the rates of DNA degradation at temperatures usually encountered by *E. coli*, it will be necessary to isolate additional temperature-sensitive alleles of the *polA* and/or *recA* genes. Since we already know what mutations eliminate survival of strains *in vivo*, it would seem judicious to isolate conditional alleles that confer cold sensitivity (CS) rather than heat sensitivity. Indeed, the use of *polA* (CS) alleles should also abolish replication and transmission of colicinigenic factor E1 (ColE1) cloning vectors at temperatures below 30° C, since ColE1 DNA replication requires a functional DNA polymerase I.

Mutations that Would Permit Replication of Cloning Vectors to be Dependent on the Host

Since we have not undertaken the development of cloning vectors, no work has been attempted in this area other than to include a *supE* allele in all strains. This gene will specifically suppress the lethal effect of the amber class of polypeptide chain termination mutations. Thus, if an amber mutation is introduced into an essential gene in a plasmid or bacteriophage cloning vector, that vector will be able to survive only in a bacterial cell containing an amber suppressor such as *supE*. The *supE* allele was therefore introduced and maintained in all strains in the expectation that most safer cloning vectors would be constructed so as to have some critical function that would be dependent upon the presence of an amber suppressor.

Mutations that Preclude or Minimize Transmission of Recombinant DNA to Other Bacteria

Mutations that cause resistance to all of the familiar specialized and generalized transducing phages of *E. coli* are well known and indeed have been introduced into some of our strains. Nevertheless, it is highly likely that alterations in the cell surface that result in resistance to these known phages will endow the strain with sensitivity to other phages that lurk in the sewers and polluted rivers. It thus may be difficult, if not impossible, to ensure that a strain could not be infected by a potential transducing phage in nature that would act to transmit a cloned DNA fragment to some other robust organism. The probabilities for such occurrences, however, would be extremely small, especially if the disarmed host were unable to synthesize DNA because of a *thyA* mutation or were otherwise in a metabolically inactive state because of growth requirements or were in the process of dying. It should be noted in this regard that *E. coli* K-12 cells suspended in phosphate buffered saline and then infected with T6 bacteriophage have been observed to yield a small burst of liberated phage several hours after infection, especially after prior starvation of the cells for four hours. Whether some of the transducing phages that are more host dependent than T6 can replicate or not is as yet unknown. This observation, coupled with our observation that *E. coli* can transfer conjugative plasmids at low frequency after long periods of starvation in nongrowth media at $37°$ C, underscores the necessity of ultimately introducing mutations that accelerate the rate of *E. coli's* death under conditions of starvation when thymineless and DAP-less death do not occur efficiently.

To block the transmission of nonconjugative plasmid cloning vectors, we have been isolating and characterizing mutants with various cell surface defects that would contribute to a conjugation-deficient (Con⁻) recipient phenotype. This approach is based on the fact that a cell possessing a nonconjugative plasmid vector harboring recombinant DNA would first have to acquire a conjugative plasmid in order to transmit that cloned DNA to some other microorganism. Although a considerable amount of work has been done on the isolation and characterization of Con⁻ recipients with regard to their ability to mate with Hfr and F' donors (8,9,17,19), there has been no previous work on the isolation of Con⁻ mutants defective in the ability to act as recipients for the 18 to 20 other conjugative plasmid types found in gram-negative enteric microorganisms. Because of the diversity of plasmid types and of mutational lesions that could potentially reduce the recipient ability of safer host strains, we have spent considerable time during the past year in isolating and characterizing Con⁻ mutants. For example, we have found that the presence of deletions of the *gal* operon reduces the recipient ability 1000 fold or more for plasmids in the M, O, W, X, and 10 incompatibility groups with smaller yet significant reductions noted for plasmids in the C, I (only some of those tested), J, L, T, and 9 incompatibility groups. The addition of *rfb* and *oms* mutations to a strain with a Δ *gal* lesion either does not alter or further reduces recipient ability for the aforementioned plasmid types but more importantly reduces recipient ability about 1000 fold for plasmids in the F and N incompatibility groups, which are very prevalent in *E. coli* strains in nature, as are I-type plasmids (10). Thus a strain such as χ 1776 with Δ *gal, rfb,* and *oms* mutations, when mated for 90 min under optimal laboratory conditions with donors possessing repressed conjugative plasmids, gives transconjugant frequencies of 10^{-4} to 10^{-5} for L, P and some I group plasmids, 10^{-5} to 10^{-6} for J, 9 and other I group plasmids and less than 10^{-7} for C, FII, H, M, N, O, W, X, and 10 group plasmids. Other mutants with different cell surface defects are currently being isolated and characterized with regard to their ability to either receive or donate the various types of conjugative plasmids.

Since little is known about plasmid transfer between gram-negative microorganisms in soil, water, and sewage, we investigated conjugational plasmid transfer at temperatures likely to occur in these environments. At 27° C we were unable, even when using wild-type Con$^+$ recipient strains of *E. coli*, to detect plasmid transfer at frequencies in excess of 10^{-8} except for transfer of plasmids in the T and P incompatibility groups. Thus, if we assume that laboratory conditions reflect those found in nature, it would be unlikely that a significant amount of mobilization of nonconjugative plasmid cloning vectors would occur in natural environments other than in the intestinal tract because of temperatures unsuitable for the expression of the donor and/or recipient phenotypes necessary for conjugative plasmid transmission. A more complete treatment of the probabilities of transmission of recombinant DNA contained on nonconjugative plasmid cloning vectors has been presented and discussed elsewhere (6,7).

Mutations that Permit Monitoring of Strains

We have introduced mutations that confer high-level resistance to nalidixic acid as a means to measure survival of the strains introduced during rat feeding experiments. This particular genetic marker is suitable for such experiments and for routine monitoring, since resistance to nalidixic acid is not common among naturally occurring gram-negative organisms and has not been observed to be plasmid mediated, and since mutations to resistance occur at a very low spontaneous frequency. In addition, nalidixic acid has a high efficiency of killing, such that it can be used to detect a single nalidixic acid resistent cell in the presence of 10^8 to 10^9 sensitive cells. Many of our strains also possess mutations to resistance to cycloserine, which also has not been observed to be plasmid mediated. It is thus possible to use double selection for resistance to cycloserine and nalidixic acid to detect very low numbers of cells when testing strains in a variety of environments in and outside the laboratory.

One of the chief concerns expressed by many who are apprehensive about the safety of recombinant DNA molecule research is the inadvertent contamination of a culture of a disarmed strain with cells of a robust transformable strain during transformation with recombinant DNA molecules. We have found that the addition of nalidixic acid and/or cycloserine to the culture during its growth prior to transformation and to the selective medium for plating transformants greatly minimizes if not precludes transformation of such contaminants. In this regard, cycloserine is somewhat preferable to nalidixic acid since the latter is difficult to dispose of in an environmentally safe way, being stable to autoclaving, whereas the former has a reasonably short half life at neutral or acidic pH's and is rapidly destroyed in nature.

Construction, Properties and Testing of χ1776

χ1776 was constructed in 13 steps from χ1276 (12) and possesses 16 different mutational lesions. Table 1 lists most of the phenotypic properties of this strain along with the responsible mutations for the designated phenotype. In summary, χ1776 and its derivative χ1876, which possesses the pSC101 nonconjugative plasmid, (i) cannot survive passage through the intestinal tract of rats, (ii) die, partially degrade their DNA, and lyse in growth media lacking DAP and thymidine, (iii) cannot grow, and die at variable rates, following drying or when suspended in water or other nongrowth media, and (iv) are unable to transmit genetic

Table 1

Phenotypic Properties of χ1776

Phenotype	Responsible Mutation(s) [a]
Requires diaminopimelic acid	dapD8 Δ29[bioH-asd]
Requires threonine	Δ29[bioH-asd]
Requires methionine	metC65 Δ29[bioH-asd]
Requires biotin	Δ40[gal-uvrB] Δ29[bioH-asd]
Requires thymidine	thyA57
Cannot use galactose for growth	Δ40[gal-uvrB]
Cannot use maltose for growth	Δ29[bioH-asd]
Cannot use glycerol for growth	Δ29[bioH-asd]
Cannot synthesize colanic acid	Δ40[gal-uvrB]
Sensitive to UV and defective in dark repair	Δ40[gal-uvrB]
Sensitive to glycerol (aerobic)	Δ29[bioH-asd]
Sensitive to bile salts, ionic detergents and antibiotics	rfb-2 oms-1 oms-2
Resistant to nalidixic acid	nalA25
Resistant to cycloserine	cycA1 cycB2
Resistant to chlorate (anaerobic)	Δ40[gal-uvrB]
Resistant to trimethoprim	thyA57
Resistant to T1, T5 and φ80	tonA53
Resistant to λ and 21	Δ29[bioH-asd]
Partially resistant to P1	rfb-2 oms-1 oms-2
Conjugation defective	Δ40[gal-uvrB] rfb-2 oms-1 oms-2
Produces minicells	minA1 minB2
Temperature sensitive at 42 °C	One mutation (oms-2) that is partially responsible for phenotype is linked to thyA
Cannot restrict foreign DNA	hsdR2
Allows for vector to be host dependent	supE42 (amber)

[a]See Bachmann et al. (1).

information contained on pSC101 to other bacteria under any of the above-described nonpermissive conditions at measurable frequencies. By factoring the various parameters necessary for a successful triparental mating to allow for receipt and then mobilization of pSC101 by a conjugative plasmid, we can predict that this series of events could occur in nature with a probability of about 10^{-22} per surviving bacterium (7).

DISCUSSION AND CONCLUDING REMARKS

Our future goals include the construction of χ1776 derivatives with additional mutations that would further (i) reduce its recipient ability in matings with donors possessing various conjugative plasmids, (ii) increase its transformability, and (iii) increase its rate of death in nonlaboratory controlled environments. At the same time, we are constructing other sublines of E. coli K-12 with defects in D-alanine metabolism as a means to block cell wall biosynthesis. The latter, together with the other mutations as described in this report, should confer the necessary attributes required for safer and more useful E. coli strains for recombinant DNA molecule research.

Since it is generally acknowledged that the cloning of DNA in bacteria, especially E. coli, is not without potential hazard, it is imperative that investigators attempting to clone DNA in these disarmed host-vector systems undertake the responsibility to verify the relevant phenotypic properties of this system immediately after isolating a clone containing recombinant DNA. With regard to χ 1776 and its derivatives, the traits to be verified should include the requirements for diaminopimelic acid and thymidine, inability to utilize or ferment galactose or maltose, inability to produce colanic acid or other capsular material, and sensitivity to UV and bile salts. These tests can be easily done either by replica plating or streak testing on appropriate agar media. Clones that are not tested or show an alteration in even one phenotypic trait should be destroyed immediately. We also believe that those who clone DNA should devote some portion of their time to more extensive tests to determine the effects of cloned DNA on the survival and transmissibility characteristics of the host-vector system. We will be more than pleased to offer advice on the design of these tests and provide any strains that might be necessary to perform them.

Soon after the discovery of recombinant DNA molecule technology using E. coli, it became obvious that there were numerous potential biohazards associated with this research. This realization was in large part based on the facts that E. coli is a normal intestinal inhabitant of humans and warm blooded animals, is sometimes a severe opportunistic pathogen, and possesses the potential to exchange genetic information, especially plasmid DNA, with representative strains of over thirty bacterial genera. It is equally obvious that the development of biological containment systems and their usage in conjunction with physical containment facilities and procedures are, therefore, imperative for continued safe recombinant DNA molecule research. Although we tend to think primarily of hazard to the human species as a direct consequence of recombinant DNA molecule research, it must be remembered that other potential biohazards in recombinant DNA molecule research can exist that could cause perturbations of the ecosystem. Also, the extent of the possible damage to the biosphere and ultimately to humans when one organism acts on another either to displace it from or interfere with its normal functioning in its ecological niche is unpredictable but potentially catastrophic. These points should therefore serve to underscore the importance for developing, testing, and using disarmed microbial host-vector systems in all experiments in which the foreign cloned DNA might contribute to such potentially biohazardous conditions. This admonition is as applicable for use of microbial hosts other than *E. coli* for recombinant DNA molecule experiments as it is when using *E. coli*. Such disarmed microbial host-vector systems should minimize if not prevent the survival and transmission of cloned DNA if the chimeric host were inadvertently released from its rigorously controlled test tube habitat.

REFERENCES

1. Bachmann, B. J., Low, K. B., and Taylor, A. L. (1976): Recalibrated linkage map of *Escherichia coli* K-12. *Bacteriol. Rev.*, **40:** 116–167.
2. Bukhari, A. T., and Taylor, A. L. (1971): Genetic analysic of diaminopimelic acid- and lysine-requiring mutants of *Escherichia coli* K-12. *J. Bacteriol.*, **105:** 844–854.
3. Burrows, T. W. (1955): The basis of virulence for mice of *Pasteurella pestis*. *Symp. Soc. Gen. Microbiol.*, **5:** 151–175.
4. Buxton, R. S. (1975): Genetic analysis of thymidine-resistant and low-thymine-requiring mutants of *Escherichia coli* K-12 induced by bacteriophage Mu-1. *J Bacteriol.*, **121:** 475–490.
5. Chang, A. C. Y., Lansman, R. A., Clayton, D. B., and Cohen, S. N. (1975): Studies of mouse mitochondrial DNA in *Escherichia coli*: structure and function of eucaryotic-procaryotic chimeric plasmids. *Cell*, **6:** 231–244.
6. Curtiss, R., III (1976): Genetic manipulation of microorganisms: potential benefits and biohazards. *Ann. Rev. Microbiol*, **30:** 507–533.
7. Curtiss, R., III, Clark, J. E., Goldschmidt, R., Hsu, J. C., Hull, S. C., Inoue, M., Maturin, L. J., Moody, R., and Pereira, D. A. (1976): Biohazard assessment of recombinant DNA molecule research. In: *Plasmids: Medical and*

Theoretical Aspects. S. Mitsuhashi, L. Rosival, and V. Kremery (eds.). Avicenum, Prague, p. 375–387.

8. Curtiss, R., III, Fenwick, R. G., Jr., Goldschmidt, R., and Falkinham, J. O. (1977): Mechanism of conjugation. In: *Transferable Drug Resistance Factor R*, ed. by S. Mitsuhashi. Univ. Park Press, Tokyo, p. 109–134.

9. Falkinham, J. O., III, and Curtiss, R., III (1976): Isolation and characterization of conjugation-deficient mutants of *Escherichia coli* K-12. *J. Bacteriol.,* **126:** 1194–1206.

10. Falkow, S. (1975): *Infectious Multiple Drug Resistance.* Pion Limited, London.

11. Frazer, A. C., and Curtiss, R., III (1973): Derepression of anthranilate synthase in purified minicells of *Escherichia coli* containing the Col-*trp* plasmid. *J. Bacteriol.,* **115:** 615–622.

12. Frazer, A. C., and Curtiss, R, III (1975): Production, properties and utility of bacterial minicells. *Curr. Top. Microbiol. Immunol.,* **69:** 1–84.

13. Freifelder, D. (1969): Single-strand breaks in bacterial DNA associated with thymine starvation. *J. Mol. Biol.,* **45:** 1–7.

14. Markovitz, A. (1976): Genetics and regulation of bacterial capsular polysaccharide biosynthesis and radiation sensitivity. In: *Surface Carbohydrates of Procaryotic Cells,* ed. by W. Sutherland. Academic Press, New York.

15. Morrow, J. F., Cohen, S. N., Chang, A. C. Y., Boyer, H. W., Goodman, H. M., and Helling, R. B. (1974): Replication and transcription of eukaryotic DNA in *Escherichia coli. Proc. Nat. Acad. Sci. USA,* **71:** 1743–1747.

16. National Institutes of Health Guidelines for Research Involving Recombinant DNA Molecules (1976): U. S. Department of Health, Education and Welfare, Public Health Service, National Institutes of Health.

17. Reiner, A. M. (1974): *Escherichia coli* females defective in conjugation and in absorption of a single-stranded deoxyribonucleic acid phage. *J. Bacteriol.,* **119:** 183–191.

18. Roozen, K. J., Fenwick, R. G., Jr., and Curtiss, R., III (1971): Synthesis of ribonucleic acid and protein in plasmid-containing minicells of *Escherichia coli* K-12. *J. Bacteriol.,* **107:** 21–33.

19. Skurray, R. A., Hancock, R. E. W., and Reeves, P. (1974): Con⁻ mutants: class of mutants in *Escherichia coli* K-12 lacking a major cell wall protein and defective in conjugation and absorption of bacteriophage. *J. Bacteriol.,* **119:** 726–735.

Acknowledgements

Research was supported by grants from the National Science Foundation (GB-37546) and the National Institutes of Health (DE-02670, AI-11456 and 5 PO2 CA 13148).

PART II

The Scientists Debate: For the Opposition

Two Lectures on
Recombinant DNA Research

Robert L. Sinsheimer

I.

First, I would like to make my position as clear as I am able. I believe the recombinant DNA issue, as it is known, will not go away. We have now recombinant DNA in viruses and bacteria. It is but a modest extrapolation to a future in which we will have recombinant DNA in plants, in invertebrates, in vertebrates, and no doubt in man.

We have come in our time to a transition point in the evolution of life on Earth and it is for precisely that reason that I believe we need think long and hard about the nature of the evolutionary process into which we would now intervene—before we may, inadvertently, wreak a great havoc.

I am not opposed to recombinant DNA research as such. I am not opposed to genetic engineering as such. I have said, and I still believe, there are some wonderful results to be derived from genetic engineering—and some that may literally be essential for the survival of our civilization. But I see also a darker potential for biological and social chaos and I would hope to maximize the former and minimize the latter.

It is for these reasons that I believe the policies to govern the recombinant DNA technology and genetic engineering need extensive

Editors' note: The first of these lectures was delivered at UCLA in November 1976, the second at the National Academy of Science Forum in March 1977.

thought and discussion, and that I believe the NIH Guidelines have been too narrowly conceived and are inadequate to the issue.

What, then, are my concerns? What would I do at this time? And, perhaps even more important, why has this issue become such a scientific contretemps—why has it been so difficult to address and resolve?

We need to understand the significance of what has now been accomplished. Recombinant DNA technology—the fruit of 25 years of extensive research in molecular genetics—makes available to us the gene pool of the planet—all of the genes developed in the varied evolutionary lines throughout the history of life—to reorder and reassemble as we see fit.

You have all seen drawings of the evolutionary tree tracing the development of each of the extant living species—each, as are we, the product of three billion years of evolution. That tree is a representation of the fact that evolution proceeds in a linear manner, by small increments, to produce gradually diverging species. Nature has, by often complex means, carefully prevented genetic interactions between species. Genes, old and new, can only reassort within species.

We can now transform that evolutionary tree into a network. We can merge genes of most diverse origin—from plant or insect, from fungus or man as we wish. Now, of course, most such combinantions will be sheer nonsense—nonviable, innocuous. A few will, by careful design, be valuable—if not, there would be little point to the whole enterprise. A few, however, may by design or inadvertence be deadly, in any of many ways.

The slow, almost measured pace of evolution permits the establishment, at any time, of quasi-equilibria among the various competing species. The balance is never static, but dynamic. Some species continue to find a suitable ecological niche; others die out. Most species that have lived have perished and have been replaced. We should perhaps recall that the giant reptiles dominated the earth for 150,000,000 years—and then perished.

Now we come, with our splendid science and our ingenuity, and we have now the power to introduce quantum jumps into the evolutionary process—but do we have the commensurate understanding to foresee the consequences to the currently established equilibria—on which, quite literally, our life support systems depend?

Organisms evolve into an ecological niche which favors and permits their survival. They are where they are and what they are because of that evolution.

Man likes to think that he is the exception—that we have made our own ecological niche. In part that *is* true—but in large part it is, at least as yet, a conceit.

We rely literally on our fellow creatures—on the plant world for our very oxygen and our food, on the microbial world to restore the planetary nitrogen and to degrade our wastes.

Our resistance to disease, our susceptibility to disease, the severity of the symptoms caused by a disease are all reflections of our evolutionary adaptation into an available niche.

There are in the United States some 25 deaths a year from botulism poisoning. Has it ever occurred to you how fortunate it is that botulism is not a contagious disease? Of course, this is not just due to good fortune. If botulism were a contagious disease, the human species could simply not be what it now is. Our ancestors would have had to find another niche.

My principal objection to the NIH Guidelines is that they do not reflect the evolutionary implications of what we are now about. They ignore the potential evolutionary consequences for ourselves *and* for our fellow creatures.

The Guidelines were conceived to cope with the perceived, immediate medical hazards of recombinant DNA research. There were, I believe, reasons for this tunnel vision to which I will return. As such, I believe their authors did a commendable job; they rank-ordered the hazards they envisioned and then in a pattern of graded risk, imposed a graded set of containment provisions commensurate with the estimated risk.

But it is clear that the authors of the Guidelines did not consider the transfer of genes across species, the introduction of quantum jumps into the evolutionary process, to be of *any* hazard, *unless* one could specifically pinpoint a gene of known toxicity.

Thus, any DNA fragment from any invertebrate can be inserted into the *E. coli* organism under P2, EK1 conditions. Any DNA fragment from any embryonic form of a cold-blooded vertebrate can be inserted into the *E. coli* organism under P2, EK1 conditions. Any DNA (from any source) that has been cloned (and is not known to code for a toxic agent) can subsequently be grown in the *E. coli* organism under P2, EK1 conditions.

Consider what is implied here. The DNA from an insect or an echinoderm can be cut with a restriction enzyme into some 20 or 30 or 50,000 fragments. Each contains some generally unknown cluster of genes. With another restriction enzyme one can produce a different set of 20 or 30 or 50,000 fragments. Any or all of these fragments can be inserted into *E. coli* and grown up into a clone. Somehow it is presumed that we know, *a priori*, that not one of those clones will be harmful to man or to our animals or to our crops or to other microbes—on which we unthinkingly rely.

I don't know that and, worse, I don't know how anyone else does.

Even more, this echinoderm DNA, for instance, may be prepared from organisms collected from Nature that live perhaps on a coastal shelf. Such organisms are surely not sterile preparations; they have their own, usually unknown, coterie of associated microbes and parasites—which can include those deposited on the coastal shelf by our waste disposal systems, as well as more indigenous forms. When the DNA of the echinoderm is prepared and cloned, one will inevitably prepare and clone, in some small proportion, the DNA of these small companions. That these small companions might include the spores of deadly bacilli, or the viruses of human waste, seems to have received scant thought.

The Guidelines reflect a view of Nature as a static and passive domain, wholly subject to our dominion. They regard our ecological niche as wholly secure, deeply insulated from potential onslaught, with no chinks or unguarded stretches of perimeter.

I cannot be so sanguine. In simple truth just one—just one—penetration of our niche could be sufficient to produce a catastrophe. I think there has been an inadequate realization of the fact that we are here concerned with potentially irreversible processes; that living organisms, if they find a suitable niche, are self-perpetuating, and even more, are subject to their own future evolution, wholly beyond our control.

This is a novel circumstance in the history of man-derived hazards. If DDT or fluorocarbons proved to be unfortunate, their manufacture can be ceased and in time they and the hazard will vanish. Once released, self-propagating organisms will be with us, potentially forever. A new pathogen need be created literally only once to cause untold harm.

In time, no doubt, we could learn to cope with new pathogens as we have with the old; but as we know, the mortality and trauma associated with the adaptation to any new disease, or new strain of an old disease, can be most costly.

In the larger sense we have need to protect not only ourselves but the entire biosphere on which we depend and which is, in a sense, increasingly in our trust. We inherited and evolved in a marvelously balanced, self-sustaining world of life. Can we in truth predict what disruptions we may introduce into that world with our extraordinary inventions—our biological innovations not derived from the historic evolutionary processes?

The concept of graded risk implicit in the Guidelines makes sense when one is concerned with large numbers at risk. If a few individuals, or even a considerable number, should perish because of our experiments, that would be tragic but not terminal. The concept of graded risk makes little sense if there is only one subject—as in one biosphere, the death of which would indeed be terminal.

But the Guidelines are also, I suggest, deficient in another respect. The recombinant DNA technology was developed by scientists to solve their own scientific problems. The Guidelines were developed by these scientists to cope with the kinds of problems and hazards they could envision as emergent from the experiments they have in mind to do—and (as we have discussed,) primarily to cope with perceived, immediate medical hazards of such experiments.

But as a technology the recombinant DNA techniques are in fact available to all sectors of society that have, or can purchase, the skills to use them. They are available to entrepreneurs, to flower fanciers, to the military, to subversives. Their application does not require vast resources. And the Guidelines have only in the most marginal way addressed the problems, the social hazards, which such applications may pose for our society. The potential for misuse is inherent in any scientific advance; it would seem to be particularly virulent here.

If, and you may judge, if I am not simply a Don Quixote, tilting at only imaginary dangers, how have we come to this parlous and perilous situation? I suggest there is a cluster of reasons which reinforce each other in such a way as to form, for many, an almost impenetrable cloak or barrier to a new and wider perspective.

There is, of course, a wholly human tendency to worry about tomorrow's woe tomorrow—an often, but not *always*, wise tendency based on the observation that many prophecies of woe are never realized.

Here I would suggest such an attitude (and I *have* heard it expressed) overlooks the potential magnitude of *this* woe and especially overlooks that uniquely irreversible character of this enterprise. We are unaccustomed to thinking about irreversible steps.

There is the very current tendency to believe that the human niche is now really quite impregnable. I suggest this is a very dangerous, if very human, conceit.

But beyond these are a set of attitudes characteristic of science and scientists—and this issue has thus far been largely within the province of scientists. These are attitudes long established within science and deeply ingrained, and if they are, as I suggest, no longer wholly appropriate, we can expect that it will take a considerable time for many to mentally and emotionally readapt.

Among these we might mention the concept of freedom of inquiry—a hard-won right, deeply treasured, earned by the trauma of our patron saints such as Galileo. If now this must, as I suggest, be at least partially restricted, acceptance will—and should—come grudgingly.

Coupled to that issue is the shock inherent in the concept that science itself—the simple pursuit of truth, the exploration of Nature— could in itself be dangerous, not merely to the experimenter, but conceivably to the entire planet. We have perhaps become accustomed to the concept that some of the products of technology, exploited on a vast scale, can be dangerous on a vast scale. But the circumstance that science itself has become so powerful that the performance of a single experiment could have, directly, vast consequence is very new. We have not needed, and are therefore unaccustomed, to incorporate elaborate measures of safety into the design of our experiments.

I referred earlier to the seemingly curious fact that the Guidelines are addressed almost exclusively to the immediate medical hazards perceived in this research. I suggest there is a reason for this limited perspective. For the Guidelines were, after all, developed under the aegis of the National Institutes of Health, whose purview is, primarily, medicine.

It is conceivable that had the Guidelines been developed under a different aegis they might have reflected a different perspective. But this circumstance reflects a deeper problem; in fact, the Guidelines *could* not have been developed under any other aegis. Essentially all the recombinant DNA research, indeed all modern biological research in the United States, has been and is supported by the National Institutes of Health.

I suggest that while the administration of the National Institutes of Health has been, truly, most enlightened, this predominant dependence upon an agency whose ultimate mission is health has distorted the science of biology in this country. It has, inadvertently I am sure, biased our values and limited our perspectives. And we are now beginning to see the cost.

Yet another aspect derives from the changed circumstances of science. These changes arise in part from factors intrinsic to science—the very success of the scientific enterprise and the great powers now thereby in the scientists' hands. They also arise, however, from the changed social perception of science.

We are now aware of the seemingly inevitable progression from scientific research to technological advance to social change. Aware of it and uncomfortable with it. And I suggest that major elements of our society are, determinedly, groping for means to reduce the apparent inevitablity of this process—to bring it more under a conscious control.

In this process, the older concept that the scientists need have no responsibility for the social consequences of their discoveries is, I suggest, becoming increasingly untenable. The convenient fiction that there is a meaningful gap between the knowledge developed by scientific discovery and the uses society may make of that knowledge is seen to be just that—a fiction. It ignores the great social cost necessary, in a free society, to maintain such a gap—if it is even possible.

And so, I suggest that scientists have—very particularly in this issue—to begin to accept the responsibility to consider the social consequences likely to emerge from their research. And this must include the novel and painful conception that science *can* make the world a more dangerous place—that scientific advance could destabilize human society and could even imperil the human future. The atomic bomb was the first breach in our innocent conviction of the beneficence of science. My concern is that genetic engineering not become the second.

This realization is especially bitter. It is potentially demoralizing, and therefore resisted. To do good science requires an intense dedication; one must live and breathe science to do it well. And to do that one must really believe that the particular science one is doing is important and ultimately beneficial. To question that belief and still preserve that essential dedication will require a very considerable maturity.

All this is a lot to swallow. These perceptions, while I believe they are true and likely essential to our very survival, bring no joy. Their recognition would, I suggest will, require a major wrench to the thinking of many if not most scientists. One could hardly expect such a transition to come quickly or easily.

Now, after all this, what would I *do?* At this time, actually not terribly much different. I would like to exploit both the scientific and practical potentials of genetic engineering with the minimum possible impact on, and interaction with, our existent biosphere. Ideally, I suppose, I would like to see such research and development done on the Moon, but more practically, I would like to see it confined to a small number of maximum containment facilities under careful and sustained supervision. And I would phase out, as soon as possible, the use of an organism such as *E. coli*, a ubiquitous species with a plethora of bacterial habitats, in favor of an organism with a very restricted range of habitats, far less likely to survive should it escape; monitorable, not indigenous to the doer of the experiments, man, etc.

To sum up, some with misplaced paternalism attribute opposition to the Guidelines to a "childish fear of the unknown." It is not fear; it is concern—concern and a realistic respect for the powers with which we toy—concern that we could, we really could blow it and thus mock all the striving, all the toil and effort, all the sweat and genius that has gone before.

I can state my objective very simply: the atomic age began with Hiroshima.

After that no one needed to be convinced that we had a problem.

We are now entering the Genetic Age; I hope we do not need a similar demonstration.

II.

The existence of an intellectual controversy is an indication of uncertainty—of a lack of knowledge that restricts our ability to make intelligent prediction of the consequences of our actions. The controversy may reflect factual uncertainty, concerning the nature of the substances or organisms involved and the general principles which control their interactions. The controversy may reflect human uncertainty, concerning the predictability of human actions and the limits of rationality. Or the controversy may reflect moral uncertainty, concerning the virtues of differing basic value sets.

All these sources of uncertainty can be seen in the controversy over recombinant DNA.

The magnitude of our uncertainty reflects the magnitude of the scientific advance which these new techniques make possible. I believe science has not taken so large a step into the unknown since Rutherford began to split atoms. The recombinors may take comfort in this analogy, for Rutherford's experiments were not in themselves disastrous. He did not, in *his* ignorance, ignite a consuming chain reaction (in a historical sense, of course, he did if we include the subsequent three decades of physics in the chain, but I would not be so deterministic).

But will we always be so fortunate in our explorations? Will Nature always be so benign and so resilient to our interventions? Are there really no evolutionary booby-traps for the unwary species?

The recombinant DNA technology brings us at one bound into a new domain with great potentials both for good and for harm—and all shrouded by our current ignorance. What are the factual uncertainties that may mask significant hazard and thus pose risks to the unwary?

One large cluster of uncertainties stems from the use of strains of *Escherichia coli* as hosts for much of this research. This organism is known to live in an intimate relationship with man and other animals. It is argued that the K12 strain is not "robust" and is unlikely to colonize the human bowel (1,2). The validity of this claim for persons on antibiotics or persons suffering from various debilitating ailments, or human infants, or other animal species, is itself uncertain.

It is proposed that we will breed mortality into the *Coli* strains to be used in the "more dangerous" experiments (3). The effectiveness of such breeding, in a variety of ecological circumstances, remains to be demonstrated. Our ability to define the "more dangerous" is arguable.

More broadly the use of *E. coli* as host organism points out that we are in considerable measure ignorant of the factors governing the ecology of the bowel. Intricate microbe-microbe and microbe-host interactions involving cross feedings of vitamins, amino acids, lactate, branched-chain acids, heme, and even hydrogen are known (4). But no one would pretend we had a full understanding of this microcosm or could predict the consequences of the introduction of novel organisms with novel capabilities.

We are largely ignorant of the normal role of the bowel flora in human nutrition, as in the production of vitamins, or in carcinogenesis, as in the controversy over the importance of bulk fiber in the diet.

We are largely ignorant of the effects of plasmids, prophages, etc. on the fitness of bacteria for bowel survival (5). We now recognize that many instances of intestinal disorder acquired in travel in foreign lands are the consequence of toxigenic plasmids, apparently endemic to the Coli strains of these countries (6,7). But we are ignorant of the details of their pathology or the factors governing the fitness of such strains.

We are ignorant of the ecology of E. coli outside the bowel, of the factors that determine its capacity to invade the intestinal wall, to sustain systemic or urinary infection, or to colonize the nasopharynx. In the environment the persistence of coliforms in uncontaminated habitats is still a matter of dispute (8). Nor is it known what factor or factors, such as bicarbonate, may limit the growth of Coli in other natural settings and how these limits might be affected in the novel organisms.

We are in large measure ignorant of the range and frequency of gene transfer throughout the prokaryotic world. E. coli is known to be capable of genetic exchange with some 40 other bacterial species, but we have little knowledge from which to estimate the rates at which novel gene constructions might spread throughout the prokaryotic world under various conditions.

We are ignorant of many aspects of the complex microbiological equilibria which truly underlie and maintain the entire world of life in its present form—quasi-equilibria which affect the bacteria which degrade our wastes and replenish the planetary nitrogen and carbon dioxide, which generate our soil and cleanse our waters—and again, we can therefore hardly estimate the consequence—short-term and long-term—of the introduction of novel microbial forms.

We are grossly ignorant of the structural gene content of the eukaryotic genomes which we introduce so blithely into this E. coli. How can we predict the consequence of the interactions of unknown gene products with the numerous macromolecules and metabolites of the Coli organism? The eukaryotic gene products themselves—whether they are the result of faithful, or partial, transcription and translation—might be toxic to a host. Because we do not know that an organism produces a toxin does not mean it may not have a gene for such. And, more generally, what is a toxin? Products analogous to human hormones, to peptide growth factors, or transmitters, or releasing factors—consider a small peptide with insulinlike activity, or somatastatinlike activity—could have grievous toxic effects.

The DNAs introduced into these strains are in no sense random sequences of nucleotides. They have been, most often, selected to code for proteins that achieve a function, often a catalysis. The action of such proteins upon indigenous components of *Coli* might split off polypeptides with unfortunate sequences, or might convert normal metabolites into undesirable products—for example, converting amino acids into catecholamines with synaptic functions.

We are ignorant of the factors affecting penetration of the intestinal epithelium. Some small proteins, such as insulin, can be transported with an efficiency of a few percent (9). At the same time very large proteins, such as the 150,000-molecular-weight botulinus toxin, appear to penetrate readily (10). With an estimated 10^{13} to 10^{14} *Coli* organisms in the human bowel, the production of quantities of insulinlike or adrenalinlike activity comparable to normal human daily syntheses (1 mg/day of insulin, 250 to 300μg/day of adrenalin) is quite possible, even allowing for limited transepithelial transport.

With respect to novel metabolites one should remember that we are as yet largely ignorant of the etiology of cancer. Does anyone imagine the roster of carcinogens or mutagens has been completed?

We are ignorant of the nature and mode of transmission of slow viruses. Could their ingredients lurk in these random bits of genome we now juggle? Ailments whose symptoms are long delayed are, of course, the most pernicious, for their causative agents could become widespread during the incubation period.

We remain largely ignorant of the factors which restrict the spread of viral species among different hosts. It is indeed fortunate that the microbial sea in which we are immersed is not to our knowledge a reservoir for human viral disease. Does this reflect differences in the potential for viral gene expression? Do prokaryotes use different promoters for RNA transcription, different recognition signals for ribosome translation (11–14)? The transferred fragments of DNA must surely carry such promoters and recognition signals. Are the initiation regions for DNA synthesis different in pro- and eukaryotes? A very appreciable proportion of the transferred fragments of eukaryotic DNA will carry the sequence for initiation of eukaryotic DNA synthesis, since these are to be found in every 30 to 40 microns of DNA (15).

Out of such interactions may Nature in time evolve the capability for prokaryote-to-eukaryote viral transfer?

Through the creation of wholly new gene combinations we are, in the broadest sense, intervening profoundly in the evolutionary process.

A plausible estimate suggests that research laboratories in the United States alone will produce some 10^{15} to 10^{16} recombinant organisms per year. Industrial production could easily exceed this by several orders of magnitude, albeit probably of more limited varieties. It is unreasonable to believe that a great many of these cells will not escape our containment provisions. Such novel strains may then in a unique development broaden the base for future planetary evolution.

Can we predict the consequences?

Except in the most general terms we are ignorant of the broad principles of evolution, of the factors which determine its rate and directions. We have no general theorems to account for the spectrum of organisms that we see and the gaps in between. In the microbial world, for one particular instance, what is the advantage of the botulinus toxin to the botulinus organism? Related strains seem to do well without it (16). Why is there no coliform which has this toxin? Did evolution simply never happen upon this path? Or was it always so lethal as to prevent the development of a successful host-parasite relation? We simply do not know.

We are ignorant of the relative importance of the various factors we currently perceive to participate in the evolutionary process. Major controversies swirl about the relative importance of neutral or advantageous mutations (17), of mutations of structural genes or mutations of control elements (18), and over factors which lead to conservation of gene order or which facilitate gene rearrangement.

We are ignorant of any absolute measure of adaptation. We are ignorant of the depth of security of our own environmental niche. How many microbes or viruses now exist which are one mutation away from human pathogenicity? Or two? Or five? Or one gene, or two? We do not know.

In this new domain into which we leap we are surrounded by terra incognita. Areas of investigation which formerly seemed of little interest are now seen, from the new perspective, to be of major import. And, of course, the new techniques provide powerful means to explore these areas. But while we reconnoiter, is not great caution advisable?

I know that some do not believe these organisms we now invent are truly novel. They postulate that Nature has experimented with the potential of eukaryotic DNA in microorganisms for a long time. This just might be true—although the evidence is very limited. There are some curious instances that may profitably be reexamined in this light. Thus the Livingstons have described an unusual microorganism that is said to produce a protein with some aspects of chorionic gonadotropic activity (19).

Indeed, the discovery of a class of serine proteinases in bacteria with major structural homologies to the trypsinlike mammalian proteins led Bryan Hartley to write in 1970, "This bacterium seems to have a mammalian gene. To finish upon a note of high and not very serious speculation I would like to suggest that an ancestor of *B. sorangium* might have acquired such a gene from an ancestor of a cow, perhaps by the accidental introduction into a mammalian cell of a lysogenic phage directed toward *B. sorangium*. Such a phage could then have lysogenized with the mammalian DNA and thereby picked up a serine proteinase gene. Returned to the soil, by the usual route, the phage would inject the mammalian gene into *B. sorangium*. In other words, the bacterium might have been infected by a cow" (20).

Obviously, we are ignorant as to whether Hartley's speculation has validity or whether this is simply an instance of convergent evolution. However, if such genetic intercourse has taken place in the past, we have no idea at what rates or in what circumstances and thus we do not know how past consequences could be compared with what we are now about.

Further, of course, one should point out that we can now create combinations of DNAs from diverse organisms such as could hardly ever, plausibly, have occurred in any natural setting.

I know that some believe we will be protected from the consequences of our ignorance by the blanket theory and workings of natural selection, which, in their view, will stifle all of our inventions. They assume in effect that in each case Nature has already achieved the highest possible level of adaptation.

I have little doubt that had they been aware of it, the buffalo and the dinosaurs would have felt protected by the same principle. I see no reason for such sanguine belief. I would add that even if Nature has indeed tried out all forms and achieved near perfectly equilibrated adaptation, that does not mean we might not introduce deeply perturbing transients (21).

Which leads me to the last unknown to add to the list. Simply, we are in the end ignorant of the extent of our ignorance.

This, then, is the substratum upon which the NIH Guidelines rest. It is crisscrossed by the faults of ignorance, the discontinuities and lacunae of our knowledge. Any one of these might fail us at any time.

Research upon novel self-perpetuating organisms is as different from prior science as was the first self-perpetuating cell from all prior abiotic chemistry.

There are other dimensions of hazard here. Let me refer briefly to the second class of uncertainty—human uncertainty.

Knowledge *is* power. As the result of the extraordinary advances in our science, biologists have become, without wanting it, the custodians of great and terrible power. It is idle to pretend otherwise. The founders of this republic understood the dangers of power, the eternal need for the restraint of power, and they embodied their understanding in the checks and balances of our government. When they drafted our Constitution they did not have in mind a nation of saints and angels who would need no laws, but rather a nation of fallible, temptable, even corruptible human beings, and they devised accordingly. As we devise our future technologies, we should profit from their example. Technology, just as much as political office, is a source of power. Let us not design a technology fit only for a rational, far-sighted, unerring, incorruptible people. We must come to accept the responsibility and restraint that must accompany the power we have fashioned—or else we really may see our world dissolve in anarchy and our science with it.

There are, equally important, uncertainties of the third kind—moral uncertainties associated with the novel questions we now confront. We are becoming creators—makers of new forms of life—creations that we cannot undo, that will live on long after us, that will evolve according to their own destiny. What are the responsibilities of creators—for our creations and for all the living world into which we bring our inventions?

A recital of risks and unknowns is lugubrious. But every risk is also a challenge and every unknown a potential for adventure. I only caution that there is a fragile line, vague and ill-marked, but fatefully real, between self-confidence and what the Greeks knew as hubris. When we are concerned with the fate not just of an individual, but of much of humanity, if not indeed our very biosphere, it is the course of wisdom to keep that line in full view and respect.

REFERENCES

1. Smith, H. W. (1975) *Nature* **255**: 500–502.
2. Anderson, E. S. (1975) *Nature* **255**: 502–504.
3. Curtiss, III, R. (1976) *Ann. Rev. Microbiol.* **30**: 507–533.
4. Bryant, M. P. (1972) *Am. J. Clin. Nutr.* **25**: 1485–1487.
5. Smith, H. W. (1976) Ciba Foundation Symposium 42, "Acute Diarrhea in Childhood," pp. 45–64.
6. Harries, J. T. (1976) Ciba Foundation Symposium 42, "Acute Diarrhea in Childhood," pp. 3–16.
7. Formal, S. B., Gemski, P., Jr., Giannella, R. A., and Takeuchi, A. (1976) Ciba Foundation Symposium 42, "Acute Diarrhea in Childhood," pp. 27–43.
8. Gray, E. A. (1975) *J. Appl. Bact.* **39**: 47–54.
9. Shichiri, M., Okada, A., Kawamori, R., Etani, N., Shimizu, Y., Hoski, M., Shigeta, Y., and Abe, H. (1973) *Endocrinology* **93**: 1371–1377.

10. Boroff, D. A., and Dasgupta, B. R. (1971) Chap. 1 in *Microbial Toxins*, Vol. IIA, ed. by S. J. Ajl, S. Kadis, and T. C. Moutie, Academic Press, pp. 1–68.
11. *Progress in Nucleic Acid Research and Molecular Biology*, Vol. 19 (1976) ed. by W. Cohn and E. Volkin, Academic Press.
12. Chamberlin, M. (1974) *Ann. Rev. Biochem.* **43:** 721–775.
13. Chambon, P. (1975) *Ann. Rev. Biochem.* **44:** 613–638.
14. Perry, R. P. (1976) *Ann. Rev. Biochem.* **45:** 605–629.
15. Huberman, J. A. (1975) *Ann. Rev. Genetics* **9:** 245–284.
16. Oguma, K. (1976) *J. Gen. Microbiol.* **92:** 67–75.
17. Crow, J. (1972) *J. Heredity* **63:** 306–315.
18. Wilson, A. C. (1976) in *Molecular Evolution*, ed. by F. J. Ayala, Sinauer Associates, pp. 225–234.
19. Livingston, V. W., and Livingston, A. M. (1974) *Trans. N.Y. Acad. Sci.* **36** (2): 569–582.
20. Hartley, B. S. (1970) *Phil. Trans. Roy. Soc. London* **B257:** 77–87.
21. Fenner, F. J. and White, D. O. (1976) *Medical Virology*, 2nd ed., Academic Press, especially Chaps. 10 and 11.

Biological, Social, and Political Issues in Genetic Engineering

Science for the People

I. INTRODUCTION

The development and application of gene-transplantation technology is a matter of concern to all members of society. The issues involved range from the generation of new environmental and health hazards to questions of how we will develop the democratic processes needed to select technologies that will benefit the whole society. Recombinant DNA techniques enable scientific workers to combine the genetic material of unrelated organisms. Typical experiments involve joining together genes of higher organisms—rats, chickens, fruit flies, and yeasts, to name a few—with the genes of a bacterium. The hybrid DNA molecules can then be incorporated into the bacterium, which will grow and reproduce itself. Such recombination events, across species barriers, have not been found in nature. The scientific debate over this type of research has focused on the risks that such hybrid novel organisms would escape from the laboratory, find a niche in the ecosystem, and cause unforeseen health hazards. Recombinant DNA technology is claimed by workers in the field to have enormous potential for the detailed examination of gene function in higher organisms (eukaryotes) as well as for the commercial production of large volumes of specific proteins and other practical, industrial, agricultural, and health-related applications. "It is only a matter of time before the race begins to exploit

Editors' note: This chapter is reprinted with permission from A. M. Chakrabarty, *Genetic Engineering,* CRC Press, West Palm Beach, Florida, 1978. Copyright© The Chemical Rubber Co., CRC Press, Inc.

these developments commercially. A new industry with untold potential is about to appear . . ." (1).

It is the public at large that will live with the potential benefits and suffer the harmful consequences resulting from the industrial and scientific exploitation of genetic manipulation techniques. The right of scientists to make essentially social decisions without public knowledge or informed public consent has become a serious issue, bringing the debate concerning recombinant DNA techniques under close public scrutiny. The power of techniques for the manipulation of genes may result in a more detailed understanding of certain aspects of life. But it is a first major step toward the ability to control human genes, and it may be the beginning of the biological destruction of our ecosystem.

A. Development of the Issue

Several scientists, whose work had led to the development of the techniques of recombinant DNA, were the first to warn publicly of the potential for health hazards. However, they did not realize the danger when they first attempted the experiments. At a workshop of viruses during the summer of 1971, a student of one of the originators of the technique described a planned experiment involving the joining of the DNA from the monkey virus SV40 with that of bacteriophage lambda, and inserting the combination into *E. coli*. SV40 is known to cause tumors in some animals, but appears not to cause tumors in humans. An instructor in the workshop raised the possibility that such a hybrid might escape from the laboratory, survive, and enable SV40 to infect humans. The experiment was abandoned (2).

Paul Berg and several others considered the potential for the new gene-splicing techniques, and during July 1974 addressed a letter to the scientific community (3), asking that there be a self-imposed moratorium on certain recombinant experiments until the risks involved had been assessed and necessary safeguards proposed. The National Academy of Sciences established a committee charged with the responsibility to assess the hazards of the research and to decide what lines of research, if any, should be followed; the question of whether or not the research should be conducted at all was never addressed. From its inception, the committee interpreted this charge to establish guidelines for carrying out recombinant DNA research safely, foregoing the conclusion that the research should be done at all.

In February 1975 a conference of experts in the field of molecular genetics was held at Asilomar, California. At the meeting, somewhat arbitrary classifications of risk were set forth and the development of safe vectors and hosts for use in these experiments was called for. There was some discussion of what organisms could be used safely as hosts. *E. coli* had the advantage of being well characterized and at the time the most convenient organism to employ because of the immense amount of research that had previously been done on its genetics. Other hosts, less well known genetically, were considered because they might not normally reside in or infect humans. Consensus emerged from the Asilomar meeting that it might be possible to manipulate strains of *E. coli* in the laboratory so that they would have a decreased probability of survival outside the laboratory, and work was begun on the construction of weakened strains. The hazards of using *E. coli*, whose pathogenicity is both extensive and poorly characterized, versus those of using microorganisms of more limited habitat and pathogenicity, were considered to be outweighed by both the convenience of using genetically well-characterized *E. coli* and the expense of the time it might take to characterize the genetics of a more limited host.

The first draft of the guidelines was prepared at Woods Hole in July 1975. The earliest criticism of the guidelines was expressed at the Cold Spring Harbor Phage Meetings in August 1975. These earliest criticisms expressed the concern that the Woods Hole draft of guidelines substantially lowered the safety standards set at Asilomar. The letter urged that the most hazardous experiments be curtailed, that mammalian DNA be cloned under more stringent conditions than those specified in the Woods Hole guidelines, and that the representation on the guidelines committee be broadened to include additional scientists not involved in the cloning experimentation as well as the general public.

By December 1975 internal compromises had been reached on many aspects of the guidelines. Some kinds of recombinant DNA research were banned outright, such as experiments with the DNA of high-risk pathogens, but the cloning of viruses known to cause cancer in animals, for which a moratorium had been asked earlier (3), was permitted under very restrictive conditions. Donald Frederickson, Director of the NIH, hoped that the guidelines for the research had found "a reasonable balance . . . between concerns to 'go slow' and those to progress rapidly" (4). In February 1976 the NIH held a hearing on the guidelines where a number of people from outside the molecular biology community came to voice opposition to the proposed safety provisions and were left with the feeling that such a voice had been denied. Senator Edward Kennedy (D.-Mass.), a leading congressional supporter of research in the health sciences, criticized the scientists' attempt at self-regulation, "because scientists alone decided to impose a moratorium, and scientists alone decided to lift it" (5).

The members of the committee that drew up the Guidelines were primarily from the field of molecular genetics, the majority of whom were personally interested in the conduct of recombinant DNA experimentation (6). As a result of pressure to have greater public input into the guidelines, a few outsiders were added to the committee after the essential decisions had been made concerning safety provisions, and even then the outsiders were observed at meetings to participate very little. Experts in the fields directly concerned with the hazards of recombinant DNA research, such as epidemiologists and microbial ecologists, were consulted during the formulation of the Guidelines but were not involved directly in the decisions affecting safety measures. The final form of the Guidelines was released on June 23, 1976. On the same day, public opposition had found a forum in the chambers of the City Council of Cambridge, Mass.

Public involvement in the recombinant DNA debate had been growing as the final version of the Guidelines was in preparation. In Ann Arbor, Michigan, during the spring of 1976, the construction of a P3 containment facility for the research at the University of Michigan was debated before the Regents of the University. A similar hearing was held by the Committee on Research Policy at Harvard University, where the construction of a containment facility in the Harvard Biological Laboratories was planned. Brought to the attention of Cambridge Mayor Alfred E. Vellucci by publicity surrounding this hearing, recombinant DNA research became an issue before the city government. After two crowded public hearings, which exposed the deep rift in the molecular biology community over the wisdom of pursuing recombinant DNA research in the city, the Council voted for a moratorium on those experiments judged to be most hazardous and established a citizen review board to examine the issue and recommend appropriate action. The Cambridge Experimentation Review Board, composed of eight city residents chosen by the City Manager, were not involved in the debate and were not molecular biologists. They arrived at the recommendations that the research could be allowed to proceed within the city, but that more stringent safety procedures should be followed than those provided by the NIH Guidelines. The Cambridge citizens' panel requested that the biological containment procedures and laboratory monitoring techniques for P3 research be strengthened. It made two other unique requests. The first was that a Cambridge Biohazards Committee be set up to oversee all recombinant DNA research in the city. The second was a petition to the Congress of the United States to establish uniform standards for recombinant work, both profit and nonprofit, and to establish a registry of all workers participating in the research for future epidemiological studies. Commenting on the role of scientists in the self-regulation of their research, the panel concluded (7):

Decisions regarding the appropriate course between the risks and benefits of potentially dangerous scientific inquiry must not be adjudicated within the inner circles of the scientific establishment. Moreover, the public's awareness of scientific results that have an important impact on society should not depend on crisis situations.

The health hazards of the research are also under consideration by the Environmental Protection Division of the New York State Attorney General's Office and by the City of San Diego, both of which have sponsored public hearings on the matter.

B. The Protection of the Environment

As an afterthought to the preparation of the Guidelines, the NIH prepared a Draft Environmental Impact Statement on recombinant DNA research. The National Environmental Policy Act requires that all major federal actions be evaluated for their effects on the environment, in order to study, develop, and describe appropriate alternatives and to recognize the worldwide and long-range character of environmental problems (8). Since its enactment in 1970, the NEPA has had a significant effect on the planning and execution of many federal and private projects. The preparation of a draft impact statement, the solicitation of comments from concerned parties, and the consideration of alternatives have guided many projects to reduce their adverse environmental effects. For this procedure to work effectively, the law requires that the predictive impact statement and consideration of alternatives be done prior to undertaking action. For the case of recombinant DNA research, the Draft Environmental Impact Statement was released well after the NIH had condoned the research and both directly and actively had been supporting its conduct in numerous laboratories.

The Draft EIS prepared by the NIH tends to rationalize the safety precautions provided, rather than to examine them critically and to suggest alternatives to the research itself. It provides little or no evidence that the NIH has undertaken, or plans to undertake, the experimental evaluation of the hazards of recombinant DNA research or of the ecology and pathogenicity of *Escherichia coli* and its plasmid vectors. The Environmental Impact Statement was prepared long after the level of hazard of various experiments had been agreed upon and the decisions had been made to employ *E. coli* as a host in this research and to support the research through direct NIH funding. The only proposal for "experimental" evaluation of the hazards has been that of the Cambridge Experimentation Review Board. They have recommended to the City of Cambridge to request funds from NIH for the health monitoring of laboratory workers engaged in the research.

Notwithstanding the amount of work that has gone into the preparation of the Guidelines, they are limited in scope and do not address the general problem of the hazards of genetic manipulation. There are many ways of manipulation of the genetic material other than those involving the specific gene-splicing techniques considered in the Guidelines. The Guidelines, moveover, are restricted in application. Applying only to research funded by the NIH, they have no jurisdiction over research conducted by industry. Industrial research, currently conducted on a much larger scale, poses greater hazards. The enforcement of the Guidelines for NIH-sponsored projects has no mechanism save the ex post facto withdrawal of funding, so the Guidelines remain no more than suggestions for the safe conduct of research. Many of their provisions are stated as suggestions or recommendations, so that research actually can be conducted according to the Guidelines with far fewer safety precautions than envisioned by the authors.*

The broader problems, both of hazards other than those anticipated with the cloning of viral or toxin genes, and of other available techniques that would produce the same resulting hybrids, have not been seriously considered.

II. HAZARDS

Several basic concerns have been expressed about the hazards of recombinant DNA research. The hybrids produced are novel, laboratory-created organisms that may never have occurred before in nature and may therefore be uniquely pathogenic to humans or other organisms. The risks are presently nonquantifiable and unpredictable because there is no way of knowing in advance the properties of these hybrid organisms. Their release and survival outside the laboratory may be irreversible because they are self-replicating. Serious doubts have been raised that the organisms can be contained by the methods presently proposed. This is a special concern, because research is planned to be undertaken extensively in industry and academia. Some of the more important objections to the containment plans for these novel microorganisms are presented below.

The strain of intestinal bacteria, *E. coli* K12, that has been chosen as a host for recombinant DNA molecules has been severely criticized as a particularly dangerous choice. Related strains of *E. coli* are normal residents of the human intestine, the pharynx (in about 15 percent of all humans), and are also human pathogens of the skin, urinary tract, and

---

Editors' note: The reader can judge for himself the adequacy of the NIH guidelines by examining the operational portion of them in Appendix II.

gastrointestinal tract (9,10). *E. coli* is one of several agents that can cause epidemic diarrhea of the newborn, a condition described as "a leading cause of infant morbidity and mortality in all parts of the world" (11). *E. coli* can cause anything from mild traveler's diarrhea to a choleralike illness and is responsible for one-third of the cases of meningitis in newly born children. Also, in recent years in a Boston hospital, *E. coli* and other gram-negative bacteria have been well documented as causing increasing sickness and death in hospital patients due to septicemia (bacteria in the blood) (12). A similar trend has been observed across the nation for hospital-acquired infections (13).

On the other hand, there is no evidence that *E. coli* K12 can cause disease in the same way as other strains of *E. coli*. *E. coli* K12 has been maintained by serial laboratory culture since being isolated from human feces 55 years ago (14). However, the transfer of a recombinant plasmid to healthier strains of *E. coli* or other gram-negative bacteria living side by side with *E. coli* poses a major threat for creating a bacterium with new or increased disease-causing potential.

This is made all the more likely by the fact that *E. coli* K12 may be able to survive and multiply outside the laboratory. Studies where K12 was fed to a very small number of human subjects, reported for eight subjects by Anderson (14) and one subject by Smith (15) demonstrated no colonization of the intestine by K12. K12 persisted in the feces for three to six days, and in two of eight subjects (14) it appeared to have multiplied during its passage through the gut. The sample size was too small to allow the conclusion that *E. coli* K12 does not colonize humans, and it is likely that populations of humans under a variety of un-examined conditions might be considerably more susceptible than the subjects used in these experiments. It is not known what factors lead to colonization of the gut; however, sheep starved for two days have been demonstrated to support K12 multiplication and plasmid transfer (16).

Following what some have termed the inevitable escape from the laboratory, it is possible that *E. coli* carrying foreign DNA could cause disease or pose a serious danger by itself. In addition, many ways exist in nature by which the foreign DNA carried by K12 could be transmitted to healthy *E. coli* or other gram-negative bacteria, which might then pose a health hazard. Foreign DNA is incorporated directly into an autono-mously replicating piece of DNA in *E. coli*, a plasmid, and could become introduced into another *E. coli* or gram-negative bacterium by, for example, infection of *E. coli* K12 by a virus or conjugational factor, which could incorporate the foreign DNA, facilitating its transmission to another cell. Recombinant DNA from dead *E. coli* could transfect other bacteria, allowing survival of the foreign genes. Although the transfor-mation of *E. coli* by *E. coli* DNA has not been well characterized *in vivo*, and its frequency cannot be as yet quantitated, it is known that the DNA of virulent *pneumococcus* bacteria injected into a mouse can transform benign *pneumococcus* to a virulent state (17). The fact that most recombinant plasmids being used at the present contain one or more drug-resistance markers increases the possibility of escape by providing potential for selective enrichment of these plasmids. The widespread use of antibiotics in our society makes this likelihood particularly serious.

Should a gram-negative bacterium carrying recombinant DNA find itself in the outside world, it would have rapid access to a whole range of ecological niches. *E. coli*, for example, is found in warm-blooded animals and some insects, in soil, in water near sewage outfalls, and in any other place contaminated by human and animal feces (10). Other experiments imply that *E. coli*, or its plasmids in which the foreign DNA is carried, can be transferred easily between domestic animals and man. For example, animals and humans living in close contact show a strong association in the patterns of antibiotic-resistant *E. coli* that can be found in them (18). Furthermore, one cannot say that selective pressure will act to remove recombinant DNA molecules from the environment before they are established in a viable organism. It has been asserted that *E. coli* carrying nonessential DNA on a plasmid would be lost from the population after many generations, owing to its inferior rate of reproduction (19). However, many of the plasmid vectors in current use, such as the bacteriophage λ, may actually increase the cell's rate of reproduction (20). One mathematical model of a situation where part of a bacterial population carries a nonselected plasmid and thus grows less efficiently shows that these plasmids will become successfully established in the bacterial population (20). Thus it is likely that recombinant-DNA-containing plasmids will survive in the environment for a substantial period. In this light it seems unwise to employ for this research an organism, *E. coli* K12, that has such an extensive host range, including humans. Given our ignorance of the habitat and ecology of *E. coli*, it is impossible to foresee how organisms carrying recombinant DNA will behave in the future.

A. Physical Containment

Will physical containment prevent the release of *E. coli* K12? The NIH guidelines describe four levels of physical containment to be used in these experiments, ranging from careful microbiological technique (P1 and P2) to special hoods, airtight rooms, and other paraphernalia used for biological warfare agents (P4) (22). Perhaps the most comprehensive study of the efficacy of physical containment is A. G. Wedum's report on the Army's biological warfare laboratory at Fort Detrick. It was commissioned by the NIH when the debate on how to contain the organisms created by recombinant DNA experiments began. No one expects containment to be perfect, and indeed, there is no question that even workers protected by the highest levels of containment could receive infectious doses of organisms against which they had been vaccinated. Under P4 conditions, with 45 personnel at risk daily, some of whom administered aerosol exposures of highly infections agents to live animals, six work-related infections were reported between 1952 and 1959 in one building, and one infection for the years 1959 to 1969 in another. In a P3-type facility where tularemia (a plaguelike organism) was studied for six years, 14 cases occurred among workers who had been vaccinated with a dead form of tularemia. The introduction of a live vaccine in 1960 prevented the appearance of further hospitalized cases, but it is clear that physical containment did not prevent occasional exposure of the workers in these laboratories. Some degree of exposure may be inevitable in all but the most severe containment facilities.

A more sensitive test of exposure is seroconversion. Paul Berg, Professor of Biochemistry at Stanford University, describes how in the P3 laboratory where SV40 (Simian Virus 40) was being studied, almost all personnel had developed substantial antibody titres to the virus after one-half to a full year in the laboratory (23). One would expect that exposure of laboratory workers to *E. coli* would be unavoidable in all but the most stringent containment facilities. Two of Wedum's recommendations at the end of his report have the aim of insuring that exposure is harmless (24).

> For research on recombinant DNA, the most effective safety measure to prevent infection of laboratory personnel is to utilize a microorganism that will not infect humans. Otherwise, the greater the required human infective dose the safer it is.

For the recombinant DNA host *E. coli* K12 there may be no firm reason to believe that such infection will not occur.

> Microorganisms for which there is an effective vaccine, and also in some cases specific effective therapy, could be considered for use in research on recombinant DNA.

This recommendation does not apply to K12, for which there is currently no effective vaccine, and it may be circumvented by both plasmid transfer and acquired or resident antibiotic resistance factors.

One underlying assumption in Wedum's report is that microepidemics may occur among those who come into contact with infectious microbes on dirty clothes, dead animals, or heavily exposed laboratory workers, but that the disease-causing agents will not spread to the general public. According to Wedum, all known cases of laboratory-caused epidemics support this assertion. Nevertheless, *E. coli-*borne recombinant DNA represents a special case. The containment achieved at Ft. Detrick represents the results of constant medical surveillance—every illness reported by a laboratory worker was assumed to have occurred on the job until shown otherwise—and this was for diseases whose symptoms were recognizable and for which vaccines were known in most cases. The bacterium harboring the recombinant DNA might not cause disease until it had passed through several hosts. Furthermore, a disease-causing bacterium might be difficult to trace from the laboratory or to recognize initially. Special hosts or environments inside or outside the laboratory, such as a person under antibiotic treatment, might allow the recombinant DNA and its host to flourish locally.

The ways in which *E. coli* can disseminate and infect humans may surprise the builders of physical containment facilities. A recent study speculated that *E. coli* could be transferred from chicken to human by air in much smaller quantities than is possible by ingestion, which usually requires 10^6 to 10^8 organisms (25). Another recent report on the spread of diarrhea-causing *E. coli* between infants in a special-care nursery suggested that the bacterium was transferred from patient to patient on the hands of hospital personnel (26). Under the right conditions, infections will follow ingestion of very low doses of *E. coli*.

B. Biological Containment

Lack of complete confidence in physical containment has led the NIH to impose additional safeguards in the form of biological containment. Crippled bacterial hosts with a one-in-10^8 chance of surviving outside the laboratory, designated EK2 and EK3, would be constructed for this work. The lowest level of biological containment, EK1, simply calls for *E. coli* K12 to be used as a host in all experiments (22). The two host-vector systems in current use that may be given EK2 status are the *E. coli* K12 strain χ 1776 of Roy Curtiss and Philip Leder's modified strain of lambda bacteriophage. At the time of this writing, no work has been published from the laboratories of either Curtiss or Leder on the safety assessment of their proposed host-vector systems, and their work consequently has not been open to public criticism. This is not the best situation in view of the widespread use of χ 1776. Some significant weaknesses of the Curtiss strain were outlined in a report by the Boston Area Recombinant DNA Group. Some of the major points of the criticism can be summarized (27).

1. "[T]here is no available plasmid cloning vector which . . . is dependent on χ1776 or some other suppressor-containing strain for its replication" (28). In other words, the plasmid could survive well in some other more healthy *E. coli* if transferred there.

2. The data presented by Curtiss concerning the survival of χ1776 do not and cannot promise the reduced survival of the same strain carrying a cloned foreign DNA sequence. Experiments performed by Struhl, Cameron, and Davis (29) have demonstrated that cloned DNA from a eukaryotic organism (*Saccharomyces cerevisiae*) can correct for genotypic deficiencies of *E. coli*, such as auxotrophy for histidine biosynthetic pathway genes.

3. The strain survives better in tap water than in certain rich environments, where the lack of specific nutrients that the cell requires, such as thymidine, causes the cells to commit suicide at a faster rate.

4. The ability of Curtiss' strain to transfer genetic information by known resistance factors is reduced only 100-fold at best for most transfer factors tested.

5. Since "markers are frequently lost during routine manipulations of a complex auxotroph," it is not certain how many of the 13 or more mutations in the strain will be lost during handling by other laboratories. Another criticism of the Curtiss strain was that of Adhya and Enquist, who comment: "It was questionable that it meets the 10^{-8} [survival rate for] containment defined by the [NIH guidelines] committee" (30).

Similar objections are likely to be valid for other current host-vector systems. None is reported to have been tested for adherence to the survival criteria of the guidelines outside the laboratories where they were originally designed.

The laboratory environment presents opportunities for genetic exchange between crippled *E. coli* K12 and healthier bacterial strains. Although the NIH guidelines ask that the phenotype of host *E. coli* K12 strains be checked regularly, wild-type bacterial and viral contaminations of lab cultures are very common occurrences in microbiological work despite the precautions taken against them. The contamination of routine well-maintained *E. coli* cultures by weaker strains carrying foreign DNA is also possible. Any of these circumstances might completely frustrate biological containment (31).

We have assumed up until this point that people would be the chief agents for the escape of *E. coli* K12 and most likely the first to suffer from the creation of a harmful organism. This human-centered perspective is shared by the guidelines, where estimates of the dangers in the research are evaluated on the basis that DNA from a higher organism might cause harm to humankind. The guidelines require considerably more strict containment for experiments using the DNA from primates than, for example, insects that are not known to cause disease. "Shotgun" experiments where the entire complement of the DNA of an insect is fragmented and inserted into a multitude of *E. coli* K12 are being carried out in P2 facilities where little more than careful microbiological technique and a sign on the door are required by the guidelines.* These are the most unpredictable of recombinant DNA experiments and may be no less dangerous than those done with organisms closer to humans in evolution. There is no way to know what harmful genes the variety of insects and the variety of parasites accompanying them may carry (32). Even though we accidentally unleash an organism that does not have harmful consequences for humans directly, can we be sure that we might not cause great damage to the environment on which we depend?

Another danger at present is that recombinant DNA research itself is proliferating rapidly and with little control in major hospitals and universities, often in large cities, in many laboratories around the world. How well regulated will these facilities be? In an academic or industrial environment disrepair of equipment, carelessness, or pressure to work faster might make the exposure of laboratory personnel to recombinant organisms much more frequent that one would judge from the Detrick experience. Certainly one or a few national centers for this research would be preferable to the ever more widespread use of the technique, especially in regions of high population density.

With the many unquantifiable difficulties of physical containment, the irreversible spread of a recombinant plasmid from the laboratory is a possibility that must be taken seriously. Complete eradication of such new, mutant, possibly pathogenic strains of bacteria might be very difficult. In the Draft EIS it is stated that "reasonable means for minimizing further dispersal [of such a pathogen] could be undertaken." No such reasonable means are mentioned and, indeed, they must be unknown to hospitals, where the incidence of *E. coli* infections has steadily been rising (33). Furthermore, the NIH has thus far failed to establish a systematic screening mechanism for monitoring the escape of recombinant organisms in laboratory personnel and the surrounding communities.

For research allowed in academic and industrial laboratories much more public involvement in the decisions about the biohazards is necessary. Local biohazard safety committees mandated by the NIH guidelines should include substantial membership from the people at

---

*Editors' note: A detailed statement of the requirements for each level of physical containment is given in Appendix II.

risk from these experiments: not only lab technicians, custodial and clerical workers, but also representatives from the local community. Some important recommendations are: (1) a rigorous course on laboratory safety should be taught; (2) biohazard reports should be included in the programs of scientific meetings, and more forums for the discussion of the social and biological impact of the research should be provided; (3) the guidelines should be made binding as law to all laboratories, industrial and academic; (4) all shotgun experiments should be limited to a few laboratories where EK2 vectors and P3 containment are used.

In conclusion, we hope that in 20 years we can look back on the way in which this research was done and say—we may have guessed right or wrong in the assessment of the dangers, but right or wrong we were not foolish.

C. Bridging the Evolutionary Gap

In the previous sections some hard, technical questions have been raised about recombinant DNA work, but other broader questions deserve some examination. Some proponents of the research have asked, "Is there anything really serious or unique to contain?" Bacteria may be constantly taking up DNA from higher organisms, from the cells shed from the lining of the intestines of mammals, for example, or from any decaying animal or vegetable (34). The assertion is not an easy one to prove, and there is no evidence to support it. No specific genes are known to be shared by *E. coli* and humans, so there is no historical evidence of such a process. Nevertheless, many have stated that with approximately 10^{22} bacteria being excreted by the human population each day, transfer of human DNA to bacteria *must* occur to some extent. The rates cannot be defined, however, and the obstacles to the process are tremendous. *E. coli* DNA can transfect *E. coli* under careful conditions in a test tube, and this is in fact a key but inefficient step in recombinant DNA technology. Given the acidic and ionic conditions of the stomach, the number of bacteria competent to take up foreign DNA would be reduced drastically. More significantly, the survival of human DNA once inside the cell would be unpredictable, given the unknown levels of internal bacterial restriction endonucleases, the lack of extensive regions of complementarity between bacterial and human DNA, and uncertainties as to why recombination may or may not occur in the cell. A barrier to genetic exchange between higher and lower organisms may effectively exist with the purpose, as some have said, of keeping the two groups evolutionarily distinct: "The recombinant DNA techniques permit the creation and likely propagation of organisms (microorganisms at this time) that could not have derived from the ordinary processes of biological evolution" (35).

As simple-minded examples of novel and also harmful organisms, imagine the potency of an organism carrying the genes for a pharmacologically active peptide or for a substance from plants or insects not previously known to be toxic to humans. Either might prove quite harmful were it lodged in a sensitive part of the human system. An entire biosynthetic pathway could be donated to *E. coli*, since foreign DNA molecules containing enough space for as much as ten different genes are commonly part of recombinant plasmids.

The question of the capability of an organism to cause disease and to survive at the same time is not well understood. What is well understood, however, is the widespread effect of a new pathogen on an innocent population, as exemplified by the suffering of the American Indians from diseases imported from the Old World. Is our ecosphere as stable as we would like to believe with respect to the presence of an unanticipated infectious agent?

More subtle changes could arise from allowing *E. coli* access to human and viral genes. An autoimmune disease might be caused by the expression of humanlike antigens on the surface of a bacterium, as in some cases of acute glomerulonephritis. Paul Berg, a leading researcher in the field, asks "whether it would be harmful to have *E. coli* in our gut carrying genes from viruses that can change normal cells into cancer cells" (36). The concern these examples raise, along with many others of which we cannot conceive, will not be resolved in the foreseeable future. It should be evident how unquantifiable the risks for this research are, with or without the most earnest schemes for containment that we may put into effect.

D. Alternatives to Recombinant DNA Techniques

Whether one is allowed to use the recombinant DNA technique is not a question of academic freedom. As Richard Goldstein, Assistant Professor of Microbiology and Molecular Genetics at Harvard Medical School, has stated, "I don't think there is any question here about the freedom of inquiry. It is a matter of freedom of manufacture of a novel microorganism" (37). Many molecular biologists feel that recombinant DNA techniques are indispensable for their work. In studying every form of genetic control from differentiation and development in an embryo to the integration of a cancer-causing virus into the DNA of its host cell, scientists are searching hard for the individual stretches of DNA involved in these processes. Many of the techniques developed for these studies were left by the wayside when recombinant DNA technology arrived.

For example, the first eukaryotic gene to be isolated, the gene for hemoglobin, was obtained by making a copy of the messenger RNA for hemoglobin without the use of a prokaryotic organism as a carrier at any point (38). Many *in vitro* techniques, such as purifying DNA sequences that code for a particular messenger RNA species, are presently within the molecular biology community's grasp and in some instances could replace recombinant DNA techniques. For example, with affinity-column techniques using messenger RNA, one can look for homologous sequences in the cell's vast supply of heteronuclear RNA (39). Similar techniques could be applied to the isolation of specific sequences of the cell's DNA as well. Other advantages presented by recombinant DNA cloning techniques such as the translation of DNA into protein, or the replication in large quantities of a purified segment of DNA, currently can be performed with measurable efficiency in cell-free systems. None of these systems appears at present to be as convenient as recombinant DNA technology, but, given the hazards of recombinant DNA research, it is the responsibility of the scientific community to pursue these other methods actively. Clearly there are experiments with recombinant DNA techniques that do not have readily apparent *in vitro* parallels. This does not mean that the questions such experiments hope to answer cannot be approached through research that does not require the manufacture of novel microorganisms. The scientists wishing to conduct this research may simply have to be more patient.

It is certain that workers in the field of recombinant DNA research could develop a number of surprising, attractive, and unusual applications of the technique for medicine and industry. One such scheme would be to fabricate an *E. coli* that produces human insulin, or that would produce the blood clotting factors that hemophiliacs lack. But recombinant DNA techniques certainly cannot promise unique solutions to these problems. Insulin is now obtained from bovine or porcine pancreas, and should it be in short supply, other methods, such as using cultured mammalian cells, which produce larger amounts of the hormone, may be simpler to use (40). In many other cases recombinant DNA technology may not provide the only solution to problems for which it has been proposed as a technique of last resort.

III. ESTIMATING BENEFITS AND RISKS

Addressing the questions of policy on recombinant DNA technology ultimately requires an assessment of benefits in relation to risks. The catalog of possible risks has been claimed to span a vast realm from the precipitation of a cancer epidemic to new kinds of overt or subtle infection and environmental catastrophe, and these risks are unpredictable in both character and likelihood. Accordingly it becomes necessary to determine whether there are broad, profound, and certain benefits arising from recombinant DNA technology that would justify the risks taken in its development. Furthermore, the benefit-to-risk estimate must be placed in a real-world context, taking account of the full impact of supposedly beneficial technologies as well as examining benefits and risks in terms of which social interests are served.

Recent experience suggests that the introduction of new technologies often has complex mixed effects, reflecting the economic and political milieu being invaded, so that simple predictions of benefit, scaled up to global operations, often are not meaningful. Nevertheless, the proponents of recombinant DNA research make such simple predictions when arguing that fundamental advances will follow in good production, drug manufacturing, an understanding of disease leading to "cures," and genetic management generally, whether related to disease, behavior, or vulnerability to environmental insult. Paul Berg, one of the Asilomar participants, explained before an NIH conference on Guidelines for Recombinant DNA Research (41):

> There are also important potential benefits to expanding the world food supply. The availability of natural and artificial fertilizers limits certain crop yields. But atmospheric nitrogen is an infinite source of ammonia, if only we could harness the microbial potential for nitrogen fixation. Experts in this field suggest that the introduction of the nitrogenase system from bacteria into plants or into symbiotic organisms is imminent and promises great rewards. . . . I have no doubt that in time the opportunities will expand as the methodology becomes more sophisticated.

Biomedical research has already begun to feel the attraction of working for genetic cures for cancer. An objective laid out in a 1974 report from the National Cancer Program Planning Conference was to "develop the means to modify individuals in order to minimize the risk to cancer development." One approach to this objective was to "alter genetic make-up or genic expression to reduce the rate of cancer development." These ideas are shared by many in the field of recombinant DNA research. Even though Nobel Laureate David Baltimore concedes that 80 percent or more of human cancer is caused by environmental agents, he implies that recombinant DNA research is essential for dealing with cancer (42):

> Even if we identify the causes of breast cancer, cervical cancer, prostate cancer, colon cancer and bladder cancer, it is unlikely that we are going to be able to design a civilization that will be acceptable to

the population and that will prevent the occurrence of these terrible diseases. We should certainly make every effort we can to understand how our lifestyle causes cancer, but we must also push forward on a more basic attack on the problem, and I believe that attack is made much easier by the ability to manipulate recombinant DNA molecules.

The Cetus Corporation, which includes on its board of advisors such eminent scientists as Joshua Lederberg, Donald A. Glaser, and Stanley N. Cohen, describes the promise of the future in glowing terms (43):

> We propose to do no less than to stitch, into the DNA of industrial microorganisms, the genes to render them capable of producing vast quantities of vitally needed human proteins. . . . This concept is so truly revolutionary to the biomedical sciences that we of Cetus predict that by the year 2000 virtually all the major human diseases will regularly succumb to treatment by disease-specific artificial proteins produced by specialized hybrid microorganisms.

Those who look forward to a future ability to intervene in the genetic control of organisms, including humans, see in recombinant DNA technology an important step toward this goal. Presently, genetic engineering remains largely limited to screening and nongenetic remedial procedures.

There are a number of fallacies behind promises of great benefits from this technology. Foremost is the premise that solutions to major problems such as insufficient world food production, or human disease, are technology-limited. We will argue that fundamental solutions are not significantly limited by existing, available technology but rather by economic and political factors. Second, there is widespread conviction that new technologies can offer striking gains, with little appreciation that they frequently have associated with them a number of negative consequences, sometimes dominant. Finally, we conclude that emphasis on "technological" solutions results in diversion or distraction from other goals that are essential for real progress.

A. Producing Enough Food

In the case of world food production, it is claimed that existing productive capacity is insufficient to feed present and future populations. However, by using recombinant DNA research it is proposed, for example, to insert nitrogen-fixing ability into nonleguminous plants, thereby expanding output greatly. But the limiting factor in food production is not capacity; rather it is insufficient market demand. As explained in a recent study prepared under the auspices of the World Bank (but not endorsed by it) (44), the large number of people throughout the world without enough food have insufficient income to purchase food that would otherwise be readily available. They find, for example, that the global caloric deficit below the recommended diet, while covering 75 percent of the population in underdeveloped countries, amounts to a mere 4 percent of world cereal production (45). Furthermore, even substantial reductions in food prices, as *might* happen with new technologies, would still not bring this population into the food marketplace. Instead, the fundamental underlying problems would continue to be unemployment, income maldistribution, land ownership, and the retention of political power by elite, minority public interests. For the large owners of agricultural resources—the relatively few who own the best land in most underdeveloped countries—and for the vertically integrated corporations involved in all aspects of global agriculture—farm machinery, fertilizers, pesticides, processing, and marketing—recombinant DNA technology offers the prospect of increased yields and higher profits.

Recent advances in the technology of agriculture, applied globally, have had adverse effects. For example, the introduction of hybrid varieties during the "Green Revolution" tended to further stratify rural class structure, favoring large landowners who could afford the costly inputs required—irrigation, fertilization, mechanization—and encouraging the further consolidation of land holdings (46,47,48). Landless and jobless peasants were forced into the cities at an even faster rate than before. There was destruction of traditional, local crops with further erosion of nutritional status, while the new crops had increased vulnerability to unexpected pests (49,50). The effects of these changes for the poorest sector in the Third World were to exacerbate the maldistribution of income and consequently the people's ability to buy, let alone grow, enough food. Presumably, new varieties created through recombinant DNA technologies will be introduced into a similar economic environment.

Accompanying the belief that advances in modern agriculture, together with population control, will provide the solution to the problems of world food production, especially through the introduction of new plant varieties, is a political planning mode that makes essential changes more remote. These changes must include a full-employment rural economy and self-sufficient small-scale industrialization in the context of rejecting the traditional, vested interests holding political leadership. "Advanced" agricultural technology requires heavy financing, which makes the credit that is vital for small farmers less available to them. It encourages government investment in irrigation, storage, and transport to suit the needs of export-oriented large-scale farms. It creates dependence on agricultural inputs from foreign rather than local sources. It strengthens the economic and political position of the large operators and weakens the rural workers, thereby further delaying basic required change in the rural economy.

B. Conquering Disease

It has been claimed that recombinant DNA techniques promise important advances in combatting disease through lower drug prices and the creation of new drugs. The possibility of the rapid, inexpensive production of human insulin by genetically engineered bacteria has often been cited in this respect. Another example frequently given is the rapid and efficient production of antibiotics by new fermentation processes. This might be achieved by implanting synthetic capabilities from slowly growing organisms such as yeasts and fungi into rapidly growing, easily cultured bacteria. However, while drug prices are indeed exorbitant, the production costs of most drugs are in most cases a small part of total cost; promotion, innovation for patent evasion or defense, regulatory requirements, distribution, and marketing comprise a far larger percent of the price (51). Projections are made for more exotic medical benefits— for example, the large-scale production of specific human proteins, such as specific metabolic enzymes, blood clotting factors, immunoglobulins, and hormones, which would have a variety of uses. One suggested direct application is to provide specific proteins that are absent or defective in persons having well-characterized genetic diseases. Rather than dwelling on technological solutions, we must recognize that the primary health problems facing people in the United States and the entire world derive directly from the conditions of their lives: working conditions, stresses, environmental quality, nutrition, and lifestyle, none of which is primarily dependent on technological innovation for amelioration, and all of which are the blunt consequences of the prevailing economic order. The resulting "civilization" or "technologic" diseases include hypertension and other cardiovascular disease, cancer, anxiety-depression, alcoholism, obesity, and others (52).

New drugs for the treatment of civilization diseases are increasingly sophisticated technologically, requiring more technical expertise for their safe and effective use and providing ample opportunity for their misuse. The proliferation and irresponsible use of antibiotics is a concrete illustration of this problem. Widespread use of antibiotics for prophylactic or improper purposes has created a selective environment favoring drug-resistant pathogens as well as compromising patients' immediate health. A number of drug-resistant strains of major pathogens are now in wide circulation, and antibiotic-resistant epidemics have been both predicted and realized (53). This trend toward technological medicine will be magnified manyfold when the products of recombinant DNA techniques begin to appear. Awesome hazards are presented by the clinical development of these therapies, an activity that will continue and increase in Third World countries where the regulation of human experimentation is less stringent. Even the routine use of therapies derived from recombinant DNA techniques will most probably be a positive advance only for those who have access to top-quality medical care—for others, harm may be a likely outcome. Aggressively marketed, difficult-to-regulate and highly complex therapies will be available within a health care system incapable of assuring their uniform and correct use.

Emphasizing a new class of wonder drugs and therapies to be made possible by recombinant DNA research not only sustains the current excessive reliance on drugs as medical panaceas, devices to make medical management more "efficient," but also diminishes the visibility of other alternatives for addressing contemporary problems, including those which look at the entire civilization and social order (e.g., pollution, working conditions) and those which increase the participation of patients in their own health management and responsibility. The funding situation in cancer research today is a graphic illustration of diversion by "high science." A small part (less than 10 percent) of the budget of the National Cancer Institute of the National Institutes of Health is budgeted for basic environmental epidemiology—identifying the agents responsible for most cancer and their modes of action (54,55). The bulk of NCI funds support programs in viral research, based wholly on the speculation that viruses are involved in human cancers, in immunology, hoping to utilize the resources of the immune system against pre-existing cancer, and in chemotherapy, attempting to poison pre-existing cancer. This is to say, cancer research is directed primarily at treating the symptoms rather than at eliminating the causes of cancer.

Much of the basic science that has been directed at the control of cancer has dealt with fundamental questions of genetic control, in part because these questions are the scientifically more "interesting" ones. The danger remains that recombinant-DNA-oriented research will sustain this bias instead of allowing emphasis to shift to vital epidemiological studies and programs aimed at eliminating exposure to carcinogens. Investment in costly containment facilities and supplies required for recombinant DNA research does nothing more than to divert funds from more directly health-oriented research that would attempt to eliminate health problems at their sources, and it may, in addition, compound already existing health problems.

C. Mastering Genetic Control

There is today a strong tendency in science and public policy to give special importance to genetic explanations of human behavior and disease. These positions are based on entirely speculative theories but derive their support from the expediencies that their conclusions allow. A current example is presented by the debate over the heritability of IQ (56) and the contention that "innate learning potential" is easily measured and catalogued using standard IQ tests. Screening for genetic factors in behavior provides a convenient method of dealing with problems whose causes are less willingly acknowledged by social planners. For example, the XYY karyotype in males was believed for a long time, on the basis of poor and eventually discredited evidence, to predispose individuals to aggressive behavior (57,58,59). Tendencies toward aggression have often been explained in this way, and more specific and subtle components of human behavior are scheduled for "elucidation" through the new pseudoscience of sociobiology (60).

In advancing recombinant DNA technologies as means to understand and deal with a great variety of problems, the proponents of these technologies are contributing further to this emphasis on genetics in social policy. This will have the consequence that programs for providing services that change the environment will be neglected in lieu of genetic assessments; attempts at understanding the interaction of environmental and genetic factors in dealing with social problems with the intent of best serving people as they are will be deferred in favor of developing systems that track or channel individuals based on their genetic "'limitations."

The civilization diseases may have some genetic components, but they remain largely undescribed. However, to focus on these undefined factors, and to pursue technological means for modifying or compensating for individual genetic variation, amounts to defending the economic, social, and political order against the people, not *vice versa*.

Genetic screening is an example of a rapidly advancing technology that under ideal circumstances could provide social services of genuine value. Experience, however, does not demonstrate this value. Genetic screening techniques have created a new realm of stigmatization and discrimination that institutions have predictably used to their own ends. Insurance companies and employers in hazardous industries have used these techniques to avoid "susceptible" people (61) while offering few clear-cut benefits to those being screened.

IV. SIGNIFICANCE OF THE RECOMBINANT
DNA DEBATE TO SCIENCE

The debate surrounding recombinant DNA technology may be directly comparable to that surrounding the issues of nuclear power development in both its gravity and its divisiveness in science. It has appeared at a time when technology in general is under suspicion and public scrutiny, when government at national and local levels is becoming increasingly uncomfortable with science policy decision making. While traditional academic freedom is being eroded by the necessity for the accountability of science to the public from which it derives its support, simultaneously some of its defenders threaten and coerce those who voice caution. A process that began as an apparently pious act of self-regulation by scientists has ignited brushfires all over the science landscape. Prominent members of the science "community" are in direct opposition over this issue—expert opinion is divided (62).

How can objective observers be in such a state of disagreement? The answer is perhaps that there is no such thing as scientific objectivity— that perceptions of "objective" reality are wholly dependent on philosophical and ideological premises as well as on other more immediate material factors in people's lives.

A. Nonobjectivity in Science

Because the benefit/risk estimate for recombinant DNA research is largely speculative, depending heavily on imagination, it is especially open to subjective valuation. As mentioned above, for example, assessing the magnitude of benefits accruing from recombinant DNA technologies is contingent on one's view of the social role of technology, an inherently political and ideological consideration. An estimation of the risks involved in this research, in view of our ignorance of the ecology of prokaryotes, is subject to the same criticism. An additional source of subjectivity arises at the personal level, involving the self-perception of one's own contribution to science and technology and the views one accepts of the role of science and technology in society.

For many people who work in science, the value of their work depends to some extent on how they see it contributing directly or indirectly to human betterment. In a society where institutions do not operate a priori to serve desirable social ends, there is created an incentive toward believing that better technology tends to shift the outcome in favor of serving those ends, that new knowledge has intrinsic positive value. Consequently, many medical researchers pursue in science answers to problems for which other solutions—that is, solutions that attempt to change social conditions—are entirely beyond their control or not even evident. Some people for this reason may have an unduly optimistic outlook in the debate over recombinant DNA research. Biases in perspective can also be observed in other scientific controversies, such as those surrounding techniques in the fields of psychosurgery, genetic screening, the funding of cancer research, and the role of computers in society.

Others in science view research issues largely from their individual, immediate vantage point, in terms of creative intellectual activity or employment. Many of these workers tend to envision limitations, constraints, and choices in scientific investigation in the same way that recognized science authority does. Many people in science have careers whose success frankly depends on the rapid exploitation of opportunities in scientific discovery, sometimes involving developments with commercial implications. The resulting advantages include publications, appointments, the realization of creative potential, esteem with family and colleagues, and recognition by institutions and officials. Further rewards may include entree into business and government circles, association with venture capitalists, and invitations to corporate boardrooms. It is clear that in situations where advances are imminent, the personal benefits and risks for some scientists, as for investors, can differ from those of the public. The influence of personal interests is no less likely a factor in determining "objective reality" for these scientists than for dedicated medical researchers or others who view themselves as protagonists of the people's interests.

B. Popular Decision Making In Science Policy

Research policy is inextricably tied to real-world politics, even though many in science would prefer it otherwise. The development of knowledge reflects the dominant social forces in a society, which in the present day United States means the interests of those who hold wealth and property. Historically, this influence is apparent in theories of biological determinism, which have been used to support the concept of social Darwinism, and are currently employed to allege fundamental, innate sex-role and racial differences as well as the inheritability of intelligence. This influence is exercised in the formulation of research objectives in government, private industry, and academia.

Public policy for science must be determined by a process based on popular awareness, organization, and control. A corresponding institutional environment would be required that does not now exist. Such an environment might include unions with strong member participation and control, with extensive internal education programs, and with active involvement in defining and enforcing both government legislation and corporate policy through collective bargaining and action. Another avenue for public control of science policy must be through community-based research and human experimentation boards, provided that they are not under the control of government bodies. They must also attract a wide participation—not only residents of the local community but also people working in the affected institutions, including hospital and university workers, technicians, and other science workers.

Without this organization, public discussion, debate, and criticism must be employed as the major vehicle to affect the existing decision-making apparatus. Few prominent scientists encourage this process; rather, the consensus among scientists remains that they alone should decide science policy. Like the issue of nuclear power, however, that of recombinant DNA technology diffuses into the public domain despite the establishment experts' attempts to retain control of the decision-making process.

How can good judgment on scientific issues be exercised by the "masses"? This, we propose, is analogous to the question: how do top government leaders and policy experts decide questions of science and technology policy? *They rely on experts whom they believe to be credible.* The public not only should gain better understanding of the sciences but also should be able to evaluate the credibility of experts. What are these experts' views on the role of technology and on specific issues bearing on the people's interests? How have they contributed to dealing with the real problems of the society, and what are their stakes in these matters? Evaluating experts is an important task for a properly functioning review board, and it was the primary concern of the Cambridge Experimentation Review Board established to assess recombinant DNA research policy. This board, through fortunate and unusual circumstances, was selected to represent a broad sample of the population.

Just as the rulers of the country can pick and choose among experts and the opinions that they espouse, so can the people.

V. SUMMARY AND CONCLUSIONS

The NIH Guidelines for Research Involving Recombinant DNA Molecules were authored to a large extent by self-interested researchers in the field of molecular genetics, who avoided the *a priori* question of whether recombinant DNA research should be pursued at all. This avoidance is revealed by the NIH Draft Environmental Impact Statement, a document released three months after the Guidelines, in direct violation of the National Environmental Policy Act of 1970. In further violation of the Act, the Draft EIP discusses neither possible alternatives to the proposed action nor the global implications of the future conduct of the research. The Guidelines are limited in applicability and establish little mechanism for enforcement, save the withdrawal of NIH-sponsored funding.

The choice of host organism for the receipt of foreign DNA, the bacterium *Escherichia coli,* is a poor one. *E. coli* is a poorly characterized pathogen that is ubiquitous in habitat. The possibility that foreign DNA implanted into *E. coli* K12 could eventually be transferred to another microorganism, resulting in the formation of a novel pathogen, cannot at present be ruled out either by experiment or by consensus within the scientific community.

Physical containment procedures established to reduce the likelihood of such an event cannot insure against the release of hybrid microorganisms generated by recombinant DNA techniques. Additional biological containment barriers do not measure up to expectations as worded in the Guidelines.

The benefits claimed as accruing from the use of these techniques are ameliorations of problems that are blunt consequences of the prevailing social and economic order. Recombinant DNA research will do little more than divert funds from much-needed social and scientific programs aimed at correcting the sources of health-related problems. This effort will reinforce the societal predisposition to correct the symptoms of such problems as cancer, while effecting no substantial progress towards the elimination of these problems.

Alternatives to the present situation include abandonment of this research program, restriction of such recombinant research to regional facilities, or the tight regulation of recombinant DNA techniques at the community level as has been sought in Cambridge, Massachusetts. The use of alternative techniques that do not involve the creation of novel microorganisms should be explored by the major federal funding agencies. Such techniques at present have the ability to supplant many of the proposed recombinant DNA experiments.

Self-regulation of science becomes untenable when that science begins to affect directly the lives of those who sponsor that societal luxury. Scientists must come to terms with the fact that they may be held accountable for the results of their work. Science must shed its guise of being "for the people" when it is not directed at eliminating the causes of social problems; at a time on this planet when the quality of the environment, and therefore life itself, is at a premium, science must redirect its goals in line with social responsibility and public accountability. Science for other ends must yield to science for the people.

REFERENCES

1. Cetus Corporation, Background material, October 1975.
2. Slesin, L., Recombinant DNA research: A chronology, Occasional Paper # 2, MIT Laboratory of Architecture and Planning, November 10, 1976.
3. Berg, P., D. Baltimore, H. W. Boyer, S. N. Cohen, R. W. Davis, D. S. Hogness, D. Nathans, R. O. Roblin, J. D. Watson, S. Weissman, and N. D. Zinder, Potential biohazards of recombinant DNA molecules, *Science* **185,** 303, 1974.
4. Fredrickson, D.S., Decision of the director, NIH, to release guidelines for research on recombinant DNA molecules, *Federal Register,* Part II, July 7, 1976.
5. Gwinne, P., S. G. Michaud, and W. J. Cook, Politics and genes, *Newsweek,* pp. 51–52, January 12, 1976.
6. Berg, P., D. Baltimore, S. Brenner, R. O. Roblin, III, and M. F. Singer, Summary statement of the Asilomar Conference on Recombinant DNA Molecules, *Proc. Nat. Acad. Sci. USA,* **72,** 1981, 1975.
7. Cambridge Experimental Review Board, Guidelines for the use of recombinant DNA molecule technology in the city of Cambridge, January 5, 1977.
8. National Environmental Policy Act.
9. Sack, R. B., Human diarrheal disease caused by enterotoxigenic *E. coli, Annual Review of Microbiology,* **29,** 333, 1975.
10. Cooke, M. E., *Escherichia Coli and Man,* Churchill Livingstone, 1974.
11. Falkow, S., *Infectious Multiple Drug Resistance,* Pion Limited, 1975.
12. McGowan, J. E., M. W. Barnes, and M. Findland, Rates of bacteremic patients at Boston city hospital during 12 selected years, 1935–1972, *J. Infect. Dis.,* **132,** 316, 1975.
13. Mejerowitz, R. L., A. A. Medeiros, and T. F. O'Brian, Recent experience with bacillemia due to gram-negative organisims, *J. Infect. Dis.,* **124,** 239, 1971.
14. Anderson, E. S., Viability of *E. coli* K12 in the human intestine, *Nature,* **255,** 502, 1975.
15. Smith, H. W., Survival of orally administered *E. coli* in the alimentary tract of man, *Nature,* **255,** 500, 1975.
16. Smith, M. G., R-factor transfer *in vivo* in sheep with *E. coli* K12, *Nature,* **261,** 348, 1976.
17. Avery, O. T., C. M. MacLeod, and Maclyn McCarty, Studies on the chemical nature of the substance inducing transformation of pneumococcal types, *J. Experimental Med.,* **79,** 137, 1944.
18. Fein, D., G. Burton, R. Tsutakawa, and D. Blenden, Matching of antibiotic resistance patterns of *Escherichia coli* of farm families and their animals, *J. Infect. Dis.,* **130,** 274, 1974.
19. Department of HEW, NIH, Recombinant DNA research guidelines, Draft Environmental Impact Statement, *Federal Register,* **41,** 176, September 9, 1976.
20. Edlin, G., L. Lin, and R. Kudrna, λ lysogens of *E. coli* reproduce more rapidly than non-lysogens, *Nature,* **255,** 735, 1975.
21. Stewart, F., and B. Levin, The population biology of bacterial plasmids, submitted to *Genetics,* 1976.

Science for the People 125

22. Recombinant DNA research guidelines, *Federal Register*, Part II, July 7, 1976.
23. Berg P., Letter to DeWitt Stetten, Deputy Director for Science, NIH, Sept. 2, 1975.
24. Wedum, A. G., The Detrick experience, in *Recombinant DNA Research*, vol. 1, Documents Relating to "NIH Guidelines for Research Involving Recombinant DNA Molecules," February 1975-June 1976, U.S. Dept. of HEW, NIH, August 1976.
25. Levy, S. B., G. B. FitzGerald, and A. B. Macone, Spread of antibiotic-resistant plasmids from chicken to chicken and from chicken to man, *Nature*, **260,** 40, 1976.
26. Ryder, R. W., I. K. Washsmuth, A. E. Buxton, D. G. Evans, H. L. DuPont, E. Mason, and F. F. Barrett, Infantile diarrhea produced by heat-stable enterotoxigenic *Escherichia coli*, *N. Engl. J. Med.*, **295,** 849, 1976.
27. Goldstein, R., C. Orrego, and P. Youderian, Analysis and critique of the Curtiss report on the *Escherichia coli* strain intended for biological containment in DNA implantation research, submitted to the NIH Recombinant DNA Molecule Program Advisory Committee, May 30, 1976.
28. Curtiss, R. III, D. Pereira, J. E. Clark, J. C. Hsu, R. Goldschmidt, S. I. Hull, R. Moody, L. Maturin, and M. Inoue, Construction, Properties and Testing of χ 1776, March 30, 1976.
29. Struhl, K., J. H. Cameron, and R. W. Davis, Functional genetic expression of eukaryotic DNA in *Escherichia coli*, *Proc. Nat. Acad. Sci. USA*, **73,** 1471, 1976.
30. Adhya, S., and L. W. Enquist, Our critique of Roy Curtiss' EK-2 host proposal, memo to W. P. Rowe, Chief, Laboratory of Viral Diseases, NIH, May 31, 1976.
31. Duncan, M., R. Goldstein, P. Primakoff, and C. Orrego, *Recombinant DNA Research*, vol. 1, p. 350, *op. cit.*
32. Nightengale, E., *Recombinant DNA Research*, vol. 1, 489, *op. cit.*
33. Falkow, p. 239, *op. cit.*
34. Davis, B., and R. H. Sinsheimer, The hazards of recombinant DNA, *Trends in Biochemical Sciences*, N178, August 1976.
35. Sinsheimer, in Davis and Sinsheimer, *op. cit.*
36. Berg, P., U.S. Department of HEW, NIH, *Proceedings of a Conference on NIH Guidelines for Recombinant DNA Molecules*, public hearings held at a meeting of the Advisory Committee to the Director, NIH, 1976.
37. Goldstein, R., *ibid.*
38. Rougeon, F., P. Kourlisky, and B. Mach, Insertion of a rabbit β-globin gene sequence into an *E. coli* plasmid, *Nucleic Acids Research*, **2,** 2365, 1975.
39. Williamson, B., Is HnRNA really pre-mRNA, *Nature*, **264,** 397, 1976.
40. Battelle Office of Corporate Communications, Membranes help insulin production, *Battelle Today*, no. 2, p. 2, November 1976.
41. Berg, P., *Proceedings of a Conference on NIH Guidelines for Recombinant DNA Molecules*, *op. cit.*
42. Baltimore, D., *ibid.*
43. Cetus Corporation, *op. cit.*
44. Reutlinger, S., and M. Selowsky, *Malnutrition and Poverty*, Johns Hopkins Univ. Press, Baltimore, 1976.

45. Wade, N., Inequality the main cause of world hunger, *Science*, **194,** 1142, 1976.
46. Cleaver, H. M., The contradictions of the green revolution, *Monthly Review*, June 1972.
47. Lappé, F. M., and J. Collins, *When More Food Means More Hunger*, Institute for Food and Development Policy, Hastings-on-Hudson, N.Y.
48. Wade, N., Green revolution (1): A just technology, often unjust in use, *Science*, **186,** 1093, 1974.
49. Cleaver, *op. cit.*
50. Wade, N., Green revolution (2): Problems of adapting a western technology, *Science*, **186,** 1186, 1974.
51. Concerned Rush Students, Turning prescriptions into profits, *Science for the People*, p. 6, January 1977.
52. Hall, R. H., *Food for Naught, The Decline of Nutrition*, p. 229, Vintage, 1976.
53. Culliton, B. J., Penicillin-resistant gonorrhea: New strain spreading worldwide, *Science*, **194,** 1395, 1976.
54. Epstein, S. E., Epidemic: The cancer producing society, *Science for the People*, p. 4, July 1976.
55. ————, The political and economic basis of cancer, *Technology Review*, p. 35, July/August 1976.
56. Felman, M. W., and R. C. Lewontin, The heritability hang-up, *Science*, **190,** 1163, 1975.
57. King, J., and J. Beckwith, The XYY syndrome: A dangerous myth, *New Scientist*, p. 474, November 14, 1974.
58. Genetic Engineering Group, The XYY controversy (continued) *Science for the People*, p. 28, July 1975.
59. Beckwith, J., and L. Miller, The XYY male: The making of a myth, *Harvard Magazine*, p. 30, October 1976.
60. Wilson, E. O., *Sociobiology: The New Synthesis*, Harvard University Press, Cambridge, Mass., 1975.
61. Miller, L., and J. Beckwith, Genetic screening: Benefits and limitations, in *The Biosciences and Society*, R. E. Munro, ed., Liverpool Univ. Press, England, in press.
62. Bennett, W., and J. Gurin, Science that frightens scientists: The great debate over DNA, *Atlantic*, p. 43, February 1977.

The Case Against
Genetic Engineering

George Wald

Recombinant DNA technology faces our society with problems unprecedented not only in the history of science, but of life on the Earth. It places in human hands the capacity to redesign living organisms, the products of some three billion years of evolution.

Such intervention must not be confused with previous intrusions upon the natural order of living organisms; animal and plant breeding, for example; or the artificial induction of mutations, as with X-rays. All such earlier procedures worked within single or closely related species. The nub of the new technology is to move genes back and forth, not only across species lines, but across any boundaries that now divide living organisms, particularly the most fundamental such boundary, that which divides prokaryotes (bacteria and bluegreen algae) from eukaryotes (those cells with a distinct nucleus in higher plants and animals). The results will be essentially new organisms, self-perpetuating and hence permanent. Once created, they cannot be recalled.

This is the transcendent issue, so basic, so vast in its implications and possible consequences that no one is as yet ready to deal with it. We can't deal with it until we know a lot more; and to learn those things we would have to venture out into this no-man's land. It is nothing like making new transuranic elements. New elements only add to the simple

Reprinted from *The Sciences,* Sept./Oct. issue (1976) by kind permission of the author and the publisher.

series of integral atomic numbers that underlie the Periodic System. Their numbers are limited and their properties highly predictable. Not so new organisms. They can be as boundless and unpredictable as life itself.

Up to now living organisms have evolved very slowly, and new forms have had plenty of time to settle in. It has taken from four to 20 million years for a single mutation, for example the change of one amino acid in the sequence of hemoglobin or cytochrome c, to establish itself as the species norm. Now whole proteins will be transposed overnight into wholly new associations, with consequences no one can foretell, either for the host organisms or their neighbors.

It is all too big, and is happening too fast. So this, the central problem, remains almost unconsidered. It presents probably the largest ethical problem that science has ever had to face. Our morality up to now has been to go ahead without restriction to learn all that we can about nature. Restructuring nature was not part of the bargain; nor was telling scientists not to venture further in certain directions. That comes hard. With some relief, most biologists turn away from so vast and uncomfortable an issue and take refuge in the still knotty but infinitely easier technical questions: not *whether* to proceed, but *how*. For going ahead in this direction may be not only unwise but dangerous. Potentially, it could breed new animal and plant diseases, new sources of cancer, novel epidemics.

We must never forget that the first intimation of these potential hazards came from workers in this field. All honor to them. Faced with unique problems, as they alone then realized, they did unprecedented things. They brought about a voluntary moratorium on certain, more clearly dangerous kinds of experiments. And now, after three years of debate, consultation and negotiation, the National Institutes of Health issued its Guidelines on June 23.

One can hardly read the Guidelines, or the careful and sensitive statement by Donald Frederickson, the Director of NIH, on releasing them, and not be impressed with the goodwill and concern that animate them. Yet there is much in this enterprise and in the Guidelines themselves that troubles me greatly.

First and foremost: the very existence of the Guidelines begs the central question, whether this kind of research should proceed at all. The experiments are quite simple and straightforward. Can they be stopped? Perhaps they can. If one could neither publish the results nor exploit them commercially there would be little incentive to do them.

As for the Guidelines themselves, the first thing to understand is the context of utter ignorance of what to expect in which they had to be formulated. The Guidelines begin by saying: "At present the hazards may be guessed at, speculated about, or voted upon, but they cannot be known absolutely in the absence of firm experimental data—and, unfortunately, the needed data were, more often than not, unavailable."

PHYSICAL CONTAINMENT

The purpose here is to keep the recombinants from escaping the laboratory. The Guidelines list four levels of containment labeled P1 to P4; but in effect there are only two levels, a lesser—P3—and a greater—P4.* This classification is itself deceptive, for it makes the prevalent P3 facility sound better than it is, three quarters of the way to the top, whereas in fact it is the lowest level of containment. P1 is just a laboratory, P2 the same laboratory with a warning sign on the door. A young woman demonstrating a P2 experiment at an open hearing before the Cambridge City Council made a point of putting on the prescribed laboratory coat: but she had long, loose, abundant hair that could have carried more bacteria or viruses than a dozen lab coats.

A P3 facility such as has just been authorized at Harvard employs various devices intended to minimize the escape of recombinants. Yet the reason proponents of the facility at Harvard gave for building it within our Biological Laboratories, close to the laboratories of prospective users—though the building is half a century old and infested with ants and cockroaches—was that workers in the facility would be the principal means of spreading contamination, and hence should have to move as short distances as possible. I think it is probably correct that the laboratory personnel will be the principal means of spreading any potential infection. But in that case, wherein lies the containment? Why the elaborate and costly precautions within the facility?—the small unit at Harvard is estimated to cost more than $800,000. And what matter whether distances between the labs are short or long? All these workers move freely throughout the building and the city; they meet with us, eat with us, and—most importantly—they teach classes of young students. I see no reason to believe that P3 containment, even if conscientiously enforced, can effectively contain.

BIOLOGICAL CONTAINMENT

One of the most unsettling aspects of present recombinant DNA research is that the host organism that receives the plasmids that carry foreign genetic material is almost always the colon bacillus, *Escherichia*

Editors' note: The specific details of each of the four levels of physical containment are described in the NIH Guidelines in Appendix II (pp. 343–378).

coli, a constant inhabitant of the human bowel. To do potentially hazardous experiments, why pick an organism that lives in us? The reason is that we know more about *E. coli* than about any other living organism. Yet what is to keep some hybridized *E. coli* turned pathogenic from infecting its conventional human hosts? Or transferring those plasmids to human cells?

Hence the stress on the assurance that all recombinant experiments with *E. coli* will use the K12 strain, which, we are told, can exist only under special laboratory conditions and neither survives nor reproduces in the human gut. The use of this strain is the "biological containment."

In this connection Stanley Falkow of Seattle, Washington, submitted to the NIH Recombinant DNA Advisory Committee a highly informative report on the ecology of *E. coli.* According to Falkow, almost innumerable serologically distinct strains of *E. coli* inhabit the human colon from time to time, the population constantly changing. The more persistent (resident) strains last several months, other (transient) strains only a few days. The statement that the K12 strain does not survive in the human bowel rests primarily on observations by E. S. Anderson and H. Williams Smith that this strain "is a poor colonizer of the human alimentary tract." Smith found a mean survival time of about three days. Anderson also found that it "multiplied to some extent in two of eight subjects." Hardly an impressive statistic! Furthermore he could detect plasmid transmission from K12 to other enteric flora when it was fed "in substantially high numbers."

Falkow confirms these observations, and adds another that is singularly important: Working with calves, he found that introducing certain plasmids into K12 increased its survival and multiplication in the gut many times over. He concludes that "it may not be too farfetched to suggest that some DNA recombinant molecules could profoundly affect the ability of this *E. coli* strain to survive and multiply in the gastrointestinal tract."

These are oddly inadequate data to carry such weight. We would like to know much more. How does K12 get along in persons whose colons are relatively empty of bacteria and hence offer it little competition—such as newborn infants, or persons who have just been treated with sulfa drugs or antibiotics? So-called biological containment seems to me as problematical as P3 physical containment.

ENFORCEMENT

The Guidelines are just that, hence wholly voluntary. The only penalty now available for simply disregarding them is the possible withholding of federal research support. Obviously this applies only to research dependent on federal funds. It leaves out completely the rapidly growing industrial exploitation of recombinant DNA technology.

BENEFITS AND RISKS

I have up to now said almost nothing of the potential benefits of this technology. I think that the most certain benefits to come out of it would be scientific: increased understanding of important biological phenomena, such as the mechanisms that turn specific gene activities on and off, that trigger cell multiplication and differentiation, that regulate cell metabolism. We are also offered the prospect of large practical benefits: teaching cereal plants to fix their own nitrogen from the air, new bacterial syntheses of drugs and hormones, the hope that increased understanding of cancer may lead to its cure. I cannot think of a single instance of such developments, scientific or practical, that does not also involve large potential risks.

Consider cancer. If indeed it turns out that recombinant DNA research will improve our understanding of cancer, that would still be far from showing us how to cure it. In spite of many statements, as vague as they are optimistic, that the cure of cancer lies in this direction, it is hard to see how that is to happen. Any such hope must be balanced against the real possibility that recombinant DNA experiments may induce new cancers. If right now I had to weigh the probabilities of either event I would guess that recombinant DNA research carries more and earlier risks of causing cancers than hope of curing them.

Add that about 80 percent of cancer in this country is now believed to be of environmental origin. The largest single cause of lung cancer is smoking, but one is free to smoke or not. About 40 percent of those environmental cancers happen in the work places, through involuntary exposure to a rapidly increasing variety of toxic materials in industrial use. If one were really concerned about cancer, there is the obvious place to attack it, with sure and immediate results.

Or consider a frankly industrial development. General Electric is reportedly trying to patent a newly assembled strain of *Pseudomonas* bacteria that can wholly digest crude oil. It was developed there by Ananda Chakrabarty by transferring plasmids from several strains, each of which could digest oil partially, into a single strain that can do the whole job. It is pointed out that this organism could be very useful for cleaning up oil spills. Very true; but how about oil that has not spilled?—oil still in the ground, or on the way, or stored? Can this organism be contained, kept from destroying oil we want to use? Or will we need to begin to pasteurize oil?

THE CORPORATE CONNECTION

As early as February 1974 *Fortune* magazine hailed the coming importance of genetic recombination in industrial developments. "The best microbes are freaks," it said and "many scientists see an important industrial role ahead for the powerful new methods of transferring genetic material from one cell to another." It named a number of them including a few who are already directing corporate activities.

The industrial exploitation of recombination technology raises special problems for in that, as any other business enterprise, the major goal is to maximize profits and frequently in the past, public and worker safety and health have been subordinated to that end. Last May representatives of about twenty drug and chemical companies met with NIH Director Frederickson to discuss the proposed Guidelines. They expressed "general support," but made three points: (1) the fear that voluntary Guidelines might lead to enforceable regulations, (2) for reasons of competition, the companies could not afford to reveal what recombinant DNA experiments they were performing, and (3) they found other features of the Guidelines onerous, for example the restrictions on large-volume experiments, which of course are less easily contained, but which they require in testing procedures for commercial feasibility.

THE DILEMMA OF THE NIH

The recombinant DNA development faces NIH with an interesting predicament. Anything I say of this is said sympathetically, for under Donald Frederickson's perceptive leadership it is doing as well as could be hoped. Yet is it possible for the same agency both to promote and regulate? The old Atomic Energy Commission, set up originally to regulate, turned instead to promoting nuclear power, and that eventually destroyed it. It has been replaced by two separate agencies, one for research and development, the other for nuclear regulation.

NIH, on the contrary, set up to promote scientific and medical research, is now being forced into regulation. Its entire impulse, as that of all other institutions concerned with research, is to avoid regulation, to maintain full freedom of inquiry. Probably that is why it can bring itself only to promulgate voluntary guidelines. Surely it recognizes the previous history of ineffectuality of voluntary self-regulation in other areas. For the NIH Guidelines to be enforced, academically and particularly industrially, they would have to become regulations, backed by legislation, with adequate provisions for licensing, inspection and supervision. The NIH would like to avoid such measures and so, as a scientist, would I. Yet this situation seems to demand them, and I fear that scientists and science will eventually have to suffer because of them.

WHAT TO DO

First, I think it essential to open a wide ranging and broadly representative discussion of the central issue: whether artificial exchanges of genetic material among widely different living organisms should be permitted.

Second, in consideration of the potential hazards and our present state of ignorance, I would confine all recombinant DNA experimentation that transcended species boundaries to one or a few national or regional laboratories where they can be adequately confined and supervised. There, every attempt should be made to define the hazards that are now only guessed at. If trouble should arise, I would expect it to involve first the workers in such laboratories and their families, whose health should be carefully monitored. Until such trials have told us better what to expect, this kind of investigation should have no place in crowded cities or educational institutions.

Third, industrial research and development in this area need most of all to be brought under control. The usual secrecy that surrounds industrial research is intolerable in a province that can involve such serious consequences and hazards. The need for licensing, inspection and supervision will probably require national legislation. Hearings in the Congress should begin at once to consider these issues.

As I write these words, they trouble me greatly. I fear for the future of science as we have known it, for human kind, for life on the Earth. My feelings are ambivalent, for the new technology excites me for its sheer virtuosity and its intellectual and practical potentialities; yet the price is high, perhaps too high. We are at the threshold of a great decision with large and permanent consequences. It needs increasing public attention here and worldwide, for it concerns all humankind. That will take time, during which we can try to learn, as safely as that can be managed, more of what to expect, of good and ill. Fortunately there is no real hurry. Let us try, with goodwill and responsibility, to work it out.

The Scientists Debate:
Arguments for the Defense

Evolution, Epidemiology, and Recombinant DNA

Bernard D. Davis

Recent developments in molecular genetics have made it possible to insert small fragments of genetic material (DNA) from any organism, including man, into tiny self-replicating units of DNA from bacteria, called plasmids. By reintroducing the altered plasmid into bacteria one can produce "hybrid" cells, and these can then be used to prepare large quantities of the inserted DNA in pure form. The prospects opened up by this technique have generated not only enthusiasm but also wide public concern, focused primarily on the presumed risk that some of the novel organisms produced may spread outside the laboratory and may then cause incalculable harm. Concern has also been expressed over a more long-term conjectural risk: that in recombining DNA from distant sources we are meddling with the natural course of evolution, and in this Promethean venture we will eventually create unforeseeable disasters.

Both these issues raise ethical questions, on which a public consensus is the ultimate arbiter. But a rational decision requires an informed public. I believe the public has unfortunately been widely misinformed, for most of the discussion has been based on two false assumptions: that we are entirely in the dark in this novel territory, and that any novel organism we may produce is likely to survive and spread if it escapes

Editors' note: This chapter is a revised version of a public lecture delivered at the Science Center at Harvard University on January 5, 1977, and published in a report of the Congressional Research Service of the Library of Congress (December 1976).

from the laboratory. In fact, the latter assumption ignores Darwin's great discovery: the dominating role of natural selection in determining what survives, multiplies, and evolves in the living world. Moreover, though the invisible organisms of the microbial world were not known to Darwin, they were soon brought into the same framework by Pasteur: in establishing that bacteria do not arise by spontaneous generation he also showed that the kinds growing out in any medium are the ones *selected* by that medium. For example, from the same mixture of contaminants from the air one regularly sees the outgrowth of one kind of organism in grape juice, producing an alcoholic fermentation, and another kind in milk, producing a lactic acid fermentation.

The contributions of Pasteur, and those of Robert Koch, also gave rise to a sophisticated methodology for identifying various microbes that cause infectious disease and for handling them in the laboratory with relatively safety. These developments then led to the science of epidemiology—a kind of applied extension of evolutionary theory concerned with analyzing the factors that promote or hinder the spread and the variation of pathogenic microbes. And our knowledge of evolutionary theory, microbiology, and epidemiology is thoroughly relevant to the problem of assessing the risks of making various kinds of recombinant DNA. It is this knowledge that I wish to review briefly. I shall also emphasize the need to consider separately three stages of risk: that a particular class of experiments may inadvertently produce a pathogenic organism; that this organism may cause disease in laboratory workers; and that it may spread outside the laboratory.

The evolutionary considerations that I shall invoke may be dismissed by skeptics as mere handwaving, since they cannot provide the hard data that we have become accustomed to in modern experimental biology. But in this light nearly all of Darwin's arguments, based on inferences about the past and not on verifiable experiments, could be (and often have been) dismissed. Indeed, such a reductionist approach is perhaps encouraged today by the fact that molecular genetics has thrown a great deal of light on one major aspect of Darwinian theory: that concerned with the origin of hereditary variation. But this theory had another major aspect, concerned with natural selection—and this set of problems cannot be reduced to the same level. For natural selection involves an altogether different set of concepts, based on the behavior of heterogeneous populations of organisms and not on the properties of homogeneous populations of molecules. And for those who would discount the value of any considerations other than direct observations on the novel organisms, I would note that Darwin's profound, unifying generalization was well established even before modern studies verified it by the direct demonstration of natural selection in a variety of organisms.

Let us review some pertinent principles from evolution and microbiology.

I. BACKGROUND

A. Microbiological and Evolutionary Principles

1. The Meaning of Species. As evolution proceeded from prokaryotes (bacteria with a single chromosome) to eukaryotes (higher organisms, with a more complex genetic apparatus), it created the mechanism of sexual reproduction. By reassorting the genes of paired parents this process provides vastly increased genetic diversity for natural selection to act on. But while diversity is necessary for evolution, unlimited recombinations from the total pool of genetic material in the living world would not be useful, for a successful organism must have a reasonably balanced set of genes. Hence the evolution of sexual reproduction was accompanied by the evolution of fertility barriers that separate organisms into species: groups that reproduce in nature only by mating with other members of the same group, and not with members of other species.

Several kinds of fertility barriers emerge as organisms diverge in evolution. First, matings between closely related species may produce hybrid offspring that are viable but not fertile (e.g., the mule); more distant crosses produce no offspring at all; and finally mating becomes physically impossible. The evolutionary value of such barriers is clear: to avoid useless production of grossly unfit, nonviable progeny.

Unlike eukaryotes, prokaryotes ordinarily reproduce by asexual cell division, which means that the genetic properties of a strain remain constant for generation after generation, except for rare mutations or rare transfer of a block of genes from one cell to another. These gene transfers, which are usually mediated by plasmids or viruses, do not show a sharp species boundary: they simply become less efficient the greater the evolutionary separation between the donor and the recipient. Prokaryotes therefore have no true species. *E. coli*, for example, is the name given to a range of strains with certain common features and also with a variety of differences—in surface molecules, nutrition, growth rate, sensitivity to inhibitors, etc. These differences determine the relative Darwinian fitness of various strains for various environments.

2. Bacterial Ecology. Every living species is adapted to a given range of habitats. The set of bacterial strains called *E. coli* thrive only in the vertebrate gut. In water they survive temporarily but quickly die out.

(Indeed, for that reason the *E. coli* count of a pond or a well is a reliable index of its continuing fecal contamination.) In the gut there is intense Darwinian competition between strains, depending on such variables as growth rate, nutritional requirements, ability to scavenge limited food supplies, adherence to the gut lining, and resistance to antimicrobial factors in the host. Hence most novel strains are quickly extinguished, in the kind of competition envisaged by Darwin for higher organisms. With bacteria the process is very rapid, because the generation time is as short as 30 minutes and the selection pressures are often intense.

It is easy to demonstrate that the environment in the gut (i.e., the type of food and physiological state) plays a decisive role in determining the distribution of organisms in its normal flora. For example, when a baby shifts from breast feeding to solid food the character of the stool changes dramatically, as lactic acid bacteria, which produce sweet-smelling products, are replaced by *E. coli* and other foul organisms. In a more experimental example, early in this century Mechnikov assumed that anything foul-smelling must be toxic and hence must limit longevity, and he attempted to displace these organisms by supplying a large number of lactic acid bacteria in the form of yogurt. The experiments were a dismal failure (except for their commercial consequences). In a third, more recent example, the administration of antibiotics frequently disturbs the bacterial population of an individual, and it has not proved possible to accelerate recovery of the normal flora by administering desired strains. From these several examples it is clear that in the gut the environment plays a dominating role in determining what strains persist.

3. Pathogenesis.   Only a very small fraction of all bacterial species can cause disease. The rest play essential roles in the geochemical cycle, in which carbon dioxide from the air is fixed in plants or bacteria by photosynthesis, plants are eaten by animals, the animals and plants return to the soil after death, and there microorganisms digest the dead organic matter and return the carbon to the atmosphere as carbon dioxide.

Various infectious bacteria differ from each other in several distinct respects: *infectivity* (i.e., the infectious dose, ranging from a few cells of the tularemia bacillus to around 10^6 cells of the cholera vibrio); specific *distribution* of the organism in the body; *virulence* (i.e., the severity of the disease once the infection has overcome natural resistance); and *communicability* from one individual host to another (including length of survival in nature). Each of these attributes, like any complex property, depends on the coordinate, balanced activity of many genes, capable of independent variation.

It is especially important to distinguish the ability to *produce* a serious disease from the ability to *spread*. For example, the tetanus bacillus produces a powerful toxin: but it is a normal, noninvasive inhabitant of the mammalian gut, and it can cause fatal illness only when it gains access (usually by trauma) to a susceptible tissue.

4. Normalizing Selection and Diversifying Selection. When an organism grows continuously in a relatively constant environment, natural selection has predominantly a normalizing or stabilizing effect, weeding out the variants that deviate too far in any direction from the well-adapted norm. But when the environment is changed, the same basic process of natural selection has another, diversifying effect: the new circumstances select for the preferential survival and reproduction of variants with increased fitness for those circumstances. This Darwinian principle, extended to bacteria, explains a feature of bacterial cultures that confused the early investigators: rapid fluctuation in various properties during prolonged cultivation. For example, when pathogenic baterial strains from infected hosts are isolated in the laboratory in artificial culture media, they face an abrupt change of environment, and on repeated transfer they quickly become better adapted to the new environment, at the expense of decreased adaptation to the old one (i.e., they lose virulence). The mechanism, which is now clear, does not involve any *directive* effect of the environment on the shifting bacterial population. Instead, rare mutants of all kinds are constantly appearing in the successive generations. (In fact, as much as 10 percent of the cells in each generation may have a change in one of the several million nucleotide units of the cell's DNA, though most of these mutations are not recognized because their effects are either too small to be seen or too large to yield a viable cell.) Among the viable mutants some may be better adapted to the new culture medium, i.e., they can grow slightly faster, or grow slightly longer with a limited food supply; and these outgrow the original strain.

An even more pertinent model is provided by epidemiological observations on the effect of the current wide use of antibiotics in man and in domesticated animals. In this case the environment being changed is the animal host, rather than a laboratory culture medium; and the result has been increased prevalence of drug resistance among some of the microbes that normally inhabit, or that occasionally infect, those hosts. Selection is again the key: the introduction of even large numbers of a variant with a specific drug resistance will not lead to its spread unless the drug is present in the environment to exert a selection pressure (or unless the variant is at least as well adapted as its competitors).

It is clear that natural selection plays an overwhelming role in evolution in bacteria, as in higher organisms. With bacteria its role was long unrecognized, for in those organisms the population shifts seemed too rapid for an undirected process; moreover, the presence of genes and mutations was not discovered until the 1940s. But now selection has become the foundation of bacterial ecology.

B. Benefits of Recombinant DNA Research

An informed decision on recombinant DNA research must take into account not only the risks but also the prospective benefits. Reviewing these very briefly, I would emphasize that this set of procedures is not just a toy to satisfy the curiosity of investigators. It is an extraordinarily powerful and simple tool for studying the structure and function of the DNA of man and other higher organisms, and it is rapidly becoming as indispensable as radioactive isotopes or the electron microscope.

While no one can foresee all the consequences of a basic discovery, the history of molecular biology assures that the new developments in handling DNA will lead to great advances in our understanding of mammalian gene regulation—the key to normal development and differentiation, and also to the defects in regulation of cell growth that result in cancer. At present we do not understand gene regulation in mammalian cells nearly as well as that in bacteria: the cells are harder to work with in many ways, and they contain DNA equivalent to several million genes, or about a thousand times the amount in a bacterium. The recombinant DNA technique can purify fragments containing a single gene and its regulatory elements, thus providing a simplification of the system that greatly facilitates detailed molecular analysis. In addition, the value of such purification of DNA fragments has recently been enormously enhanced by the development, by W. Gilbert and F. Sanger, of simple techniques for determining the sequence of short pieces of DNA, up to 200 bases long. In two days one can now completely determine such a sequence, which previously took two years: hence we can anticipate rapid progress in determining the chemical structure of innumerable mammalian genes.

One can also safely predict the use of recombinant bacteria for producing medically valuable human cell products. Insulin and other protein hormones are the most immediate candidates, but we can also anticipate the manufacture of specific antibodies to replace the deficiencies that are being rapidly revealed by new developments in immunogenetics. More distant prospects include specific antigens for immunization against tumors, and the specific genes or their products that may ultimately be used to treat hereditary enzyme deficiencies.

Quite apart from these practical benefits, I would also emphasize the enormous cultural importance of continuing to encourage free inquiry: the potential loss to society from a precedent of curtailing such inquiry is huge. And here I am not setting up a demand for absolute freedom. It has always been clear that the right to freedom of inquiry has limits, just like the right to freedom of expression. One limit is cruelty; another is unacceptable hazard, whether to individuals, to the population, or to the environment. The question is therefore not whether restraints are acceptable in principle but whether any proposed restraints relate sensibly to the actual hazards. Perhaps the closest model is research with radioactive isotopes. Here investigators uniformly accept regulations and licensing requirements, which they recognize as reasonable responses to real hazards. Severe restrictions based on conjecture, however, will present a different problem, and a demand for absolute freedom from risk would be a prescription for paralysis.

In turning now to the risks, I would note that they are often not directly commensurable with benefits (i.e., expressible in similar units), unlike costs compared with benefits. For this reason a particular activity must be judged for acceptability not only in terms of such a comparison but also in terms of its probable increment to the risks that we already live with. I would further emphasize that it is easy to draw up a scary hypothetical scenario if one's imagination need not be limited by considerations of probability. But any realistic discussion must consider probabilities. And as I mentioned earlier, we must consider three probabilities: that experiments with a given kind of DNA will produce a dangerous organism, that that organism will infect a laboratory worker, and that the organism will escape and spread in the community or the environment.

II. HAZARDS

A. Risk of Producing a Harmful Organism

There is no doubt that molecular recombination *in vitro* could produce pathogenic derivatives of *E. coli*. For example, if a strain carrying the gene for a potent bacterial toxin multiplied enough in the host, or even without multiplication if it were taken up in a large enough dose to produce a harmful amount of the toxin, it could cause disease. A strain carrying the genome of a tumor virus might also be hazardous, but its effect would be less certain, for unlike a toxin producer it would require more than the normal function of the foreign DNA within the bacterial carrier: it would require the release of the viral DNA from the bacterial cell and its infection of animal host cells. While that probability may be very low, we cannot assume that it is negligible. Both these kinds of experiments are appropriately prohibited in the NIH Guidelines today.

I would like to concentrate on a kind of experiment that is allowed, but that is causing great concern and is restricted to quite special facilities: the so-called "shotgun" experiment, in which one transfers random fragments of DNA from mammalian cells. It is clear that with this procedure the probability of isolating a strain with a gene for a toxic product, or with the genes of a tumor virus, is exceedingly low. In addition, I would seriously question whether the kind of novelty that we fear in the products of such experiments is real: for there are strong reasons to believe that recombinants of this kind have been appearing in nature ever since higher animals evolved.

The reasons are the following. It is known that bacteria can take up naked DNA from solution. In fact, two different strains of pneumococcus have been shown to be able to produce a third, recombinant strain in an animal body, by release of DNA from a lysed cell of one strain and its uptake by an intact cell of the other. Moreover, bacteria in the gut are constantly exposed to fragments of host DNA, released by death of the cells lining the gut; while bacteria growing in carcasses have a veritable feast. The efficiency of such uptake of mammalian DNA by bacteria (especially the kinds found in the gut) is very low. However, the scale of the exposure in nature is extraordinarily large—around 10^{20} to 10^{22} bacteria are produced collectively in the human species per day. (To place this figure in perspective we might note that the weight of the earth is about 10^{27} grams.) Hence it seems virtually certain the recombinants of this general class have been formed innumerable times over millions of years, and thus have been tested in the crucible of natural selection. Moreover, such organisms are undoubtedly also being formed in nature today. If they had high survival value, we would be recognizing short stretches of mammalian DNA in *E. coli*. We do not. On the other hand, naturally occurring recombinants might be appearing and even causing transient epidemics, which are escaping our attention. But then we would have to ask how much our laboratories could add, performing experiments on the scale of 10^{10} to 10^{12} bacteria.

B. Risk of Laboratory Infection

Having considered the probability of inadvertently producing a harmful organism, we must now consider the probability that such an organism would cause a laboratory infection. Let us assume the worst case: an *E. coli* strain producing a potent toxin absorbable from the gut, such as botulinus toxin. (Production of such a strain is at present prohibited.) The danger of harm from a laboratory infection with such a strain would be real. However, there are a number of reasons to expect it to be less than the danger encountered with the pathogens that are handled every day in diagnostic and research laboratories.

(a) Though in the history of microbiology about 6000 instances of laboratory infection have been recorded, the rate has dropped markedly since safety cabinets were introduced in the 1940s. Moreover, these cases were largely due to various agents of respiratory infection, spread by droplets. And in contrast to such respiratory infections, enteric infections arise through swallowing of contaminated food or other material; hence even the most virulent enteric pathogens are relatively safe to handle in the laboratory with simple precautions, such as not putting food or a cigarette on the laboratory bench.

(b) Strain K12, used in almost all genetic work with *E. coli* (including current work with recombinant DNA), has been transferred in the laboratory for over 50 years, during which it has become much better adapted to artificial media than to the human gut. In fact, recent tests showed that after a large dose in man (much larger than what one would expect from a laboratory accident) this strain disappeared from the stools within a few days. Its problems of survival outside the laboratory are clearly analogous to those of a delicate hothouse plant thrown out to compete with the weeds in a field.

(c) The addition of a block of foreign DNA to a bacterial strain will ordinarily increase its adaptation to survival in nature. For one thing, the mere replication of such additional DNA exacts a metabolic price for an organism; and if the DNA is active, its products are likely further to disturb the cell's metabolic balance.

(d) A very large safety factor is added by the provision in the present Guidelines for biological containment. All work with mammalian DNA must be carried out only in mutant strains (the class called EK2) that have a drastically impaired ability to multiply or to transfer an introduced plasmid except under very special conditions provided in the laboratory. For example, the presently certified EK2 strain has several stable mutational defects in which genes essential for the survival of this organism in nature have actually been deleted from the DNA. These include loss of the ability to synthesize an essential wall component (diaminopimelate), which can be supplied in laboratory cultures but is not available in the gut. An infecting dose therefore could not multiply in the gut. But the protection goes much farther, reaching a degree unprecedented in the annals of man's exploration of new materials: this material has been coded for self-destruction. For without the required diaminopimelate such mutant cells can continue to grow and expand, but without forming more wall, and so they quickly burst. Accordingly, under conditions similar to those in the gut, such an EK2 strain not only fails to multiply but less than 1 in 10^8 cells survives after 24 hours—and it would be an extraordinarily sloppy laboratory accident that would result in ingestion of as many as 10^8 cells. In addition, while the cells are dying off in the absence of diaminopimelate, they are severely impaired in their ability to transfer plasmids to other, well-adapted cells (which is the important point for the danger of spreading harmful genes). Finally, the plasmids being used to carry recombinant genes are also weakened mutant derivatives, selected for severe impairment of their ability to be transmitted from the host cell to another cell.

We thus see that even with a strain known to carry the gene for a potent toxin the production of disease in a laboratory worker would require the compounding of two low probabilities: that the strain will initiate an infection; and that it will survive long enough to cause harm despite its multiple disadvantages. These disadvantages include being a laboratory-adapted strain, carrying the burden of foreign DNA, carrying the very large burden of being a suicidal EK2 strain, and carrying the gene on a highly defective, infertile plasmid. With recombinants from shotgun experiments we have an additional, very low probability, already mentioned: that of having picked up a dangerous gene from mammalian tissue.

I conclude that with the kinds of recombinants now permitted (which exclude a known gene for a potent toxin or a known tumor virus) the danger of a significant laboratory infection is vanishingly small compared with the dangers encountered every day by medical microbiologists working with virulent pathogens. And such dangers must ultimately be balanced against the potential benefits. In the United States, up to 1961, of the 2400 recorded cases of laboratory infections 107 were fatal—over half of these from diagnostic laboratories. On the other side, millions of lives have undoubtedly been saved by bacteriological research and diagnosis.

But even if the risks in recombinant DNA research are much smaller than the public has been led to believe, it is important to keep all the probabilities low. In particular, even though a toxin-producing EK2 strain would survive only very briefly in the gut, a large enough dose might meanwhile produce enough toxin to cause disease. Hence it is important for molecular biologists working in this area to learn, and to use, the standard techniques of medical microbiology, at least until we have acquired much more experience with the organisms. Indeed, the main benefit from the current discussion might be the enforcement of such practices.

C. Risk of Spread

I now come to the most important point of all, from the point of view of the public: the enormous difference between the danger of causing a laboratory infection and the further danger of unleashing an epidemic. In our government's bacteriological warfare laboratories at Camp Detrick, working for 25 years on the most communicable and virulent pathogens known, 423 laboratory infections were seen. Moreover, most of these infections occurred via respiratory transmission, over which control is very imperfect. Nevertheless, *there was only one probable case of secondary spread* to a member of the family or to any person outside the laboratory. Similarly, in the Communicable Disease Center of the U.S. Public Health Service 150 laboratory infections were recorded, with one case of transmission to a family member. Elsewhere in the world there have been about two dozen laboratory-based microepidemics recorded— and each involved at most a few outsiders.

With enteric pathogens the danger of secondary cases is minimal, for with this class of agents modern sanitation provides infinitely better control than we can provide for respiratory infection: the appearance of typhoid, in contrast to that of influenza, does not lead to an epidemic. Enteric epidemics appear only when sanitation is poor or has broken down, or when a symptom-free carrier with filthy personal habits serves as a food handler; and such epidemics are always small (except when sewage freely enters the water supply).

This epidemiological information is clearly pertinent to the recombinants that we are discussing. For while there has been widespread apprehension over the presumed production of biparental chimeras, with totally unknown properties, the fact is that the recombinants envisaged are all genetically 99.9 percent *E. coli*, with about 0.1 percent foreign DNA added. It is exceedingly improbable that such an organism could have a radically expanded habitat, no longer confined to the gut. It is even harder to see that the organism would be more communicable, or more virulent, than our worst enteric pathogens, which cause typhoid or dysentery. The Andromeda Strain remains entertaining science fiction.

I conclude that if by remote chance a recombinant strain should be pathogenic, and if it (or a recipient of its plasmid) should cause a laboratory infection, that infection would give an early warning, which would decrease the chance of spread. Moreover, if a case should appear outside the laboratory, the enteric habitat of *E. coli* provides powerful protection, in a country with modern sanitation, against the chain of transmission required for an epidemic.

We must therefore ask whether the problem really merits deep concern by the general public. We have seen that to produce a laboratory infection by introducing random fragments of mammalian DNA into *E. coli* would require the compounding of several low probabilities. By any reasonable analysis the risk seems very much less than that from pathogens that are being cultivated in laboratories all the time. As with known pathogens, the risk should concern those involved, at various levels, in the investigation. However, the risk of producing a serious epidemic seems vanishingly small. I therefore see *no realistic basis for public anxiety over this issue,* any more than over the way diagnostic work or research on known pathogens is conducted.

Apart from the danger of inadvertent epidemics, it has been suggested that terrorists might deliberately create harmful recombinant bacteria as a powerful new tool. But it is hard to see why a terrorist would be interested in an *E. coli* strain containing, say, a gene for botulinus toxin, when that gene is already housed in the naturally occurring *Clostridium botulinum*—an organism from which the toxin can be obtained without the necessity of a time-consuming, expensive, and uncertain effort to produce a new species of bacterium having no discernible advantage.

D. Tumor Viruses

Tumor viruses present a special problem. Unlike other viruses, whose entry in an adequate dose regularly causes disease in a susceptible host, tumor viruses do not cause a tumor regularly after infection but require special circumstances. Indeed, it is their frequent presence in apparently normal animal tissues that has given rise to fear of "shotgun" experiments. Moreover, whether they make any contribution to human cancer is still quite unknown. On the other hand, if they should do so it would be after a latent period of years. Hence any conceivable infection by a bacterium containing a tumor virus genome would lack the early warning that would be seen with a bacterium producing a potent toxin.

However, all other aspects of the problem remain the same for the two groups. And this loss of one protective feature is balanced by the fact that viruses, by definition, have their own means of spread. Indeed, in general the natural spread of viruses is even more effective than that of bacteria, for each infected animal cell produces thousands of infectious virus particles, while each bacterium produces two daughter cells. Moreover, since viral DNA in a bacterium would have to get out of its host cell and get into human cells, through an extremely inefficient process, it is hard to imagine that such DNA in a bacterium would be more hazardous than the same DNA in its own infectious, viral coat, adapted by evolution for entering animal cells.

Indeed, if we fear the danger of such indirect uptake of unrecognized tumor virus DNA from normal mammalian tissue, via a bacterial vector, we must ask whether the direct ingestion of such tissue, as in a steak, may not present at least as great a danger. Or, in another hypothetical scenario, we might consider the risk from ordinary blood transfusions. Since tumors cannot be detected until they reach a substantial size, the average transfusion may have as high a probability as 0.1 or 1 percent of having come from an early tumor patient. Since cancers can spread to distant organs via the bloodstream, a transfusion from a person with undetected cancer has significant probability of containing cancer cells—which are almost certainly more hazardous than naked tumor virus DNA in the gut.

I am not suggesting that the danger of acquiring a cancer by eating rare meat or by receiving a transfusion is something to be concerned about. I am suggesting only that the danger of using recombinant DNA to study tumor viruses must be judged against that background, as well as against the background of the virus's own inherent ability to spread. I would also ask whether there is greater danger if we use the recombinant DNA technology to help us to understand tumor viruses, or if we presumably play safe and inhibit that research. For if we choose the latter we meanwhile allow the tumor viruses to spread as they presently do in nature, under circumstances where we really do not understand their relation to human tumors at all.

E. The NIH Guidelines

In the face of the alleged dangers that have been widely discussed I cannot blame the public for having a high level of anxiety; and I would regard the present Guidelines as a reasonable response to that anxiety. On the other hand, in the light of the technical realities that I have discussed above I would regard these Guidelines as excessively conservative.

The Guidelines contain a provision for periodic revision; and since these revisions (or the nature of any future legislation) will depend on public attitudes as well as on the results of actual experience with the organisms, there is need for a great deal of public education. In this connection I would criticize the *New York Times* for the article by L. Cavalieri on recombinant DNA in its Sunday Magazine (August 1976). Though the writer is a molecular biologist whose official credentials would lead the public to assume a reasonable degree of objectivity, the article not only was inflammatory but it exhibited extraordinarily little understanding of either microbiology or evolution. In discussing *E. coli* as though it were a standard, uniformly distributed organism, which would carry with it through the world any additional genes that one might insert, the writer ignored the most important factor of all: natural selection. He also made the remarkable statement that the insertion of tumor viruses into bacteria may make them infectious—as though viruses were not infectious. And he suggested that scientists working in this field may produce yet another Andromeda strain—as though the first strain existed in fact rather than fancy.

Given the present level of public anxiety, scientists in this field seem quite willing to accept the Guidelines. But I hope it will not be too long before these rules are modified in the light of further experience. For since the technique is potentially useful for a large number of investigators, a requirement for excessively elaborate facilities will add up to a very large expense and will inevitably inhibit desirable experiments. The principle of erring on the side of caution is laudable up to a point—but if it is pushed too far it can end up being paralytic.

III. INTERVENTION IN EVOLUTION

A. The Prokaryote-Eukaryote Barrier

The hazard that we have been discussing—that of creating novel, dangerous organisms—is a legitimate cause for public concern: there is no question about society's right to limit an individual's activities that may harm others. However, when we ask whether our increasing power to manipulate genetic material creates long-term evolutionary dangers, we are in quite a different area, involving the concept of dangerous knowledge rather than dangerous actions. The most prominent exponent of this view is Robert Sinsheimer of The University of California at Santa Cruz. Perhaps we can clarify the issue by trying to translate into more specific terms some of the general sources of apprehension that he has expressed in various publications.*

*See page 85, "Two Lectures on Recombinant DNA Research," by Dr. Sinsheimer.

1. Dr. Sinsheimer questions our moral right to breach the barrier between prokaryotes and eukaryotes, since we simply cannot foresee the consequences. Apart from the lack of demonstration of a greater barrier between these groups than between distant animals, this argument seems to turn evolutionary principles through 180 degrees. Evolution is concerned with selection for fitness, in the Darwinian sense; and since fitness depends upon a balanced genome, it is not surprising that evolution proceeds in small steps, which will not excessively unbalance the genome in one respect while improving its adaptation in another. The barriers that evolution has established between species are accordingly designed to avoid wasteful matings, i.e., matings whose products would be monstrosities, unable to survive, rather than monsters, able to take over. It is therefore exceedingly unlikely that artificial transfers of genes between the most distant organisms—man and prokaryotes—would pass the test of Darwinian fitness.

2. "This is the beginning of synthetic biology." I wonder whether this statement can really be defended, considering that man has been meddling with evolution by domesticating animals and plants by selective breeding since neolithic times. He has also been cloning and grafting plants.

3. "The power to change the evolutionary process is as significant as cracking the atom." But atoms are not subject to extinction by Darwinian selection: stores of nuclear weapons are likely to be more permanent than any dangerous organism that might reach the world from a laboratory working with recombinant DNA. The statement by George Wald that "a living organism is forever" is dramatic, but it disregards two powerful evolutionary predictions: first, that natural selection will rapidly extinguish all evolutionary departures except for the infinitesimal fraction that have improved their adaptive fitness; and second, that the recombination of genes from distant sources has an exceedingly small probability of providing such improved fitness.

4. "We no longer have the absolute right to free inquiry." But we never had: visibly dangerous procedures have always been subject to social limitations. On the other hand, to invoke dimly foreseen, undefined dangers as a basis for limitation seems to be starting on the slippery slope of excluding dangerous ideas rather than dangerous actions.

5. A further push in this direction may be seen in the statement that power over nucleic acids, as over the atomic nucleus, "might drive us too swiftly toward some unseen chasm. We should not

thrust inquiry too far beyond our perception of its consequences." I would paraphrase this statement and suggest that we
should not thrust our limitations on research too far beyond our
perception of its hazards. Otherwise we will find ourselves reenact-
ing the drama of Galileo and Urban VIII. And the analogy is
uncomfortably close: for the mystical quality of the current argument suggests that at its core the issue is whether man's possible
interference with evolution is not blasphemous.

B. Genetic Engineering in Man

Perhaps the most significant of Sinsheimer's statements is his suggestion that the study of recombinant DNA in bacteria is the beginning of a
genetic engineering that will ultimately extend to man. Here, in contrast
to the vagueness of the preceding propositions, we finally come to
something concrete that one can wrestle with.

I would suggest that concern over genetic engineering in man is
utterly irrelevant to the question of the danger of creating an epidemic;
hence it is irrelevant to Sinsheimer's recommendation that all research
on recombinant DNA be presently restricted to a few maximal security
federal facilites. This concern also seems irrelevant to the question of
breaching the prokaryote-eukaryote barrier; for while gene transfers at
the cellular level across this border, in either direction, are of great
scientific interest, it is hard to envisage any reason to try to introduce
into man genetic material from the opposite end of the evolutionary
spectrum.

Nevertheless, vague concern over possible extensions of gene manipulation to man, even more than concern over epidemics or over
meddling with evolution in general, may lie at the heart of much of the
current uneasiness over recombinant DNA research. And because of
the enormous publicity given to our new power to splice blocks of DNA
into plasmids, we have perhaps lost sight of the fact that this development is no more radical a step toward genetic engineering in man than
are many other steps, which have aroused no such public terror. These
include the isolation of a gene, its chemical synthesis, the cultivation of
human cells, the use of viruses to incorporate genes into those cells, and
the achievement of genetic recombination in vitro between human cells
and other animal cells.

The promise and the threat of genetic engineering in man received
extensive discussion in 1970, which then subsided; but the topic has
been reactivated by the very different question of genetic engineering in
bacteria. It cannot be considered in detail here, but I would like to make
a few brief points.

First, the medical aim of genetic engineering in man is gene therapy for well-defined hereditary diseases, i.e., those due to single defective genes, with a well-defined chemistry. For this purpose we would have to be able to introduce the appropriate DNA in a reliable, controlled way, in the right cells: and I believe we are still a long way from that goal. But even if this guess is wrong, and if we succeed in genetically curing such diseases as phenylketonuria or cystic fibrosis, it is clear that this success would still leave us very far from being able to manipulate in any useful way the large number of genes, all still undefined, that specifically direct the development and the function of the brain. Moreover, in a developed organism, with an already formed brain, no conceivable manipulation of DNA could reorganize the wiring diagram of that brain. Hence the possibility that a tyrant could use genetic engineering to manipulate personalities seems still too remote to justify present concern. Finally, even if we could use genetic technology to manipulate human personalities I would question whether the technological imperative would necessarily (or even likely) lead us to do so. For the simple but effective techniques of selective breeding and artificial insemination are not being used to influence the human gene pool, though their efficacy has been established for thousands of years in animal husbandry. It is therefore not clear what motivation would impel society to try to use the much more elaborate techniques that might emerge from the current research.

Philosophical questions about the effects of science and technology on man's fate go back to Galileo—and the history of Italy's failure to build on that early start should give us pause. For better or worse, we cannot unlearn the scientific method: and if we restrict it in one country it will turn up in another. To be sure, our world has only recently come to realize how large (and often unexpected) is the price for various aspects of technology, how finite our terrestrial resources, and how clumsy our responses to the need to limit the size of our population and its demands on those resources. Faced with these crushing problems, it is only too easy to take the benefits of science and technology for granted and to object to the new problems that they are raising. But in the long run it is difficult to see how we can plot a more prudent course than to try to recognize the hazards and the costs of specific possible applications *as soon as they become visible;* to seek a reasonable balance between the demand for freedom of action and the demand for protection from excessive risks and costs; and to avoid arousing unwarranted public anxiety, while we seek orderly and responsible methods for involving the public in matters that so deeply affect its interests.

I share Sinsheimer's concern for the future, and his passionate advocacy of vigilance. But the vigilance must be directed at specific, definable applications of knowledge. Vigilance concerning new knowledge that *might* someday be misused is a threat to freedom of inquiry, and I believe a threat to human welfare. We may conceivably be entering dangerous territory in exploring recombinant DNA—but we are surely entering dangerous territory if we start to limit this exploration on the basis of our incapacity to foresee its consequences.

Real and Imagined Dangers
of Recombinant DNA Technology:
The Need for Expert Evaluation

Rolf Freter

1. INTRODUCTION: THE CURRENT STATE
OF THE DEBATE (MARCH 1977)

The present volume is an impressive witness not only to the uniqueness of recombinant DNA technology as such, but also to the uniqueness of the debate its advent has triggered. In dealing with a subject that is said to be *sui generis*, important, and threatening, one should perhaps not be surprised if some aspects of the worldwide debate have transcended the bounds of uniqueness and entered the realm of the bizarre. In fact, the speculative nature of most of the arguments available for discussion has on occasion reduced what should have been a series of public scientific inquiries into a form of intellectual Russian roulette, in which a number of participants could not be sure whether the big guns they carried to the rostrum were actually loaded and could penetrate reality (1). This is not the kind of situation cherished by the scientist nor indeed by any thinking person. In my opinion, these disagreeable circumstances are a major cause of one of the most bizarre aspects of this debate, namely the fact that the group of scientists who have participated in it have been largely specialists in genetics and molecular biology—scientists who use the new technology and are directly affected by the present inquiry, but who are not necessarily the best experts to discuss certain points that are of immediate public interest and concern. However, because of the various undesirable features of this debate, not many scientists with expertise in other areas have volunteered their contributions and, apparently, not many have been asked to participate.

Imagine! Here is a debate triggered by serious concern about the potential dangers of a new technology that as some envision, might create monstrous microorganisms capable of exterminating much of the human race, of spreading throughout the environment and thereby threatening animal and plant life and the entire delicately balanced ecology of our planet. Here we are in the mid-1970s, well into the age of the academic specialist! One would assume therefore that this issue would be debated by those with expertise in infectious diseases, in microbial pathogenicity, in the microbial ecology of sewage, water, soil, and related areas. One would assume further that the National Institutes of Health (the agency that finances most medical and biological research in this country and that rightly refuses to give a penny of support to anyone who cannot demonstrate extensive expertise in his area of research) would have convened an advisory committee consisting of recognized experts in these various relevant disciplines—with, of course, a sizable assortment of molecular biologists to act as consultants.

None of this has happened. Neither the proponents of the new technology nor those who counsel caution nor those who demand a moratorium because of the imagined dangers nor the U.S. National Academy of Science (e.g., their "Forum," March 7–9, 1977) have been willing or able to muster a representative contingent of specialists with expertise in those very areas where most of the suspected dangers are said to lie. Likewise, the National Institutes of Health have convened an advisory committee to devise guidelines for work with recombinant DNA molecules that, during some very important periods of its history, did not include on its roster a single member who would claim to speak with authority on matters of microbial pathogenicity and ecology. Unusual? Bizarre? Be this as it may, it seems to me that the recombinant DNA technology has already presented future historians, psychologists, and sociologists with a wealth of material for study and, if it achieves nothing else, this alone will justify the efforts invested in it.

Having managed in a few lines to affront and impugn the judgment of so many, I must hasten to make amends for this discourtesy by making some equally unflattering remarks about the potential contributions that microbiologists in my own and related areas of specialization might have made to the problem of defining and containing the conjectural dangers inherent in the new technology. To come right out with it: I do not believe that the current guidelines issued by the National Institutes of Health would have been substantially different in concept if they had been drawn up by individuals whose research interests include the manner in which microbes cause disease, in which they spread among people and animals and disseminate throughout our environment. This statement might serve to increase the apprehensions of those who hold that all expert scientific advice should be heavily discounted when it comes to making important decisions on public policy. I prefer an alternative interpretation: that, in this never-never-land of conjecture and imagined dangers, no one who wants to avoid the fate suffered by Don Quixote at the hands of the windmills could do more than to offer a set of conceptually simple and straightforward precautions. That is what the current guidelines essentially represent.

This is not to say that input by a representative group of experts into the current debate and into the formulation of the official guidelines might not be beneficial at all. The basic shortcoming of some important segments of the current debate lies in the fact that isolated statements have not infrequently been made in order to support a given viewpoint, rather than to explore the problem. For example: "*E. coli* is the safest possible bacterium for recombinant DNA work because our knowledge of its genetics is the most advanced," or, conversely, "*E. coli* is a commensal* and also a common human pathogen and is therefore one of the more dangerous bacteria one could choose for such work." All too frequently, however, such statements have been made alone and have not been supported and tempered at the same time by other relevant evidence, a shortcoming that most speakers or authors would probably not have tolerated in debates of problems relating more closely to their own area of interest. For example, a discussion in which the above statements had been made would be seriously deficient if it did not also include a consideration of the following points: "Our extensive knowledge of *E. coli* genetics is of little practical use, because our current understanding of the genetics of bacterial pathogenicity is, at best, fragmentary," or "The nature of the known pathogenic potential of *E. coli* virtually assures that this species cannot become the cause of global epidemics." I believe, therefore, that a thorough examination of what is known about the mechanisms by which microorganisms cause disease and by which they spread among humans and throughout the environment would contribute noticeably toward sharpening the debate and would allow it to mature and to concentrate on those aspects of the problem which deserve closer attention. A penetrating debate of this nature would perhaps be tiring in its attention to detail and for this reason not well suited to public forum-type meetings, but it should go a long way to allay the fears of those who still contemplate some of the

---

*The term *commensal* is used here to denote a bacterium that is a part of the body's normal microbial flora. For example, *E. coli* is found in the normal intestine of most people.

more outlandish consequences of the new technology. It should also facilitate an early experimental resolution of those remaining problems which cannot be fully evaluated at the present time.

As a brief and necessarily imperfect illustration I shall present in the remainder of this chapter my own view of two relevant problems, namely the questions of (a) whether novel, highly pathogenic bacteria are likely to be created by the new technology, and (b) whether a microorganism such as *E. coli* carrying foreign DNA may become implanted in the human gut and eventually transfer this DNA to man. These views are, of course, those of a single individual and, knowing the nature of scientific debates, one may predict that some of my colleagues will perceive misinterpretations and omissions in what I have to say. I should remind the reader therefore that I have advocated in the above paragraphs a *thorough* discussion by *representative groups* of experts in various fields of medicine and science. I have further suggested that this, and only this kind of extensive analysis would have a significant effect on the direction of the current discussion. Unfortunately, a talk by one or two individuals buried in the midst of a three-day symposium, the specific advice of one or two consultants to some committee, or single contributions such as the present one, while perhaps useful, simply cannot be expected to do full justice to a problem as complex as the one before us.

II. CAN *E. COLI* BECOME A SUPER PATHOGEN CAPABLE OF SUDDENLY WIPING OUT LARGE SEGMENTS OF THE HUMAN POPULATION?

The main argument on this question is quite straightforward and can be stated in a single paragraph:

The characteristics that a bacterial pathogen must have in order to cause disease in man or animals are exceedingly complex. While much remains unknown in this area, it is clear that a given bacterium must possess a very large number of attributes before it can colonize man and cause disease. For example, a typical pathogenic bacterium must be able to (i) survive in the environment (water, sewage, air, etc.) in order to spread from one host to the next. It must then (ii) be able to multiply on the body surface (skin, nose, throat) or in the intestine, or lung. It must (iii) have some mechanism for penetrating into the body, for (iv) spreading throughout the body, and for (v) resisting the numerous defense mechanisms of the mammalian tissue. Pathogenic bacteria must finally (vi) be able to produce a toxin or to otherwise interfere with the host's physiology to bring about the symptoms that we recognize as disease. Each of these various stages requires that the bacterium possess a discrete and complex set of characteristics in order to survive. If only one of these characteristics is missing, the chain is broken; the bacterium will be stopped in its progress and will be unable to cause disease. For this reason, and fortunately for us, only very few of the many thousand bacterial species are capable of causing human disease, because only those few microbes possess the complete set of the essential characteristics. The new recombinant DNA technology makes it possible to insert into a microorganism a relatively small piece of DNA that will add *one or a few* new characteristics to its new microbial host. If new genes are inserted into an initially apathogenic, or into a specially "enfeebled" microorganism (as required in the NIH guidelines), it is exceedingly improbable that these newly introduced characteristics will include every single number of the large set of characteristics required to convert the enfeebled microorganism into a functional pathogen.

As always in biology, there are exceptions to this generalization. Pathogens such as the diphtheria bacillus, cholera vibrio, and the pathogenic *E. coli* strains do not need to go through the whole sequence, but settle on the surface of the throat or the intestinal wall and produce a toxin at that location. This toxin, in turn, will enter the tissues of the body and cause disease. But, even with these types of pathogens, the apparent simplicity of the manner in which they cause disease is highly deceptive. In the third section of this chapter we will examine some of the various parameters involved in determining the implantation of bacteria into the gut, and it will become apparent that this seemingly uncomplicated process has so far defied analysis because of its complexity. In fact, there is *no* bacterial pathogen at all for which we can state all the relevant characteristics that make it different from related nonpathogenic species.

Textbooks of medical bacteriology often mislead the beginning student by showing lists of "virulence factors"* for each pathogen, which the poor fellow may even have to memorize. After doing so, the student still does not understand virulence because the known (and more often only "suspected") virulence factors represent only the tip of the proverbial iceberg of pathogenicity. For example, "the" virulence factor of the diphtheria bacillus is thought to be the well-known toxin. However, the bacterial strain that is most active in producing this toxin (and which is therefore used in industry for this purpose) is not highly virulent, simply because it lacks other attributes and therefore cannot multiply or produce toxin in the mammalian body (2). Closer to our problem is the example of the enteropathogenic *E. coli* Strains.

---

*Virulence factors are those characteristics which enable a pathogenic microorganism to cause disease. The terms *virulence* and *pathogenicity* are used here synonymously and refer to the ability to cause disease.

It has been shown that such strains must have distinct virulence factors in order to cause diarrheal disease. One such factor, a toxin, is the agent that induces the diarrhea. However, the K12 strain of E. *coli* was unable to produce diarrheal disease in the animal gut even when it carried the gene coding for this toxin, again obviously because it lacked other equally important but as yet unidentified virulence factors that would enable it to survive in the intestine (3). The reader should note here that the E. *coli* strains to be used for recombinant DNA work are all derivatives of this K12 strain.

When applying for admission to the bar, lawyers often fail to mention that their heart is capable of pumping blood, that their kidney can purify the blood, and that their stomach can digest their breakfast. This is surprising because, if their organs did not have these capabilities, the applicants would be confined to a hospital, which would have to supply a heart-lung machine, an artificial kidney, and an intravenous feeding device. Consequently such applicants could not practice law with the degree of efficiency to which their clients would be entitled. Nevertheless, lawyers' applications are seldom rejected on the grounds of having omitted such vital information for the simple reason that "it goes without saying" that *any* specialized activity of man must be supported by the myriad of physiologic, metabolic, and other functions essential to the life of the human body. Analogously, when medical microbiologists discuss "virulence" mechanisms of pathogenic bacteria, they rarely point out that these mechanisms include all the myriad of metabolic functions and structural features that *every* bacterial cell must have in order to survive. This fact is seldom emphasized because it is not very exciting and because it is so obvious that it "goes without saying." Consequently, the term "virulence factor" is usually applied to only those characteristics of pathogenic bacteria that distinguish them from the common and harmless nonpathogenic microorganisms. Nevertheless, it is obvious that the number of characteristics a bacterium must have in order to cause disease is rather enormous and is by no means confined to those traits commonly listed under the heading of "virulence factors."

This consideration is important in the evaluation of the so-called EK2 and EK3 ("enfeebled" or "disarmed") strains of E. *coli*, which are to be used in most recombinant DNA research. These micoorganisms have been deliberately deprived of a number of those metabolic or structural features which are required for the normal functions of *any* common bacterium. For this reason EK2 and EK3 strains of E. *coli* can grow only under those special laboratory conditions which provide the bacterial equivalent of our lawyer's heart-lung machine, artificial kidney, intravenous feeding device, etc. Thus, each deletion of an ordinary metabolic or structural trait from an "enfeebled" E. *coli* strain is at the same time also a deletion of an indispensable link in the long chain of virulence factors this strain would have to possess if it were to become a serious pathogen.

As the reader will recall, our original argument states that E. *coli* is a common and usually harmless bacterium, which is not a highly dangerous pathogen because it lacks a large number of virulence factors, and that, consequently, the inadvertent introduction by recombinant DNA methods of genes coding for only one or a few of these factors would not be sufficient to restore the entire complement of traits necessary for virulence. But, one may ask, how can we be sure that E. *coli* really lacks a large number of virulence factors? Is it not possible that E. *coli* may in fact possess all but one of the factors required to turn it into a pathogen capable of causing serious epidemics among human populations? This is a reasonable question, and there are indeed precedents. For example, certain bacteria that often inhabit the normal human throat are close relatives of the diphtheria bacillus and some of these, as far as we know, differ from this pathogen only in that they lack the gene for toxin production. Moreover, when two-week-old chicks were injected intravenously with a rather enormous quantity of E. *coli* (approximately a billion cells, enough to overcome most body defense mechanisms by sheer bacterial numbers), the introduction of a limited amount of genetic information allowed these bacteria to kill the animals more efficiently (4). It is therefore possible that the introduction of a limited amount of genetic material into E. *coli* may *augment those disease-causing properties which these bacteria already possess.*

It is also true that some strains of E. *coli* cause diarrheal disease in humans, and others are common causes of urinary tract infections. Nevertheless, I would consider it highly unlikely, but certainly not entirely impossible, that the introduction of a single gene into an E. *coli* strain might convert it into a more efficient causative agent of infectious diarrheal or urinary tract disease. However, urinary tract infections by E. *coli* are not contagious at all, and enteric infections by special strains of E. *coli* are self-limiting and not life threatening (except perhaps in newborns) and do not give rise to widespread epidemics in countries with adequate sanitation, because the route of dissemination of this type of disease (from the anus of an infected patient to the mouth of a prospective victim) can be effectively disrupted by established practices of common-sense hygiene. Only the respiratory route of transmission of infectious agents is difficult to control in countries like the U.S., and the ability to travel this route is not within the arsenal of pathogenic strains of E. *coli*. Thus, unfortunate as diarrheal diseases and urinary tract infections may be for the individual involved, these conditions are definitely not candidates for life-threatening global epidemics. *E. coli* (as well as many other normally inocuous bacteria) may also cause serious infections of organs and blood in newborns and in adults whose defense mechanisms against infections have been seriously impaired by other underlying diseases (e.g., leukemia). For obvious reasons, none of these conditions can give rise to widespread epidemics among the general population.

In order for *E. coli* to cause a worldwide epidemic of fatal infection it would have to acquire most or all of the following traits, each of which is likely to be complex in itself and therefore to require the presence of several genes for its expression. (i) It must be able to spread via the respiratory route of infection, which, in turn, means that it must be able to infect the adult (i.e., other than newborn) human lung and must survive in droplets of sputum, which dry out in the air and which are exposed to lethal ultraviolet radiation from the sun. (ii) It must be able to actively penetrate through skin or the epithelial linings that cover the body's tissues. (iii) It must be able to resist engulfment or otherwise avoid being killed by the phagocytic cells of the body. (iv) It must resist the effects of antibodies and other antibacterial substances in the body fluids. (v) It must be capable of producing powerful iron chelators in order to compete with the body for this metal. (vi) It must be resistant to the action of all clinically useful antibiotic drugs. Certainly, it will take more than the introduction of one or a few genes into *E. coli* to give it all the properties needed to turn it into a global menace, especially when an "enfeebled" EK2 or EK3 strain is used.

As a final thought it is pertinent here to point out that modern medicine has been able to conquer all those highly pathogenic bacteria that were the ancient scourges of mankind. The once-dreaded worldwide epidemics of highly fatal plague or cholera are no more, the large numbers of deaths from tuberculosis of pneumococcal pneumonia have disappeared from the statistics. Infections are still a major cause of disease and death in the United States, but the nature of the causative agents has shifted from "superpathogens" to bacteria of relatively low virulence (including *E. coli*), which often are highly resistant to most clinically useful antibiotics and which cause serious infections in patients whose body defenses against bacteria have been impaired by other underlying diseases (5). It is apparent therefore that, if some terrorist were indeed able to deliberately create a new bacterial pathogen of the same caliber as say, the plague bacillus, this would be the very kind of infectious agent that modern medicine is best equipped to handle.

In summary, then, it is not possible to definitely exclude on theoretical grounds the possibility that the inadvertent introduction of one or a few genes into wild-type (i.e., not "enfeebled") *E. coli* strain may augment those disease-causing properties which some *E. coli* strains already possess. None of these diseases can, by any stretch of the imagination, be expected to result in worldwide epidemics. Current knowledge of the manner in which bacteria cause disease allows us to exclude the possibility of an inadvertent creation from *E. coli* of a superpathogen by recombinant DNA techniques. Even if the deliberate introduction by some terrorist of a large number of genes into *E. coli* were to succeed in creating a pathogen that rivals in virulence the ancient scourges of mankind (and it should be remembered that some countries have spent large sums of money on such projects with little apparent success), modern medicine may be expected to deal swiftly with such an emergency.

III. CAN *E. COLI* IMPLANT INTO THE INTESTINAL TRACT OF MAN AND RELEASE CLONED DNA INTO THE HUMAN BODY?

E. coli is a commensal in the human intestinal flora and sometimes also in the throat. It seems therefore reasonable to inquire whether this bacterium, after accidental escape from a laboratory culture, could contaminate humans, become a part of their intestinal flora, and proceed from this source to implant itself into the normal body flora of a large segment of the human population. This process is, of course, the second in the above-described series of steps that a bacterial pathogen must be able to execute in order to cause disease, and we will now examine it in some more detail.

By definition, commensals differ from pathogens in that they will not advance through subsequent steps toward causing disease, but will remain a part of the intestine's normal bacterial flora for periods ranging from a few days to several years. For purposes of this discussion we will assume that the newly inplanted commensal bacterium carries a segment of cloned DNA that would be harmful if it entered certain cells of the human body. For this reason we will examine whether there exists a possible pathway for a commensal bacterium (with its cloned DNA) to enter the human body.

The study of implantation and growth of enteric bacteria in the gastrointestinal tract and the transfer from human to human of these microorganisms constitutes an area in which a relatively large volume of literature is available. Unfortunately, in spite of all this published work, practically nothing definite is known about the attributes that a bacterium, such as *E. coli* K12, must have in order to be easily transmitted and implanted into the intestinal tract of humans (6). Consequently, there is no way to predict which characteristics should be deleted from a bacterial genome in order to make the bacterium less "implantable." Furthermore, there is no way of testing the safety of prospective bacterial hosts, because it is not possible at the moment to devise an experiment that could reliably determine the degree of "implantability" into man's normal body flora of any given bacterium.

To be sure, the recent literature reports a number of widely quoted attempts to implant prospective host bacteria, such as *E. coli* K12, into the intestine of human volunteers by feeding large doses of these microorganisms (e.g., ref. 7,8). Such data regularly show that it is almost impossible to transmit *E. coli* in this manner, and the inoculum usually survives only a few days in the host's intestine. To the casual observer this kind of evidence may appear to be quite pertinent and conclusive. Unfortunately, this is not the case: these experiments do not tell us much that is new about the ability of bacteria to grow in the human intestine. It has been known for many decades that feeding of bacteria, especially those grown under the usual laboratory conditions, rarely results in implantation, because the normal intestinal flora of the recipient antagonizes the growth of invaders (much of this had been reviewed by Wiedemann and Knothe, ref. 9; see also, as another example, ref. 10).

The data available in the older literature may be summarized (and dramatized) in terms of the following hypothetical situation: If one were to isolate the predominant *E. coli* strain from volunteer A (i.e., an indigenous strain that obviously is well capable of intestinal growth) and were to feed it to volunteer B, it usually would not implant permanently in B. Conversely, one would not expect B's *E. coli* to implant in A.

Returning to actual published data, it is quite clear from the older literature that feeding experiments in humans usually show a lack of permanent implantation even for indigenous *E. coli* strains. Consequently, there is little point in repeating the same exercise with prospective host bacteria: the fact that these do not implant permanently does NOT indicate very much about the ability of such strains to grow in the human intestine. While the recently reported feeding experiments are very interesting as far as they go (they show, for example, that K12 has *no greater* ability to establish itself in the human intestine than indigenous strains), one can certainly not assume that such experiments could be utilized as "tests" for the "implantability" of prospective *E. coli* hosts, even if sufficient numbers of human volunteers were available. (In fairness to the original authors I should make it clear that they themselves have not claimed such exaggerated significance for their work. I am emphasizing the points above because a reader who is less familiar with the area of intestinal growth of bacteria could easily draw erroneous conclusions.)

Unfortunately, the evidence above does not allow the conclusion that E. coli strains cannot be transmitted to humans at all. On the contrary, while most human subjects studied in the past by various authors carried one or a few E. coli strains for long periods, other individuals are known to change their E. coli strains frequently and therefore are presumably more susceptible to implantation of E. coli (9). Moreover, it is well known that E. coli strains do spread among human populations that presumably were not exposed to antibiotics, as witnessed for example by the frequency of R-factor*-carrying E. coli in the normal population (e.g., review in ref. 11; also ref. 12). Apparently, then, transfer of E. coli to humans does occur in nature, and one must assume that the mechanisms involved in such transfers differ from those which are observed in simple feeding experiments. There is consequently no assurance that differences between E. coli strains that can be observed in feeding experiments (e.g., survival in the intestinal tract of volunteers for ten days vs. three days) have any relation to those bacterial or host characteristics which would be important in permanent implantation.

As mentioned above, nothing definitive is known about the mechanisms that inhibit or promote intestinal colonization by bacteria, even though there are numerous theories. Adhesion to mucosa appears to be important for cholera vibrios and enteropathogenic E. coli, but whether this factor is necessary for indigenous E. coli is not known. There is evidence to indicate that one reason why feeding experiments do not usually result in implantation is simply that the bacterial inoculum is not in the proper physiologic state, and that feeding would result in at least improved implantation if the inoculum had been grown under conditions (of substrate, pH, Eh, etc.) resembling those in the intestine. Two laboratories (13,14) have shown that the exclusion of invading microorganisms from an established flora by bacterial antagonism may become entirely inoperative when the invading bacteria come from an inoculum that is already adapted to growth in the environment it is about to enter. Thus, bacterial antagonism exerted by an indigenous flora may prevent the implantation of other bacteria or, conversely, may have little or no effect, depending on the physiologic state of the invading bacteria. This evidence suggests that an escaped E. coli strain may have less difficulty in spreading among people once it has succeeded in adapting itself to the intestinal environment in at least one individual.

Before environmental mechanisms affecting bacterial growth can be understood, we need some knowledge of the characteristics of the environment in which this growth takes place, and especially of the substrates available for growth. Again, almost nothing is known about this

---

*See the glossary for a description of R-factors.

subject. Several authors have speculated that intestinal bacteria, such as *E. coli*, live a "feast and famine" existence caused by periodic food uptake by their hosts, in which their habitat is suddenly swamped with nutrients and subsequently depleted after a meal has been digested and absorbed (15,16). However, it has been observed for some years that enterobacteria, including *E. coli*, may actually reach higher populations in a stressed or starved host (e.g., ref. 17) a phenomenon recently studied in more detail by Morishita and Ogata (18) and by Tannock and Savage (19). One may wonder, therefore, whether and to what extent intestinal *E. coli* actually depend on diet-derived substrates and whether perhaps the main support for the growth of *E. coli* and some other microorganisms may come from a more or less steady flow of host-derived substances such as effete epithelial cells, secreted mucus and enzymes, etc. In contrast, some data are available showing that high dietary intake of a given substance, e.g., lactose, may select for lactose-utilizing bacteria in the intestine. Unquestionably, diet can at times affect the intestinal flora (cf. review in ref. 20). However, as the above examples suggest, the question whether or not any dietary substance constitutes an important substrate for a given intestinal bacterium, such as *E. coli*, probably depends to a large extent on the nature of the other bacteria present—i.e., their ability to compete for or to modify this substrate—and may therefore differ under various circumstances and in different individuals, populations, or animal colonies.

It should be pointed out, finally, that the control of intestinal *E. coli* appears to be predominantly a function of the "normal" flora of the large intestine, which consists mainly of strictly anaerobic bacteria (e.g., ref. 21). *E. coli* rapidly populates the intestine of the germfree animal,* and the implantation of a large number of different anaerobic species is required before the intestinal *E. coli* population is reduced to the relatively low levels found in conventional animals (21). Implantation of intestinal anaerobes is often difficult, and some species do not implant in the germfree animal at all unless the animal has been previously associated with other bacteria. Moreover, a number of intestinal bacteria have never been cultivated *in vitro*, but have been observed by electron microscopy. It is apparent, therefore, that the intestinal environment of conventional animals, and presumably of man, embodies features that cannot be duplicated at present either by *in vitro* methods or in the germfree animal, a fact that greatly enhances the experimental difficulties inherent in the study of intestinal microecology.

To summarize our analysis of the current state of the art:

---

*A germfree animal has been raised under sterile conditions and has never come in contact with live bacteria. As a result, a germfree animal is exquisitely sensitive to the implantation of most kinds of bacteria into its body.

1. While it is reassuring to know that bacteria are difficult to implant into the normal flora of humans or conventional animals when the microorganisms are simply fed by mouth, tests based on such a procedure cannot be relied upon to measure the ability of bacterial strains to proliferate in the intestinal environment. Epidemiological studies have shown that dissemination of certain *E. coli* strains from person to person can indeed occur.
2. The nutritional and other factors that govern the growth of bacterial populations in the intestine are essentially unknown.
3. The intestinal environment has certain features that are demonstrably important for bacterial growth. Some of these features cannot be duplicated at present under laboratory conditions or in the germfree animal and therefore can be studied only with difficulty.

It is apparent from the discussion above that those who wish to construct "enfeebled" EK2 and EK3 strains of *E. coli*, strains that are unable to implant into the normal body flora of humans or animals, cannot follow the most rational approach, namely that of deleting those traits which are essential for bacterial survival in the gut. Those traits have not yet been identified. Obviously a large amount of further study is required to resolve this issue satisfactorily. As related earlier in this chapter, attempts are being made to construct "enfeebled" strains of *E. coli* that lack several of the common metabolic or structural traits required for bacterial proliferation in all specially formulated laboratory media. This strategy is also highly promising in preventing the implantation of such *E. coli* strains into the gut. However, careful testing is required because of our ignorance of the characteristics of the intestinal environment. For example, a bacterium unable to synthesize an essential cell wall component (diaminopimelic acid), which does not occur in mammals, grew unexpectedly well in the intestine of germfree mice even though it failed to grow in laboratory media lacking this substance (unpublished data). Obviously, the intestinal milieu must have provided a substrate that enabled the bacterium to multiply in spite of its biosynthetic deficiency.

The paragraph above contains an apparent contradition. Having first stated that rational testing for the "implantability" of *E. coli* strains into the intestine is not yet feasible, I nevertheless advocate in a subsquent sentence the careful testing of all prospective "enfeebled" EK2 and EK3 strains. This difficulty can be avoided for practical purposes by carrying out the tests for "implantability" under conditions so stringent that one can be virtually certain that any enteric microorganism that passes this test will not be able to multiply in the intestine of any mam-

malian species, including humans. I believe that such stringent conditions obtain in germfree animals. Such animals, born and raised in an
enviroment free of microbes, have not developed most of the defense
mechanisms that protect the rest of us against bacterial colonization and
invasion. If one or a few cells of any member of the family Enterobacteriaceae (the large taxonomic group to which *E. coli* belongs) are introduced into the gut of such animals, they will multiply overnight to form
a population of approximately 1 to 10 billion bacteria in the intestine.
Even those enterobacteria which are not members of the normal bacterial flora of a given animal species will grow in germfree individuals of
that species. In other words, an enterobacterium that did not colonize
the conventional mouse would still be able to grow to large microbial
populations in the gut of a germfree mouse.

A test requiring the inability of a bacterium to grow in the gut of
germfree animals would therefore be excessively stringent; i.e., many a
putative "enfeebled" *E. coli* strain would fail this test because it could
multiply in a germfree animal's gut while, in fact, it might have been
totally unable to do so in the normal human intestine. The stringency of
this test is also illustrated by the fact that, with one exception, I have
never encountered in my own work, or heard about, a strain of *E. coli* or
a related species that would fail to multiply in the gut of germfree animals. Thus, the inability of a given *E. coli* strain to grow in the gut of a
germfree animal would, in my opinion, virtually guarantee that it would
also be unable to multiply in the gastrointestinal tract of normal human
beings.

The one exception mentioned above is the *E. coli* strain χ1776.*
When I presented to germfree mice a billion organisms of this strain in
the drinking water, not a single viable bacterium could be recovered on
subsequent days from the gut of these animals. This strain is, of course,
the only one currently certified by the National Institutes of Health for
use as an EK2 bacterium in recombinant DNA work (its inability to grow
in germfree mice was, however, not a factor considered in the certification process). Obviously, then, it is not impossible for geneticists to
construct *E. coli* strains that pass this excessively stringent test for "implantability" into the gastrointestinal tract. In view of this, I would
recommend that the inability to multiply in the germfree intestine be
added by the National Institutes of Health to the list of requirements for
certification of putative "enfeebled" EK2 and EK3 strains of *E. coli* that
are to be used in research with recombinant DNA molecules. Until this
is done, however, we can rejoice in the fact that the one *E. coli* strain that
has already been certified for this purpose has somehow managed to

---

*See page 67, "Biological Containment: The Construction of Safer *E. coli* strains" for the
characteristics of strain χ1776.

pass this stringent test even without benefit of official coercion. As more becomes known about the factors that control the growth of E. *coli* and other bacteria in the human intestine, this test may be replaced with more precise (i.e., less excessively stringent) procedures.

If, for purposes of this discussion, we assume that an E. *coli* strain carrying a segment of cloned DNA has succeeded in implanting itself into the intestinal flora of humans, what is the chance that this DNA can enter the body and subsequently enter cells of the body in intact form? (I will not discuss in this chapter the measures that have been taken to reduce the possibility that the DNA might be transferred in the gut to another, more virulent bacterium.) It is well known that bacteria from the human mouth frequently enter the body and may then be found in the blood. This is a normal and everyday process, which may be triggered by, for example, the simple act of chewing (22). It has been shown in humans, mice, rats, and dogs (23–25) that particles such as starch granules or bacteria from the intestinal tract regularly invade the body. The number of bacteria involved is very small; rather crude estimates made in my laboratory suggest that one in a billion bacteria present in the mouse gut may invade and reach the mesenteric lymph nodes during a period of several hours. Since the number of E. *coli* in the intestine rarely exceeds 10^6 per gram of intestinal contents, the actual number of E. *coli* cells reaching the body's tissues must be very small (i.e., 1 per 1000 grams).

Moreover, when bacterial cells of species such as E. *coli* invade the healthy body, they are quickly phagocytosed by leukocytes. Inside the leukocytes the bacteria are sequestered in vacuoles, killed, and subsequently broken up by the enzymes released into the vacuoles. Since these enzymes include those which degrade DNA, it is unlikely that intact DNA would escape into the general circulation of the body, even in the event that leukocytes killed the bacterium but subsequently either egested the contents of the vacuole or were totally unable to lyse certain structural components of the bacterial cell (26). Alternately, invading bacteria may be killed and lysed by the antibacterial substances in the body fluids. In this case, any DNA released by the disintegrating bacterium would be quickly destroyed by the DNA degrading enzymes in these fluids, probably long before it had a chance to enter a cell. It is apparent, therefore, that, even if an E. *coli* strain succeeded in implanting itself into the gut, the chance that its DNA would enter a cell of the human body in intact form is indeed exceedingly small. However, on theoretical grounds (the only tool we have in this chapter) we cannot exclude this possibility with the same degree of certainty as we are able to reach in discussing the possibility of creating superpathogens from E. *coli* or implanting into humans an E. *coli* strain that cannot multiply in a germfree animal.

It seems prudent, therefore, to direct biological containment efforts in recombinant DNA technology toward the development of microorganisms that, if they should escape from the laboratory, cannot perpetuate themselves on or in the human body or elsewhere in the environment. These are, of course, the EK2 and EK3 strains or "enfeebled" microbes, which were conceived for the first time in the discussions that culminated in the development of the now notorious NIH guidelines.

IV. GENERAL CONCLUSIONS

This chapter has, it is hoped, demonstrated to the reader that one of the various arguments implying terrifying consequences of recombinant DNA research does not stand up under thorough examination of facts that are well known to those who study infectious bacteria. It appears furthermore that problems concerning the possibility of implanting harmful hybrid bacteria in the human gut can be overcome satisfactorily if, but only if, precautionary measures are taken that are based on *all* the data currently available to medicine and science. I do not advocate that the public or other scientists accept such conclusions on the strength of a single article. However, if a representative group of specialists were to study these same issues and were to arrive at generally similar conclusions (as I believe they would), this should certainly go a long way toward channeling the increasing public concern into more rewarding avenues.* I am convinced that this same approach—which, after all, is a part of the enigmatic "scientific method"—would also prove beneficial in dealing with the many other concerns raised during the current debate. I expect strongly that many a scary scenario, such as the one in which oil-digesting bacteria leave the oil spill into which they were implanted and gobble up the earth, would be proven groundless when examined in the light of what is known about ecological niches and enzyme specificity. Scenarios such as this may seem to many scientists so obviously foolish that they consider it beneath their professional dignity even to discuss the matter. One must consider, however, that these same imagined dangers must appear very real to the intelligent layman and that it therefore behooves the specialists, who after all are supported by the tax dollars of these same layman, to do their best to clarify such matters of general public concern.

Editors' note: A meeting of specialists in microbial ecology and infectious diseases has subsequently considered these issues and has in fact concluded that the potential for the conversion of *E. coli* K12 into an epidemic pathogen is nil. The proceedings of this meeting have been published in J. Infec. Dis. **137**: 709 (1978).

As pointed out earlier in this chapter I consider the paucity of input from experts other than geneticists and molecular biologists to be a major cause of the lack of progress made so far in the public debate. Understandably, laymen often equate the term *scientist* with *expert*. For this reason the public is gaining the impression that the new recombinant DNA technology is an area of grave concern because "even the experts disagree" as to its inherent dangers. In fact, the experts have so far remained strangely silent in the discussion of many important issues that have been raised. It is becoming quite obvious that the debate, because of these defects, is contributing materially to public anxiety over goings-on in science in general. If nothing else, this unfortunate trend should provide sufficient motivation for experts in areas such as medicine and ecology to volunteer their services (even if they are not specifically asked for their advice) and to help in the separation of imaginary dangers of recombinant DNA work from those problems whose features may be less apt to excite our instinctive fears, but that nevertheless may require some attention.

NOTES

1. Cohen, S. N. 1977. Recombinant DNA: Fact and fiction. *Science 195*: 654–657.
2. Gary, W. E. 1917. Virulence and toxin production in *B. diphtheriae*. *J. Infec. Dis* **20**: 244–271.
3. Smith, H. W., and S. Halls, 1968. The transmissible nature of the genetic factor in *Escherichia coli* that controls enterotoxin production. *J. Gen Microbiol.* **52**: 319–334.
4. Smith, H. W. 1974. A search for transmissible pathogenic characters in invasive strains of *Escherichia coli*: the discovery of plasmid-controlled toxin and a plasmid-controlled lethal character closely associated with colicine V. *J. Microbiol.* **83**: 95–111.
5. Finland, M. 1970. Changing ecology of bacterial infections as related to antibacterial therapy. *J. Infect. Dis.* **122**: 419–431.
   McGowan, J. E., Jr., M. W. Barnes, and M. Finland, 1975. Bacteremia at Boston City Hospital: Occurrence and mortality during 12 selected years (1935–1972) with special reference to hospital-acquired cases. *J. Infect. Dis.* **132**: 316–335.
6. Freter, R. 1976. Factors controlling the composition of intestinal microflora. In: H. M. Stiles, W. J. Loesche, and T. C. O'Brien (eds.), *Microbial Aspects of Dental Caries* Information Retrieval, Inc., Washington, D.C.
7. Smith, H. W. 1975. Survival of orally adminstered *E. coli* K12 in alimentary tract of man. *Nature* **255**: 500–502.
8. Anderson, E. S. 1975. Viability of, and transfer of a plasmid from, *E. coli* K12 in the human intestine. *Nature* **225**: 502–504.
9. Wiedemann, B., and H. Knothe. 1969. Untersuchungen über die Stabilität der Koliflora des gesunden Menschen. 1. Mitteilung: Über das Vorkommen

permanenter und passanter Typen. *Arch. Hyg.* **153**: 342–348. 2. Mitteilung: Über den induzierten Wechsel permanenter Stämme. *ibid.* **255**: 349–352.

10. Anderson, J. D., W. A. Gillespie, and M. H. Richmond. 1973. Chemotherapy and antibiotic-resistance transfer between enterobacteria in the human gastro-intestinal tract. *J. Med.. Microbiol.* **6**: 461–473.

11. Richmond, M. H. 1974. R factors in man and his enviroment. In: *Microbiology 1974*. American Society for Microbiology, Washington, D.C., pp. 27–35.

12. Levy, S. B., G. B. FitzGerald, and A. B. Macone. 1976. Spread of antibiotic-resistant plasmids from chicken to chicken and from chicken to man. *Nature* **260**: 40–42.

13. Ozawa, A., and R. Freter. 1964. Ecological mechanism controlling growth of *Escherichia coli* in continuous flow cultures and in the mouse intestine. *J. Infect. Dis.* **114**: 235–242.

14. Anthony, B. F., and L. W. Wannamaker, 1967. Bacterial interference in experimental burns. *J. Exptl. Med.* **125**: 319–336.

15. Koch, A. L. 1971. The adaptive responses of *Escherichia coli* to a feast and famine existence. *Adv. Micro. Physiol.* **6**: 147–217.

16. Savageau, M. A. 1974. Genetic regulatory mechanisms and the ecological niche of *Escherichia coli*. *Proc. Natl. Acad. Sci. USA* **71**: 2453–2455.

17. Rolle, M., and. H. Mayer. 1954. Untersuchungen über die Darmflora des Meerschweinchens. *Arch. f. Hyg. und Bakt.* **138**: 505–510.

18. Morishita, Y., and M. Ogata. 1970. Studies on the alimentary flora of the pigs. V. Influence of starvation on the microbial flora. *Jap. J. Vet. Sci.* **32**: 19–24.

19. Tannock, G. W., and D. C. Savage. 1974. Influences of dietary and environmental stress on microbial populations in the murine gastrointestinal tract. *Infect. Immun.* **9**: 591–598.

20. Drasar, B. S., and M. J. Hill. 1974. *Human intestinal flora*. Academic Press, London.

21. Syed, S. A., Abrams, G. D., and Freter, R. 1970. The efficiency of various intestinal bacteria in assuming normal functions of enteric flora after association with germfree mice. *Infect. Immun.* **2**: 376–386.

22. Cobe, H. M. 1954. Transitory bacteremia. *Oral Surg. Oral Med. Oral Pathol.* **7**: 609–615.

23. Volkheimer, G., and F. H. Schulz. 1968. The phenomenon of persorption. *Digestion* **1**: 213–218.

24. Gordon, L. E., D. Ruml, H. J. Hahne, and C. P. Miller. 1955. Studies on susceptibility to infection following ionizing radiation. *J. Exper. Med.* **102**: 413–424.

25. Wolochow, H., G. J. Hildebrand, and C. Lamanna. 1966. Translocation of microorganisms across the intestinal wall of the rat: effect of microbial size and concentration. *J. Infect. Dis.* **116**: 523–528.

26. Ginsburg, I., N. Neeman, R. Gallily, and M. Lahav. 1976. Degradation and survival of bacteria in sites of allergic inflammation. In: D. C. Dumonde (ed.): *Infection and immunity in the rheumatic diseases*. Blackwell Scientific, Oxford.

DNA Splicing: Will Fear Rob Us of Its Benefits?

Joshua Lederberg

Although our theoretical understanding of the cell has been completely transformed in the last 30 years, there has not yet been a corresponding advance in the practical application of our knowledge to medicine. Indeed, very little in the practice of medicine (even of clinical genetics) is directly related to the fundamental knowledge that DNA has a bihelical structure.

Nonetheless, our faith remains steadfast that further theoretical understanding of viruses, the neoplastic cell, the aging cell, the immune response mechanism, and the aberrant chromosome will bring far-reaching changes to medicine. The human benefit from such understanding will someday surely match the theoretical impact that DNA study has already made on cell biology.

These expectations for a possibly long-delayed future benefit have been heightened and accelerated by new findings that give us much greater technical ability to manipulate microbial DNA. New methods of DNA splicing have already opened up many lines of investigation into the structure of eukaryotic (higher life form) chromosomes.

We can now fragment animal or human DNA into perhaps a million segments and transfer a single segment to a bacterial host for study in a microcosm or for production of large quantities of a specific DNA segment. This allows more elaborate analysis than has ever been possible with the enormously complex original unfragmented source material.

Editors' note: This chapter is reprinted with permission from *Prism* magazine, November 1975. Copyright © by the American Medical Association, November 1975.

This technique of gene implantation can also be used to transfer the genetic information for a given product from the cell of one species to that of another; and this is the direction, in my own view, that will lead to a technology of untold importance in diagnostic and therapeutic medicine: the ready production of an unlimited variety of human proteins. Analogous applications may be foreseen in fermentation processes for the cheap manufacture of essential nutrients and in the improvement of microbes for the production of antibiotics and special industrial chemicals.

In the face of such a revolution, the primary concern of researchers in the field has been the public hazards that such a technology may create. While we may indeed inherit a Promethean dilemma, public policy decision can lead to social good only if we are equally well informed about the potential risks and benefits of further work on DNA splicing. If substantial risk can be identified, there is no doubt of the need for ethical and operational safety standards; the only question must be whether the form and implementation of such standards are adequate.

Too often, the "easy" way to handle such a problem is to invoke a formal regulatory statute, ignoring how well the actual bureaucratic enforcement or policing of the rules meets the intended balance of risks and benefits. Before elaborating on the policy issues, it may be well for me to outline what is currently being done in DNA splicing, some promising applications, and also the risks of further work in this field.

DNA recombination, as the ultimate purpose of the sexual form of reproduction, is, of course, one of the major happenings in the natural world. Among higher life forms, DNA exchange is almost always limited to members of the same or closely related species. Bacteria and viruses, however, exhibit many exceptions to this rule, which perhaps reflects the fragility of the concept "species" when applied to these life forms.

For example, the entire group of enteric bacteria, including such forms as *Shigella, Escherichia coli, Proteus,* and *Serratia,* can exchange genetic fragments without special intervention. Our own experiments in genetic exchange would not seriously increase the risks already latent in that natural process.

CONVENIENT TOOLS

An especially interesting and important level of genetic organization in bacteria is the plasmid: a bit of circular DNA that behaves like an extra chromosome and seems to survive in nature by virtue of its easy transmissibility from one bacterial strain to another. Many different kinds of plasmids are known; in medicine, the most prominent are those which confer transmissible antibiotic resistance on human pathogens, notably staphylococci and some enteric pathogens such as *Shigella*.

These plasmids are a by-product of the evolution of their host organisms: the spread of antibiotic-resistance plasmids is the most formidable bacterial response yet to our widespread use of antibiotics. Other plasmids are undoubtedly involved in altering the pathogenicity and host-specificity of various bacteria; therefore, in simple self-defense, we must learn all we can about them, without delay.

Plasmids have also achieved special prominence for a technical reason—they are especially convenient tools for DNA splicing and for the transmission of DNA segments from one species to another, particularly in conjunction with another elegant tool: the R-(for restriction) enzyme. (The R-nucleases are widely distributed among cell types; they may be an important mechanism by which a cell fends off any "foreign" DNA while protecting its own.)

Stanley N. Cohen, M.D., of the Department of Medicine, Stanford University, has used an R-enzyme to simplify a naturally occurring plasmid to the point where it consists of a small circle of DNA, embracing the minimum amount of genetic information needed to replicate, plus a single R-enzyme recognition site.

This artificial plasmid, pSC-101, has been an important tool in DNA splicing research. When exposed to R-enzyme, the circle is cut into a single open length with sticky ends. It is then possible to insert other sticky-ended pieces of DNA from diverse sources into the plasmid, and finally close it up with another enzyme, ligase. This process is the key to the convenient design and construction of new DNA molecules, which subsequently can be transferred to a bacterial host.

One important aspect of this research is that the new DNA does *not* have to come from the same bacterial species. For example, Dr. Cohen and his collaborators have already reported the successful transfer of DNA from a toad, *Xenopus*, to *E. coli* with evidence of the production of toad-like ribosomal nucleic acids in the modified bacteria.

In addition to these plasmids, bacterial viruses are being used in a similar fashion. Less elegantly, perhaps, segments of DNA from intact bacteria may also be used both for insertions and as the acceptors. So far, all of these techniques depend on the innate (and poorly understood) ability of bacterial cells to incorporate DNA furnished from without. There have been many published claims of similar phenomena with plant and animal cell acceptors, but to date the claims are unconfirmed.

The special power of the enzyme transfer techniques is that they depend on the basic chemical structure of DNA rather than on biological adaptation. Thus laboratory manipulation may produce constructs that occur rarely, if ever, in the natural world. Most of these constructs would resemble hothouse plants, and be poorly adapted to competitive survival in the world outside the laboratory. But some, by chance, might be harbingers of new diseases or the source of ecological upsets difficult to control—like the mongoose in Hawaii or the crabgrass in your lawn.

R-enzymes, mixed DNA, and acceptor bacteria surely bring about some DNA segment transfers in nature. Our knowledge of the extent of natural plasmid transmission among "unrelated" life forms was widened by recent discoveries of plasmids with extraordinarily broad host ranges. It is difficult, however, to assess just what *can* or *cannot* occur in nature.

RAPID ADVANCEMENT

DNA splicing is, however, merely the most powerful of several artificial techniques which bring together more-or-less natural assemblages of DNA. Indeed, it may prove to be less powerful than older methods (sexual crossing, transduction with bacteriophage, DNA-mediated transformation) for special constructions involving larger complexes than the segments yielded by R-enzymes.

These methods, in turn, are an extension of the artificial breeding of domestic animals and plants. In any event, the most efficient application of DNA splicing requires intimate knowledge of the genetic structure of both the donor and the acceptor strains, for which breeding methods are important if not indispensable.

Perhaps the single most important conclusion is that this technology is just in its infancy but has already advanced far—and that it is simple enough to be applied in any laboratory which can handle pure bacterial cultures. But it is just this simplicity, which makes for great convenience and speed of development, that has raised concern about the proliferation of such methods in the hands of people with perhaps less-than-mature professional and ethical judgment, and with insufficient skill to contain bacterial cultures in the laboratory.

Now that we have put the dangers of DNA splicing research into perspective, let us examine the promise that it holds. DNA segmentation and splicing is certain to play a vital role in the further domestication of microbes for such uses as the development of new antibiotics and the production of high-quality food protein supplements. However, the unique strength of this procedure is that it allows the large-scale production of gene products of a less easily domesticated species: man.

Human proteins already play a substantial role in medicine but a role which is hindered by scarce supply. Today, the most attractive candidates for such large-scale production are the human antibody globulins. Compared to the rare genetic defects in other proteins (as in the case of hemophilia), failure of error in the production of antibody globulin is quite prevalent and is known to play a major role in the breakdown of the body's defense against infectious disease, in autoimmune and allergic disease, and perhaps also in cancer.

The most comprehensive use for biosynthetic proteins would be in passive immunization against infectious disease. (Animal antisera were once used but had to be abandoned because of the anti-animal antibody that they provoked in man.) With wholesale production of biosynthetic proteins, passive globulin therapy could be targeted at those diseases for which either technical or social factors may bring about gaps in the protection provided by active immunization. Included in that group of diseases are influenza, hepatitis, smallpox, encephalitis, rubella, herpes, rabies and perhaps also trypanosomiasis, malaria, schistosomiasis, tuberculosis, leprosy, and many others.

NEED FOR A READY DEFENSE

There is reason for special urgency in the development of a backup capability in passive immunization. Complacency about active immunization against diseases such as polio and the technical inadequacy of such vaccines as rubella and hepatitis have weakened our general posture of defense against viral pandemic. We have no assurance that the next influenza epidemic, slightly more virulent than the last one, will not take a million lives for lack of a ready defense.

A broader need for biosynthetic proteins lies in polyvalent prophylaxis for infants. The principal medical argument for breast feeding is that human milk provides the infant with colostrum and a continuing supply of maternal mixed globulins. In the future there might be a huge demand for polyvalent gamma globulin supplements for infants both in industrialized and in poorer countries. And an analogous veterinary use could bring about greater efficiency in livestock production.

Specific antibodies, of course, are already widely used as diagnostic reagents of high specificity and selectivity. But in sufficient quantity, blocking antibodies might also play a useful role in helping protect transplanted tissues and organs from immunological attack by the new host. Conversely, tissue-specific ligating antibodies, although not necessarily cytotoxic themselves, may be useful in enhancing the cell-specific toxicity of certain cancer drugs. Cell-specific reagents would also be invaluable in diagnosis and in the specific separation of human cell types for either diagnostic or therapeutic applications.

Besides the specific antibody globulins, a number of important, but less specific, proteins (complement, properdin) play a major part in defense against infection. Fibrinolysin (plasmin) and urokinase (plasminogen-activator) represent a group of enzymes that experimentally have shown promise in the control of embolism. Besides these human proteins, many human hormones are also discouragingly scarce for use in clinical trials. The list of such bioproducts could be extended substantially. And perhaps the most important products are those that remain to be discovered.

Of course, microbial biosynthesis may well be supplemented by organic synthesis in human and hybrid somatic cell cultures and by cell-free ribosomal synthesis with m-RNA extracted from natural sources or synthesized. Each of these methods has its own peculiar difficulties and hazards, and the whole field will be advanced most rapidly by using the best available methods for any given problem.

At present, perhaps a half-dozen bacterial species are well enough understood to serve as prime vehicles in laboratory studies of DNA splicing. For safety and convenience, investigators have preferred not to use pathogenic forms. Yet many scientists are primarily concerned that DNA splicing may inadvertently generate a new pathogen inimical to man or to some other species important to man's ecology. The most likely, but not necessarily the only, sources of such pathogenic genes are the organisms that most urgently need further study—the subtle and insidious killers not now amenable to medical treatment. These include slow virus infection that may be involved in a wide range of chronic diseases, including cancer, and more familiar viruses, such as herpes, for which satisfactory vaccines are not available.

SPECULATING THE HAZARDS

The public debate over DNA splicing has focused on the possible hazards of new microorganisms, and away from their utilitarian prospects. The most urgent concern has been the danger of introducing potentially cancer-causing DNA into common bacteria. While this hazard is clearly speculative, the general territory is so poorly understood that no one can argue against the need for cautious laboratory procedures. A number of workers—particularly those whose special experience or training has been in fields other than medical microbiology—have confessed giving almost no thought in the past to safety; some of them are now among the most zealous in demanding tighter regulation of such research. And that zeal has spread to create a sincere, almost frantic effort to ferret out and identify the remote, conceivable hazards.

Viewed as a rather public soul-searching and self-education, these discussions are invaluable. The main danger is that some political imperative may forge these tentative questions into iron-clad regulations which will be with us long after their origins have been forgotten. After all, similar questions can be raised about the widest range of human activities: should it be lawful to keep domestic cats when we suspect that they harbor toxoplasmosis, and possibly leukemia as well? Similarly, what assurance do we have that artificial pollination will not produce a weed that could ruin the wheat crop a decade from now? Closer to home, should we forbid international travel simply because our quarantine procedures do not guarantee that exotic diseases will be kept out?

For each of these cases, and many more, the apparently innocuous doctrine, "As long as there is any risk, don't do it!" can only bring a loss to human welfare. We must instead make every feasible effort to assess both the risks and the benefits of a given course of action—only then will we be able to find the optimal balance. But individuals can hardly determine the best policy about their own future—including their expectations for what medicine will offer for the infirmities of their own later years—without expert assessment.

Such assessments are difficult, problematical, and controversial. But a committee of the National Academy of Sciences has made some headway in trying to classify different categories of hazard. Where such hazard is reasonably predictable, the committee has recommended laboratory containment precautions akin to those appropriate for known pathogens. This applies, for example, to experiments in the recombination of known tumor virus DNA with bacterial plasmids.

For more conjectural hazards, such as the introduction of antibiotic resistance into common, non-pathogenic species, the high security requirements recommended by the committee may be an inordinate burden for laboratories (who, in fact, will pay for them?) in relation to the prospective gains. The best strategy in such a case seems to be the development of safe vectors: plasmids and bacteria engineered so that they have little chance of surviving outside the laboratory. In fact, in the long run this is a safer procedure than relying upon the uncertainty of human compliance with fixed rules and regulations.

Remaining controversies in this area center upon rather complicated analyses of the most remote risks. Given some additional time, most research institutions will work out their own reasonable plans, based on the national guidelines. A premature imposition of external regulation will not only frustrate useful research, but will also hinder that research which is needed to more accurately assess the dangers. Those who consider themselves guardians of the public safety must count the costs to the public health of *impeding* research, as well as the speculative *hazards* of research.

SOCIETY'S CONSENT

This partly voluntary approach will not assure absolutely that no foolish experiment is ever attempted. But the history of human institutions should suffice to show that *no* system of sanctions can achieve such a goal. The human species is inevitably attended by contaminating and parasitic microbes—the person suffering from an enteric infection who fails to wash his hands or the influenza victim who insists on going to work is behaving unethically and to the peril of his fellows. But we would scarcely invoke serious regulatory sanctions in preference to public education, except where there is an unusual public risk with some attendant evidence that an enforced quarantine would be effective.

Senator Edward Kennedy (D-Mass.) has remarked that society must give its informed consent to technological innovation. The power of the purse is enough to enforce that doctrine; nor can there be any quarrel with it on ethical grounds. Informed consent surely includes knowing the hazards of saying no to the prospects of significant medical advances. DNA splicing research, far from being an idle scientific toy or the basis for expensive and specialized aid to the privileged few, promises some of the most pervasive benefits for the public health since the discovery and promulgation of antibiotics.

PART III

Philosophical, Legal, and Social Issues

The Recombinant DNA Debate: Some Philosophical Considerations

Stephen P. Stich

The debate over recombinant DNA research is a unique event, perhaps a turning point, in the history of science. For the first time in modern history there has been widespread public discussion about whether and how a promising though potentially dangerous line of research shall be pursued. At root the questions being debated are *moral* questions: What sorts of experiments *should* be permitted and which *should* be banned? What sorts of controls *ought* to be imposed? Like most moral debates, this one requires factual input at crucial stages in the argument. A good deal of the controversy over recombinant DNA research arises because the facts simply are not in. There are many empirical questions we would like to have answered before coming to a decision—questions about the reliability of proposed containment facilities, about the viability of enfeebled strains of *E. coli*, about the ways in which pathogenic organisms do their unwelcome work, and much more. But all decisions cannot wait until the facts are in; some must be made now. It is to be expected that people with different hunches about what the facts will turn out to be will urge different decisions on how recombinant DNA research should be regulated. However, differing expectations about the facts has not been the only fuel for the controversy. A significant part of the current debate can be traced to differences over *moral* principles. Also, unfortunately, there has been much unnecessary heat generated by careless moral reasoning and a failure to attend to the logical structure of some of the moral arguments that have been advanced.

183

My goal in this essay is to help sharpen our perception of the moral issues that underlie the controversy over recombinant DNA research. My strategy will be to begin with some brief observations on the workings of moral arguments in general. This will provide us with the analytical tools needed to clear away frivolous arguments that have deflected attention from more serious issues. With that done, we can take a careful look at the problems involved in deciding whether the potential benefits of recombinant DNA research justify pursuing the research despite the risks that it poses.

I. SOME PATTERNS OF MORAL REASONING

To begin, let me note the handy distinction between *particular moral judgments* and *general moral principles*. Under the former heading I include claims about what should be done by a specific person or group of people in a specific historical situation. By contrast, general moral principles make more general claims about what should be done in *every* case of a certain sort or by *everyone* in a certain sort of situation. The claim: "Congress should enact legislation placing a three-year moratorium on all recombinant DNA experiments," is an example of a particular moral judgment. Traditional moral codes such as the Ten Commandments provide many examples of general moral principles, though not all such principles need be quite so global in their scope.

Using the terminology of the previous paragraph, we can isolate a feature of defensible moral views that proves central to the process of rationally supporting or attacking a moral position. The feature is this: If a particular moral judgment is defensible, then it must be *supportable* by a defensible general moral principle. A general moral principle *supports* a particular moral judgment if the principle (or more often the principle supplemented by some true factual claims) logically entails the particular moral judgment. There is much discussion in the philosophical literature on just why defensible particular moral judgments must be supportable by a defensible moral principle (1), but there is relatively little dispute that some such constraint is built into our moral concepts. Thus if a person advances a particular moral judgment and is unable to produce a supporting principle when challenged, his position will generally be considerably weakened. And if a person recognizes no need to support his particular moral judgments by appeal to moral principles, we may well conclude that he simply does not understand how moral concepts are used. To put the point in a rather different way, a person who recognizes no need to support particular judgments by principles is likely using moral words (such as 'should' and 'ought') with a meaning very different from their common meaning.

Moral dialogue can be complex and varied; it is often highly elliptical, with necessary premises left out. However, a good deal of what goes on in rational moral dialogue can be interpreted in light of the requirement that particular moral judgments be supportable by defensible general moral principles. Frequently the proponent of a particular moral judgment will try to defend her view by citing a general moral principle and some facts, which together entail the particular moral judgment she is advocating. Debate may then center on a number of points. If both parties to the dispute are prepared to grant the general moral principle, then debate will likely focus on the factual claims. Since the factual claims needed to complete the pattern of support are often exceptionally difficult to evaluate, much of the effort in rational ethical deliberation will be aimed at getting straight on matters of empirical fact.

A moral argument may take quite a different turn, however, if one party to the dispute is unwilling to grant the general moral principle that the other party uses to support her particular moral judgment. When this occurs, the debate may well turn on whether the general moral principle is itself defensible. There is a range of commonly used rational strategies for supporting a general moral principle whose acceptibility has been challenged. One strategy is to appeal to a still more general principle that (alone or with the addition of some factual claims) entails the principle needing defense. To be persuasive, of course, the higher principle must itself be one our opponent accepts. Quite a different idea for supporting a moral principle is to show that the principle, along with relevant facts, entails a significant number of particular moral judgments that our opponent accepts. Of course, this sort of argument, even if it can be made to stick, does not yield a conclusive defense of the disputed principle. For some quite different principle may do as well or better at entailing a range of particular moral judgments acceptable to our opponent. However, if we can show an opponent that a proposed principle does in fact entail a fair number of particular moral judgments that he accepts, we have at least succeeded in shifting the burden of the argument to him. If he is to reject the principle we are defending, he had best have an alternative that does at least as good a job in supporting his particular moral judgments.

Perhaps the most frequently used strategy for *attacking* a moral principle is to find one or more particular moral judgments that the principle's advocate firmly rejects, and then argue that the principle plus some factual claims entail the unacceptable particular moral judgments. If this can be shown, then the defender of the principle is left with the choice of either abandoning the principle or changing her view about the particular moral judgments that the principle has been shown to support.

The foregoing discussion of the mechanics of moral arguments is intended only as a brief introducion. I think, however, that we now have enough of the anatomy of moral argument to venture a dissection of some living specimens.

II. THREE BAD ARGUMENTS

In the current section I will scout a collection of untenable arguments, each of which has surfaced with some frequency in the public debate over recombinant DNA research. Their recurrent appearance has diverted attention from more serious issues, and it is high time they were laid to rest.

1. The first argument on my list has as its conclusion that recombinant DNA research should not be controlled or restricted. The central premise of the argument is that scientists should have full and unqualified freedom to pursue whatever inquiries they might choose to pursue. The claim appeared repeatedly in petitions and letters to the editor during the height of the public debate over recombinant DNA research in the University of Michigan community (2). The general moral principle that is the central premise of the argument plainly entails the particular moral judgment that investigators using recombinant DNA technology should be allowed to pursue their research as they see fit. However, we need only consider a few examples to see that the principle being invoked in this "freedom of inquiry" argument is utterly indefensible. No matter how sincere a researcher's interest may be in investigating the conjugal behavior of American university professors, very few would be willing to grant him the right to pursue his research in my bedroom without my consent; and no matter how interested a researcher may be in investigating the effects of massive doses of bomb-grade plutonium on preschool children, it is hard to imagine that anyone thinks he should be allowed to do so. Yet the "free inquiry" principle, if accepted, would allow both of these projects and countless other Dr. Strangelove projects as well. So plainly the simplistic "free inquiry" principle is indefensible (3). It would, however, be a mistake to conclude that freedom of inquiry ought not to be protected. A better conclusion is that the right of free inquiry is a qualified right and must sometimes yield to conflicting rights and to the demands of conflicting moral principles. The task of articulating a properly qualified principle of free inquiry is subtle and challenging. Carl Cohen's essay in the following chapter of this volume takes up this challenge.

2. The second argument I want to examine aims at establishing just the opposite conclusion from the first. The particular moral judgment being defended is that there should be a total ban on recombinant DNA research. The argument begins with the observation that even in so-called low-risk recombinant DNA experiments there is at least a possibility of catastrophic consequences. We are, after all, dealing with a relatively new and unexplored technology. Thus it is at least possible that a bacterial culture whose genetic makeup has been altered in the course of a recombinant DNA experiment may exhibit completely unexpected pathogenic characteristics. Indeed, it is not impossible that we could find ourselves confronted with a killer strain of, say, E. *coli*, and worse a strain against which humans can marshall no natural defense. Now if this is possible, if we cannot say with assurance that the probability of its happening is zero, then, the argument continues, all recombinant research should be halted. For the negative utility of the imagined catastrophe is so enormous, resulting as it would in the destruction of our society and perhaps even of our species, that no work that could possibly lead to this result would be worth the risk.

The argument just sketched, which might be called the "doomsday scenario" argument, began with a premise that no informed person would be inclined to deny. It is indeed *possible* that even a low-risk recombinant DNA experiment might lead to totally catastrophic results. No ironclad guarantee can be offered that this will not happen. And while the probability of such an unanticipated catastrophe is surely not large, there is no serious argument that the probability is zero. Still, I think the argument is a sophistry. For the moral principle needed to go from the undeniable premise that recombinant DNA research might possibly result in unthinkable catastrophe to the conclusion that such research should be banned is that *all* endeavours that might possibly result in such a catastrophe should be prohibited. Once the principle has been stated, it is hard to believe that anyone would take it at all seriously. For the principle entails not only that recombinant DNA research should be prohibited, but also that almost all scientific research should be prohibited, along with many other commonplace activities having little to do with science. It is, after all, at least logically possible that the next new compound synthesized in an ongoing chemical research program will turn out to be an uncontainable carcinogen many orders of magnitude more dangerous than aerosol plutonium. And, to vary the example, there is a nonzero probability that experiments in artificial pollination will produce a weed that will, a decade from now, ruin the world's food grain harvests (4).

As someone who earns his bread and butter teaching introductory philosophy, I cannot resist noting that the principle invoked in the doomsday scenario argument is not new. Pascal used an entirely parallel argument to show that it is in our own best interests to believe in God. For though the probability of God's existence may be very low, if He nonetheless should exist, the disutility that would accrue to the disbeliever would be catastrophic—an eternity in Hell. But, as introductory philosophy students should all know, Pascal's argument looks persuasive only if we take our options to be just two: Christianity or atheism. A third possibility is belief in a jealous non-Christian God who will see to our damnation if and only if we *are* Christians. The probability of such a deity's existing is again very small, but nonzero. So it looks as if Pascal's argument is of no help in deciding whether or not to accept Christianity. For we are damned if we do, and damned if we don't.

I mention Pascal's difficulty because there is a direct parallel in the doomsday scenario argument against recombinant DNA research. Just as there is a nonzero probability that unforeseen consequences of recombinant DNA research will lead to disaster, so there is a nonzero probability that unforeseen consequences of *failing* to pursue the research will lead to disaster. There may, for example, come a time when, because of natural or man-induced climatic change, the capacity to alter quickly the genetic situation of agricultural plants will be necessary to forestall catastrophic famine. And if we fail to pursue recombinant DNA research now, our lack of knowledge in the future may have consequences as dire as any foreseen in the doomsday scenario argument.

3. The third argument I want to consider provides a striking illustration of how important it is, in normative thinking, to make clear the moral *principles* being invoked. The argument I have in mind begins with a factual claim about recombinant DNA research, and it concludes that stringent restrictions, perhaps even a moratorium, should be imposed. However, advocates of the argument are generally silent on the normative principle(s) linking premise and conclusion. The gap thus created can be filled in a variety of ways, resulting in very different arguments.

The empirical observation that gets the argument going is that recombinant DNA methods enable scientists to move genes back and forth across natural barriers, "particularly the most fundamental such barrier, that which divides prokaryotes from eukaryotes. The results will be essentially new organisms, self-perpetuating and hence permanent" (5). For this reason it is concluded that severe restrictions are in order. Plainly this argument is an enthymeme; a central premise has been left unstated. What sort of moral principle is being tacitly assumed?

The principle that comes first to mind is simply that natural barriers should not be breached, or perhaps that "essentially new organisms" should not be created. The principle has an almost theological ring to it, and perhaps there are some people who would be prepared to defend it on theological grounds. But short of a theological argument, it is hard to see why anyone would hold the view that breaching natural barriers or creating new organisms is *intrinsically* wrong. For if a person were to advocate such a principle, he would have to condemn the creation of new bacterial strains capable of, say, synthesizing human clotting factor or insulin, *even if* creating the new organism generated *no unwelcome side effects.*

There is quite a different way of unpacking the "natural barriers" argument that avoids appeal to the dubious principles just scouted. As an alternative, this second reading of the argument ties premise to conclusion with a second factual claim and a quite different normative premise. The added factual claim is that at present our knowledge of the consequences of creating new forms of life is severely limited, and thus we cannot know with any assurance that the probability of disastrous consequences is very low. The moral principle needed to mesh with the two factual premises would be something like the following:

> If we do not know with considerable assurance that the probability
> of an activity's leading to disastrous consequences is very low, then
> we should not allow the activity to continue.

Now this principle, unlike those marshalled in the first interpretation of the natural barriers argument, is not lightly dismissed. It is, to be sure, quite a conservative principle, and it has the odd feature of focusing entirely on the dangers an activity poses while ignoring its potential benefits (6). Still, the principle may have a certain attraction in light of recent history, which has increasingly been marked by catastrophies attributable to technology's unanticipated side effects.

I will not attempt a full-scale evaluation of this principle just now. For the principle raises, albeit in a rather extreme way, the question of how risks and benefits are to be weighed against each other. By my lights, that is the really crucial moral question raised by recombinant DNA research. It is a question that bristles with problems. In the section that follows I will take a look at some of these problems and make a few tentative steps toward some solutions. While picking our way through the problems we will have another opportunity to examine the principle just cited.

III. RISKS AND BENEFITS

At first glance it might be thought that the issue of risks and benefits is quite straightforward, at least in principle. What we want to know is whether the potential benefits of recombinant DNA research justify the risks involved. To find out we need only determine the probabilities of the various dangers and benefits. And while some of the empirical facts—the probabilities—may require considerable ingenuity and effort to uncover, the assessment poses no particularly difficult normative or conceptual problems. Unfortunately, this sanguine view does not survive much more than a first glance. A closer look at the task of balancing the risks and benefits of recombinant DNA research reveals a quagmire of sticky conceptual problems and simmering moral disputes. In the next few pages I will try to catalogue and comment on some of these moral disputes. I wish I could also promise solutions to all of them, but to do so would be false advertising.

1. Problems about Probabilities

In trying to assess costs and benefits, a familiar first step is to set down a list of possible actions and possible outcomes. Next, we assign some measure of desirability to each possible outcome, and for each action we estimate the conditional probability of each outcome given that the action is performed. When we attempt to apply this decision-making strategy to the case of recombinant DNA research, the assignment of probabilities poses some perplexing problems. Some of the outcomes whose probabilities we want to know can be approached using standard empirical techniques. Thus, for example, we may want to know what the probability is of a specific enfeebled host *E. coli* strain surviving passage through the human intestinal system, should it be accidentally ingested. Or we may want to know what the probability is that a host organism will escape from a P-4 laboratory. In such cases, while there may be technical difficulties to be overcome, we have a reasonably clear idea of the sort of data needed to get a fix on the required probabilities. But there are other possible outcomes whose probabilities cannot be determined by experiment. It is important, for example, to know what the probability is that recombinant DNA research will lead to a method for developing nitrogen-fixing strains of corn and wheat. And it is important to know how likely it is that recombinant DNA research will lead to techniques for effectively treating or preventing various types of cancer. Yet no experiment we can perform nor any data we can gather will enable us to *empirically* estimate these probabilities.

Nor are these the most problematic probabilities we may want to know. A possibility that weighs heavily on the minds of many is that recombinant DNA research may lead to negative consequences for human health or for the environment *that have not yet even been thought of*. The history of technology during the last half century surely demonstrates that this is not a quixotic concern. Yet here again there would appear to be no data we can gather that would help much in estimating the probability of such potential outcomes.

It should be stressed that the problems just sketched are not to be traced simply to a paucity of data. Rather, they are conceptual problems; it is doubtful whether there is *any clear empirical sense* to be made of objective probability assignments to contingencies such as those we are considering.

Theorists in the Bayesian tradition may be unmoved by the difficulties we have noted. On their view all probability claims are reports of subjective probabilities. And, a Bayesian might quite properly note, there is no special problem about assigning *subjective* probabilities to outcomes like those that worried us. But even for the radical Bayesian, there remains the problem of *whose* subjective probabilities ought to be employed in making a *social* or *political* decision. The problem is a pressing one, since the subjective probabilities assigned to potential dangers and benefits of recombinant DNA research would appear to vary considerably even among reasonably well-informed members of the scientific community.

The difficulties we have been surveying are serious ones. Some might feel they are so serious that they render rational assessment of the risks and benefits of recombinant DNA research all but impossible. I am inclined to be rather more optimistic. Almost all the perils posed by recombinant DNA research require the occurrence of a sequence of separate events. For a chimerical bacterial strain created in a recombinant DNA experiment to cause a serious epidemic, for example, at least the following events must occur:

1. A pathogenic bacterium must be synthesized.
2. The chimerical bacteria must escape from the laboratory.
3. The strain must be viable in nature.
4. The strain must compete successfully with other microorganisms which are themselves the product of intense natural selection (7).

Since *all* these must occur, the probability of the potential epidemic is the product of the probabilities of each individual contingency. And there are at least two items on the list, events 2 and 3, whose probabilities are amenable to reasonably straightforward empirical assessment. Thus the product of these two individual probabilities places an upper bound on the probability of the epidemic. For the remaining two probabilities, we must rely on subjective probability assessments of informed scientists. No doubt there will be considerable variability, yet even here the variability will be limited. In the case of event 4, as an example, the available knowledge about microbial natural selection provides no precise way of estimating the probability that a chimerical strain of enfeebled *E. coli* will compete successfully outside the laboratory. But no serious scientist would urge that the probability is *high*. We can then use the highest responsible subjective estimate of the probabilities of events 1 and 4 in calculating the "worst-case" estimate of the risk of epidemic. If, when we use this worst-case estimate, our assessment yields the result that benefits outweigh risks, then lower estimates of the same probabilities will, of course, yield the same conclusion. Thus it may well be that the problems about probabilities we have reviewed will not pose insuperable obstacles to a rational assessment of risks and benefits.

2. Weighing Harms and Benefits

A second cluster of problems turns on the assignment of a measure of desirability to the various possible outcomes. Suppose we have a list of the various harms and benefits that might possibly result from pursuing recombinant DNA research. The list will include such "benefits" as development of an inexpensive way to synthesize human clotting factor, development of a strain of nitrogen-fixing wheat, and such "harms" as release of a new antibiotic-resistant strain of pathogenic bacteria, release of a strain of *E. coli* carrying tumor viruses capable of causing cancer in humans. Plainly, it is possible that pursuing a given policy will result in more than one benefit and more than one harm. If we are to assess the potential impact of various policies or courses of action, we must assign some index of desirability to the possible *total outcomes* of each policy, outcomes which may well include a mix of benefits and harms. To do this we must confront head-on a tangle of normative problems that are as vexing and difficult as any we are likely to face.

What we must do is *compare* the moral desirabilities of various harms and benefits. The task is particularly troublesome when the harms and benefits to be compared are of very different kinds. Thus, for example, some of the attractive potential benefits of recombinant DNA research are economic; we may learn to recover small amounts of valuable metals in an economically feasible way, or we may be able to synthesize insulin and other drugs inexpensively. By contrast, many of the risks of recombinant DNA research are risks to human life or health. So if we are to take the idea of cost-benefit analysis seriously, we must at some point decide how human lives are to be weighed against economic benefits.

Some contend that the need to make such decisions indicates the moral bankruptcy of attempting to employ risk-benefit analyses when human lives are at stake. On the critics' view, we cannot reckon the possible loss of a human life as just another negative outcome, albeit a grave and heavily weighted one. To do so, it is urged, is morally repugnant and reflects a callous lack of respect for the sacredness of human life.

On my view, this sort of critique of the very idea of using risk-benefit analyses is ultimately untenable. It is simply a fact about the human condition, lamentable as it is inescapable, that in many human activities we run the risk of inadvertently causing the death of a human being. We run such a risk each time we drive a car, allow a dam to be built, or allow a plane to take off. Moreover, in making social and individual decisions, we cannot escape weighing economic consequences against the risk to human life. A building code in the midwest will typically mandate fewer precautions against earthquakes than a building code in certain parts of California. Yet earthquakes are not impossible in the midwest. If we elect not to require precautions, then surely a major reason must be that it would simply be too expensive. In this judgment, as in countless others, there is no escaping the need to balance conomic costs against possible loss of life. To deny that we must and do balance economic costs against risks to human life is to assume the posture of a moral ostrich.

The point I have been urging is that it is not *morally objectionable* to try to balance economic concerns against risks to human life. But if such judgments are unobjectionable, indeed necessary, they also surely are among the most difficult any of us has to face. It is hard to imagine a morally sensitive person not feeling extremely uncomfortable when confronted with the need to put a dollar value on human lives. It might be thought that the moral dilemmas posed by the need to balance such radically different costs and benefits pose insuperable practical obstacles for a rational resolution of the recombinant DNA debate. But here, as in the case of problems with probabilities, I am more sanguine. For while some of the risks and potential benefits of recombinant DNA research are all but morally incommensurable, the most salient risks and benefits are easier to compare. The major risks, as we have noted, are to human life and health. However, the major potential benefits are *also* to human life and health. The potential economic benefits of recombinant DNA research pale in significance when set against the potential for major breakthroughs in our understanding and ability to treat a broad range of conditions, from birth defects to cancer. Those of us, and I confess I am among them, who despair of deciding how lives and economic benefits are to be compared can nonetheless hope to settle our views about recombinant DNA research by comparing the potential risks to life and health with the potential benefits to life and health. Here we are comparing plainly commensurable outcomes. If the balance comes out favorable, then we need not worry about factoring in potential economic benefits.

There is a certain irony in the fact that we may well be able to ignore economic factors entirely in coming to a decision about recombinant DNA research. For I suspect that a good deal of the apprehension about recombinant DNA research on the part of the public at large is rooted in the fear that (once again) economic benefits will be weighed much too heavily and potential damage to health and the environment will be weighed much too lightly. The fear is hardly an irrational one. In case after well-publicized case, we have seen the squalid consequences of decisions in which private or corporate gain took precedence over clear and serious threats to health and to the environment. It is the profit motive that led a giant chemical firm to conceal the deadly consequences of the chemical that now threatens to poison the James River and perhaps all of Chesapeake Bay. For the same reason, the citizens of Duluth drank water laced with a known carcinogen. And the ozone layer that protects us all was eroded while regulatory agencies and legislators fussed over the loss of profits in the spray deodorant industry. Yet while public opinion about recombinant DNA research is colored by a growing awareness of these incidents and dozens of others, the case of recombinant DNA research is fundamentally different in a crucial respect. The important projected benefits that must be set against the risk of recombinant DNA research are not economic at all, they are medical and environmental.

3. Problems about Principles

The third problem area I want to consider focuses on the following question: Once we have assessed the potential harms and benefits of recombinant DNA research, how should we use this information in coming to a decision? It might be thought that the answer is trivially obvious. To assess the harms and benefits is, after all, just to compute, for each of the various policies that we are considering, what might be called its *expected utility*. The expected utility of a given policy is found by first multiplying the desirability of each possible total outcome by the probability that the policy in question will lead to that total outcome, and then adding the numbers obtained. As we have lately seen, finding the needed probabilities and assigning the required desirabilities will not be easy. But once we know the expected utility of each policy, is it not obvious that we should choose that policy whose expected utility is highest? The answer, unfortunately, is no, it is not at all obvious.

Let us call the principle that we should adopt the policy with the highest expected utility the *utilitarian principle.* The following example should make it clear that, far from being trivial or tautological, the utilitarian principle is a substantive and controversial moral principle. Suppose the decision that confronts us is whether or not to adopt policy A. What is more, suppose we know there is a probability close to 1 that 100,000 lives will be saved if we adopt A; however, we also know that there is a probability close to 1 that 1000 will die as a direct result of our adopting policy A, and these people would survive if we did not adopt A. Finally, suppose that the other possible consequences of adopting A are relatively inconsequential and can be ignored. (For concreteness, we might take A to be the establishment of a mass vaccination program, using a relatively risky vaccine.) Now plainly if we take the moral desirability of saving a life to be exactly offset by the moral undesirability of causing a death, then the utilitarian principle dictates that we adopt policy A. But many people feel uncomfortable with this result, the discomfort increasing with the number of deaths that would result from A. If, to change the example, the choice that confronts us is saving 100,000 lives while causing the deaths of 50,000 others, a significant number of people are inclined to think that the morally right thing to do is to refrain from doing A, and "let nature take its course."

If we reject policy A, the likely reason is that we also reject the utilitarian principle. Perhaps the most plausible reason for rejecting the utilitarian principle is the view that our obligation to *avoid doing harm* is stronger than our obligation to do good. Many examples, some considerably more compelling than the one we have been discussing, seem to illustrate that in a broad range of cases we do feel that our obligation to avoid doing harm is greater than our obligation to do good (8). Suppose, to take but one example, that the child of my indigent neighbor is struck by a car on the very day that I win $1500 in the state lottery. If the injured child requires an operation costing $1500 to restore her to good health, surely I have no strong obligation to provide the $1500, and it would be a supererogatory gesture if I were to do so. By contrast, if I am offered $1500 to hit a neighbor's child with my car, I have the very strongest obligation to refuse. My obligation to avoid harming the child (even though it may leave me $1500 poorer) is much stronger than my obligation to help her (at a cost to me of $1500).

Suppose that these examples and others convince us that we cannot adopt the utilitarian principle, at least not in its most general form where it purports to be applicable to all moral decisions. What are the alternatives? One cluster of alternative principles would urge that in some or all cases we weigh the harm a contemplated action will cause more heavily than we weigh the good it will do. The extreme form of such a principle would dictate that we ignore the benefits entirely and opt for that action or policy that produces the *least* expected harm. (This principle, or a close relation, emerged in the second reading of the "natural barriers" argument discussed in II, 3, above.) A more plausible variant would allow us to count both benefits and harms in our deliberations, but would specify how much more heavily harms were to count.

On my view, some moderate version of a "harm-weighted" principle is preferable to the utilitarian principle in a considerable range of cases. However, the recombinant DNA issue is not one of these cases. Indeed, when we try to apply a harm-weighted principle to the recombinant DNA case we run head on into a conceptual problem of considerable difficulty. The distinction between doing good and doing harm presupposes a notion of the normal or expectable course of events. Roughly, if my action causes you to be worse off then you would have been in the normal course of events, then I have harmed you; if my action causes you to be better off than in the normal course of events, then I have done you some good; and if my action leaves you just as you would be in the normal course of events, then I have done neither. In many cases, the normal course of events is intuitively quite obvious. Thus for the case of the neighbor's child, in the expected course of events I would neither intentionally hit the child with my car nor would I pay for the child's operation. Thus I am doing good if I pay for the operation, and doing harm if I hit her. But in other cases, including the recombinant DNA case, it is not at all intuitively obvious what constitutes the "expected course of events," and thus it is not at all obvious what to count as a harm.

To see this, suppose that as a matter of fact many more deaths and illnesses will be prevented as a result of pursuing recombinant DNA research than will be caused by pursuing it. But suppose that there *will* be at least some people who become ill or die because recombinant DNA research was pursued. If these are the facts, then who would be harmed by imposing a ban on recombinant DNA research? That depends on what we take to be the "normal course of events." Presumably, if we do not impose a ban, then the research will continue and the lives will be saved. If this is the normal course of events, then if we impose a ban we have *harmed* those people who would be saved. But it is equally natural to take as the normal course of events the situation in which recombinant DNA research is not pursued. And if *that* is the normal course of events, then those who would have been saved are not harmed by a ban, for they are no worse off than they would be in the normal course of events. However, on this reading of "the normal course of events," if we *fail* to impose a ban, then we have harmed those people who will ultimately become ill or die as a result of recombinant DNA research, since as a result of not imposing a ban they are worse off than they would have been in the normal course of events. I conclude that, in the absence of a theory detailing how we are to recognize the normal course of events, "harm-weighted" principles have no clear application to the case of recombinant DNA research.

Harm-weighted principles are not the only alternatives to the utilitarian principle. Another cluster of alternatives take off in a quite a different direction. These principles urge that in deciding which policy to pursue there is a strong presumption in favor of policies that adhere to certain formal moral principles (i.e., principles that do not deal with the *consequences* of our policies.) Thus, to take the example most directly relevant to the recombinant DNA case, it might be urged that there is a strong presumption in favor of a policy that preserves freedom in inquiry. In its extreme form, this principle would protect freedom of inquiry *no matter what the consequences;* and as we saw in II, 1, this extreme position is exceptionally implausible. A much more plausible principle would urge that freedom of inquiry be protected until the balance of negative over positive consequences reaches a certain specified amount, at which point we would revert to the utilitarian principle. On such a view, if the expected utility of banning recombinant DNA research is a bit higher than the expected utility of allowing it to continue, then we would nonetheless allow it to continue. But if the expected utility of a ban is enormously higher than the expected utility of continuation, banning is the policy to be preferred. In the next chapter of this volume, Carl Cohen defends this sort of limited protection of the formal free inquiry principle over a straight application of the utilitarian principle.

IV. LONG-TERM RISKS

Thus far in our discussion, the risks that have occupied us have been what might be termed "short-term" risks, such as the release of a new pathogen. The negative effects of these events, though they might be very long lasting indeed, would be upon us relatively quickly. However, some of those who are concerned about recombinant DNA research think there are longer-term dangers that are at least as worrisome. The dangers they have in mind stem not from the accidental release of harmful substances in the course of recombinant DNA research, but rather from the unwise use of the *knowledge* we will likely gain in pursuing the research. The scenarios most often proposed are nightmarish variations on the theme of human genetic engineering. With the knowledge we acquire, it is conjectured, some future tyrant may have people built to order, perhaps creating a whole class of people who willingly and cheaply do the society's dirty or dangerous work, as in Huxley's *Brave New World.* Though the proposed scenarios clearly are science fiction, they are not to be lightly dismissed. For if the technology they conjure is not demonstrably achievable, neither is it demonstrably impossible. And if only a bit of the science fiction turns to fact, the dangers could be beyond reckoning.

Granting that potential misuse of the knowledge gained in recombinant DNA research is a legitimate topic of concern, how ought we to guard ourselves against this misuse? One commonly aired thought is to try to prevent the acquisition of such knowledge by banning or curtailing recombinant DNA research now. Let us cast this proposal in the form of an explicit moral argument. The conclusion is that recombinant DNA research should be curtailed, and the reason given for the conclusion is that recombinant DNA research could possibly produce knowledge that might be misused with disastrous consequences.. To complete the argument we need a moral principle, and the one that seems to be needed is something like this:

> If a line of research can lead to the discovery of knowledge that might be disastrously misused, then that line of research should be curtailed.

Once it has been made explicit, I think relatively few people would be willing to endorse this principle. For recombinant DNA research is hardly alone in potentially leading to knowledge that might be disastrously abused. Indeed, it is hard to think of an area of scientific research that could *not* lead to the discovery of potentially dangerous knowledge. So if the principle were accepted, it would entail that almost all scientific research should be curtailed or abandoned.

It might be thought that we could avoid the extreme consequences just cited by retreating to a more moderate moral principle. The moderate principle would urge only that we should curtail those areas of research where the probability of producing dangerous knowledge is comparatively high. Unfortunately, this more moderate principle is of little help in avoiding the unwelcome consequences of the stronger principle. The problem is that the history of science is simply too unpredictable to enable us to say with any assurance which lines of research will produce which sorts of knowledge or technology. There is a convenient illustration of the point in the recent history of molecular genetics. The idea of recombining DNA molecules has been around for some time. However, early efforts proved unsuccessful. As it happened, the crucial step in making recombinant DNA technology possible was provided by research on restriction enzymes, research that was undertaken with no thought of recombinant DNA technology. Indeed, until it was realized that restriction enzymes provided the key to recombining DNA molecules, the research on restriction enzymes was regarded as a rather unexciting (and certainly uncontroversial) scientific backwater (9). In an entirely analogous way, crucial pieces of information that may one day enable us to manipulate the human genome may come from just about any branch of molecular biology. To guard against the discovery of that knowledge we should have to curtail not merely recombinant DNA research, but all molecular biology.

Before leaving the topic of long-range risks, we would do well to note the profound pessimism reflected in the attitude of those who would stop recombinant DNA research because it might lead to knowledge that could be abused. It is, after all, granted on all sides that the knowledge resulting from recombinant DNA research will have both good and evil potential uses. So it would seem the sensible strategy would be to try to prevent the improper uses of this knowledge rather than trying to prevent the knowledge from ever being uncovered. Those who would take the more extreme step of trying to stop the knowledge from being uncovered presumably feel that its improper use is all but inevitable, that our political and social institutions are incapable of preventing morally abhorrent applications of the knowledge while encouraging beneficial applications. On my view, this pessimism is unwarranted; indeed, it is all but inconsistent. The historical record gives us no reason to believe that what is technologically possible will be done, no matter what the moral price. Indeed, in the area of human genetic manipulation, the record points in quite the *opposite* direction. We have long known that the very same techniques that work so successfully in animal breeding can be applied to humans as well. Yet there is no evidence of a "technological imperative" impelling our society to breed people as we breed dairy cattle, simply because we know that it can be done. Finally, it is odd that those who express no confidence in the ability of our institutions to forestall such monstrous applications of technology are not equally pessimistic about the ability of the same institutions to impose an effective ban on the uncovering of dangerous knowledge. If our institutions are incapable of restraining the application of technology when those applications would be plainly morally abhorrent, one would think they would be even more impotent in attempting to restrain a line of research that promises major gains in human welfare.

V. SOME THOUGHTS ON THE CURRENT DEBATE

Before concluding, I want to offer a few observations about a problem posed by the current debate over recombinant DNA. The scientists who planned the Asilomar conference took the unprecedented step of calling attention to the potential dangers of their own research and inviting a broader dialogue on how these dangers should be controlled. Their actions have been widely and justifiably praised as an exemplar of moral responsibility in the scientific community. Yet as I write this essay, two years after the Asilomar conference, it is an open secret that a number of the scientists who were responsible for Asilomar are disillusioned, and perhaps a bit bitter. They feel, with obvious justification, that in inviting broad-based discussion and input they have opened a Pandora's box filled with unanticipated problems. By "going public" the Asilomar scientists succeeded in focusing the attention of many serious and responsible people on the complex problem of how work with recombinant DNA molecules could be safely pursued. But they also succeeded in making recombinant DNA research the focus of attention of a broad spectrum of political ideologues and opportunists. Some of the idealogue critics admit privately (indeed, sometimes publicly) that their real concern is not recombinant DNA research and the problems unique to it, but rather the entire system of research support and review in the United States. Others are concerned with even more general issues such as the availability of health care to the poor in this country, or perhaps even the system of distribution of wealth in the western world. Now these issues are surely important ones, and it is good that they be discussed widely and from diverse perspectives. However, using public concern over recombinant DNA research as a lever in the debate over these broader issues is unfair to those scientists who want to pursue their experiments in this area safely and responsibly; it is also potentially very dangerous. Let me elaborate on each of these points.

When concern about recombinant DNA research is used as a convenient device for generating public skepticism about the direction of government-supported research in general, a heavy and unfair burden is placed upon those scientists involved with recombinant DNA research. In the last several years, many of these researchers have been forced to spend enormous amounts of time away from their laboratories attempting to correct public misconceptions about their work, spawned carelessly (or, worse, perhaps deliberately) by political idealogues whose interest in recombinant DNA research was at best peripheral. Privately, a number of the scientists who participated in the Asilomar conference now say that if they had it to do over again, they would have "kept their mouths shut" and never brought their concerns about safety to the attention of a wider audience. It is just here that I see a profound danger. For one of the crucial needs of our society in the last years of this century is to develop ways of anticipating and forestalling the untoward consequences of scientific advances and new technologies. The job of monitoring research and technology will obviously be enormously simplified if researchers themselves cooperate fully, coming forward with their concerns as the Asilomar organizers did. We must, in short, find some way to encourage the fullest cooperation of scientists themselves in the process of locating potential problem areas well before the potential problems become actual ones. But the lesson of Asilomar's aftermath has not been lost on the scientific community. The current politicized debate with its entourage of idealogues and political opportunists is plainly the wrong model if we hope to encourage openness and cooperation on the part of scientists in other areas whose research may harbor future hazards. Unfortunately, I cannot end on a constructive note. For though the present ad hoc politicized review process is surely inadequate, I have no better system to propose.

NOTES

1. See, for example, R. B. Brandt, *Ethical Theory* (Englewood Cliffs, N.J.: Prentice-Hall, 1959), pp. 19-35; also Marcus G. Singer, *Generalization in Ethics* (New York: Alfred A. Knopf, 1961).
2. For example, from a widely circulated petition signed by both faculty and community people: "The most important challenge may be a confrontation with one of our ancient assumptions—that there must be absolute and unqualified freedom to pursue scientific inquiries. We will soon begin to wonder what meaning this freedom has if it leads to the destruction or demoralization of human beings, the only life forms able to exercise it." And from a letter to the editor written by a professor of engineering humanities: "Is science beyond social and human controls, so that freedom of inquiry implies the absence of usual social restrictions which we all, as citizens, obey, respecting the social contract?"
3. It is interesting to note that the "freedom of inquiry" argument is rarely proposed by defenders of recombinant DNA research. Rather, it is proposed, then attacked, by those who are opposed to research with recombinant molecules. Their motivation, I would conjecture, is to discredit the proponents of recombinant DNA research by attributing a foolish argument to them, then demonstrating that it is indeed a foolish argument.
4. Professor Lederberg alludes to the "doomsday scenario" argument, or a close relation, in the essay reprinted in this volume. He writes of the "zeal [which] has spread to create a sincere, almost frantic effort to ferret out and identify the most remote, conceivable hazard." and he disparages "the apparently innocuous doctrine, 'As long as there is any risk, don't do it.' " Unfortunately the doomsday scenario argument is *not* a straw man conjured only by those who would refute it. Consider, for example, the remarks of Anthony Mazzocchi, spokesman for the Oil, Chemical and Atomic Workers International Union, reported in *Science News*, March 19, 1977 (p.181): "When scientists argue over safe or unsafe, we ought to be very prudent. . . . If critics are correct and the Andromeda scenario has *even the smallest possibility* of occur-

ring, we must assume it will occur on the basis of our experience." (Emphasis added.)

5. The quote is from Professor Wald's essay, "The Case Against Genetic Engineering," reprinted in this volume.

6. Note, however, that the principle is considerably less conservative, and correspondingly more plausible, than the principle invoked in the doomsday scenario argument.

7. For an elaboration of this point, see Bernard D. Davis, "Evolution, Epidemiology, and Recombinant DNA," in this volume.

8. For an interesting discussion of these cases, see J. O. Urmson, "Saints and Heros," in *Essays in Moral Philosophy*, ed. by A. I. Melden (Seattle: University of Washington Press, 1958).

9. I am indebted for both the argument and the illustration to Professor Ethel Jackson.

When May Research Be Stopped?

Carl Cohen

The uses and possible misuses of recombinant DNA are so threatening, some believe, that research into that technology should now be stopped. But reasons good enough to justify prohibition of research in this sphere must, in fairness, apply equally to other spheres, if threats of similar gravity arise there.

I ask, therefore: What are the alternative principles that, if adopted, might reasonably justify prohibition of research in a given sphere? Which of these alternative principles should be rejected, and which accepted in some form or part? And to the extent that any one of these principles is acceptable, what bearing does it have upon continued research with recombinant DNA?

My answers to these questions rest upon two fundamental propositions, very generally agreed upon. First: freedom of inquiry is a value of such profound importance that it must not be abridged without the most compelling reasons. This proposition, true generally, carries great weight in a society holding liberty as a paramount ideal; it carries extraordinary weight in universities and research institutions committed explicitly to the enlargement of knowledge. Second: some research undertakings should properly be restricted. Everyone may not agree upon particular cases, but it will be agreed that a rational commitment to

Reprinted, by permission, from the *New England Journal of Medicine* **296**, 1203–1210 (1977).

freedom of inquiry does not protect every research enterprise in every circumstance.

The task is to characterize the enterprises and circumstances in which prohibition may prove defensible or even obligatory.

ALTERNATIVE PRINCIPLES OF PROHIBITION

To justify prohibition, some would present a practical syllogism in this form: Major premise: research having certain identifiable features (p, q, r, . . .) may (or must) be stopped. Minor premise: this research (in recombinant DNA, or in nuclear fission, or . . .) has precisely those features. Hence, this research may (or must) be stopped. My first objective is to formulate alternative, plausible major premises of such syllogisms.

Principles of prohibition must pertain either to the product or to the process of the research in question. The line between the two may be hard to draw, but under one or the other can be listed all the general principles that seem remotely tenable.

Set A: By-product (#1–#3)

1. Research should not be permitted when it aims at (or is likely to result in) the discovery of knowledge that is wrong for human beings to possess.
2. Research should not be permitted when it aims at (or is likely to result in) the discovery of knowledge that is not wise to place in human hands.
   2a. When there is any probability that the knowledge developed will be used with very injurious consequences.
   2b. When there is moderate probability that the knowledge developed will be used with very injurious consequences.
   2c. When there is high probability that the knowledge developed will be used with very injurious consequences.
3. Research should not be permitted when it aims specifically at the development or perfection of instruments for killing or injuring human beings.

Set B: By-process (#4–#6)

4. Research should not be permitted when it is not conducted openly, for all to examine the ongoing process and results.

5. Research should not be permitted when its continuation is unfair, either to the subjects of the experimentation or to those otherwise involved in the research. Research is unfair when, through coercion or deceit, or in some other way, the rights of those involved are not respected.

6. Research should not be permitted when its conduct (as distinct from its product) presents risks so great as clearly to outweigh the benefits reasonably anticipated. Risks here must be understood to include all hazards that that process of inquiry entails, of which there are two large categories: risks of "misfire" (i.e., achieving results different from and more dangerous than those sought) and risks of "accident" (i.e., unforeseen mishap during the process).

   6*a*. When the risks are essentially to persons involved in the research.

   6*a*1. the subjects of experimentation;

   6*a*2. the researchers and their associates.

   6*b*. Where the risks are essentially to others than those involved in the research.

These exhaust the alternative principles that are at all reasonable, or arguably tenable, for the prohibition of research. Phrasing (and the degree of specificity) may vary, but every plausible candidate, I contend, will fall under one or another of the six kinds distinguished.

As an illustration: Robert Sinsheimer, Chancellor of the University of California at Santa Cruz and an acute critic of recombinant DNA research, has suggested some possible answers to the question, "For what specific purposes might we wish to limit inquiry?" He proposes, or at least entertains, several candidates (1).

(a) To preserve human dignity. "We should not do experiments that involuntarily make of man a means rather than an end." (1). But, of course, human subjects in medical experiments are means. Sinsheimer surely intends to emphasize, with Kant, that one must not treat human beings as means only, but also as ends—for which reason research committees do rigorously insist that their participation be truly voluntary and informed. This is but a more elaborate statement of principle 5, above, demanding fairness.

(b) To avoid "involuntary physical or biological hazard" (1). Sinsheimer recognizes, of course, that one cannot avoid all such hazards. He wants us all to be very sensitive to the level of danger. This is but another formulation, less precise, of principle 6, above, addressing the balance of risk and benefit.

(c)   Cost. "[One] asks if the primary consequences of the inquiry, that is, the knowledge to be gained, is worth the expenditures of talent, time and resources" (1). But refusing to support research is one thing, prohibiting it is another. When the research enterprise involves heavy expense, one is right to insist that the worth of the knowledge to be gained be very carefully estimated. Protecting free inquiry does not entail irrationality in the expenditure of resources. And it is true that institutional refusal to fund a research project often has the effect of blocking that project in that context. Many research activities, however (work with recombinant DNA technology being one important example), do not require very large investments of institutional or governmental funds. Although one may conclude, therefore, that for some projects the object does not justify the expenditure, it is essential to see that cost cannot serve as grounds for prohibition. Some research is simply not worth doing, but reasons for not troubling to do certain things ourselves must not be taken as reasons for keeping other persons from doing them.

Some principles, very different from those stated above, are procedural, requiring special machinery for the approval of research protocols in certain areas. Approval (some say) must be given by the appropriate bodies, with appropriate membership, deliberating with appropriate care. Others say that the decision to permit research of certain kinds may be made only by some larger community (city or state) through some democratic voting procedure. Such regulation, although awkward, is being tried in some quarters. But, even when feasible and fitting, those procedural requirements are not germane here. One seeks to discover reasons that may serve to deny approval. Whatever the decision-making machinery, the individual users of that machinery need grounds for concluding yea or nay. These grounds—not the system of their application—are what I am concerned with here.

Finally, I note that the six alternative principles are not mutually exclusive; one could rationally hold (say) both 3 [on killing], and some variant of 6 [risks over benefits]. They may overlap, relying upon the same feature of a given inquiry—for example, what is objectionable under 4 [openness] may also be objectionable under 5 [unfairness].

It remains to determine which (if any) of these principles should be adopted, and which (if any) apply to the sphere of recombinant DNA research, and to identify the restrictions (if any) that properly follow therefrom.

What follows is a critical appraisal of the six alternative principles listed above. The conclusions reached unavoidably rely upon some personal judgments, but are put forward for general agreement.

Principles Based upon the Product of Research (#1–#3)

1. "Research should not be permitted when it aims at (or is likely to result in) the discovery of knowledge that is wrong for humans to possess."

This principle should be rejected utterly. There is no body of knowledge, or item of knowledge, that is intrinsically wrong to possess. The conviction that there are forbidden precincts, that there is an intellectual sanctum sanctorum into which all entry is sinful, must rely upon some claim of special revelation, or some other nonrational restriction that has no rightful authority to limit inquiry in a university or research institution.

Principles of privacy, it is true, may render certain sorts of knowledge about individuals not suitable for public scrutiny. But there is a vast difference between the claim that some knowledge is not properly public and the claim that some knowledge is intrinsically unfit for human acquisition. Individuals are free, of course, to limit themselves if they honestly hold beliefs of the latter sort. The search here, however, is not for principles that some persons may cling to but for the principles research institutions ought to defend.

Note that principle 1 is widely attractive. Much of the anxiety that attends DNA research, I submit, flows from vague, unformulated doubts about whether this probe into "the code of life itself" might not be a form a human presumption, a playing of God. One is understandably awed by the cumulative powers of human intelligence; those powers can be (and often have been) misused. But fear of that misuse does not give rational warrant for closing the avenues of exploration. If one did believe that there are domains in which human knowing is taboo, molecular genetics might indeed be one of them. Inquiry into nuclear fusion or celestial exploration might then be equally taboo, as might be the study of theories of relativity, or the development of contraceptive technics. The penetration of every intellectual frontier threatens deeply held convictions. Every striking advance and appears to a few as the profane invasion of the holy of holies. The difficulty lies not in discriminating between the real holy of holies and those only mistakenly supposed; it lies in the unwarranted assumption that there are any spheres of knowledge to which ingress is forbidden. This first principle is wholly untenable; a fortiori it is not tenable as applied to the biochemistry of genetics.

2. "Research should not be permitted when it aims at (or is likely to result in) the discovery of knowledge that is not wise to place in human hands."

Of this principle I distinguish three varieties, depending on the degree of probability with which great injury may be anticipated as a result of the possession of that knowledge. Of course, there will be argument about what constitutes great injury, about how such probabilities are to be calculated or estimated, and about what the probabilities are in a particular case. But supposing rough agreement is reached on these matters, the rightness (or wrongness) of the principle here formulated remains to be determined, and is of profound importance.

Principle 2*a* [that prohibition of research is justified when there is any probability of very injurious consequences] may be rejected categorically; it verges on the absurd. On that principle one ought not rise from bed.

Principle 2*b* [that prohibition of research is justified when there is moderate probability of very injurious consequences] must be more seriously considered—but it too deserves rejection in the end. It is true, of course that the results of inquiry will often be such that there is some moderate likelihood that their use will prove very injurious, even disastrous. But the ubiquity of such possibilities renders 2*b* [as it does 2*a*] so sweeping as to entail the cessation of a great deal of the research—both in biologic and in physical sciences—best calculated to improve the human condition.

Nevertheless, serious thinkers have urged the adoption of some such principle. Two especially—one a historian, and the other a biophysicist—deserve response.

Shaw Livermore, professor of history at the University of Michigan, in concluding that his university should not proceed with the development of recombinant DNA technology, argues as follows (2):

(i) Research on recombinant DNA promises the development of a special, elemental capability: that of overcoming the natural genetic barriers that separate the species, and (ultimately) of uniting genetic components of different species to produce new forms of life.

(ii) This capability would be so great, so overwhelming, as to exceed the capacity of society to direct and control it.

Therefore, (iii) success in this research "will bring with it a train of awesome and possibly disastrous consequences" (2), and (iv) decisions demanded by this technology may well have effects that are "unintended but irreversible." (2)

Livermore suggests, in effect, that scientists here are like the Sorcerer's Apprentice, conjuring into existence what they do not fully understand and none of them can control. Although appealing, the argument fails upon careful test. Consider its elements in reverse order.

(iv) That decisions regarding DNA will sometimes have unintended and irreversible effects is surely true, but not very weighty. That is precisely the case for every important human enterprise in research and development—it leaves the world in an irreversibly different condition from that in which it was before, and has consequences that could not have been foreseen, and therefore could not have been intended. Many of the discoveries that have proved the greatest boons to mankind have arisen from basic research in ways that were—when that research was first pursued—wholly unforeseen and unintended. That some consequences (bad and good) of any major inquiry will be irreversible and unintended is evident, and cannot reasonably be taken as grounds for the prohibition of inquiry.

(iii) But the consequences of this inquiry may be awesome—possibly disastrous. Awesomeness, again, may be for good as well as evil. One should bear in mind, in weighing arguments like these, that recombinant DNA technology also opens possibilities for monumental improvements in the human condition. Both sides must be weighed. But it is the "possibly disastrous" results that are the nub of this complaint—a complaint that cuts either not at all or entirely too well. There are no grounds for supposing that the likelihood of disaster is greater in this arena than in other research arenas that one would not seriously think of foreclosing. There is "moderate probability," I suppose, that the results of research into nuclear fusion will one day be put to malevolent uses—but one would not on this ground seriously suggest a prohibition of inquiry into the ways in which the nuclei of atoms may combine. In any sphere knowledge may be put to devilish use; that is a poor reason for prohibiting its acquisition.

(ii) It is suggested that the acquisition of certain awesome capabilities be barred because the capacity to direct them, once acquired, is lacking. But do we lack that power? The supposition is very doubtful. Many reflective historians and philosophers would insist that we have the capacity for the direction and control of the products of research. Whether we will sharpen such capacities as finely as we ought remains to be seen. Recent self-imposed restraints, followed by extended public deliberation, precisely in this sphere of recombinant DNA, strongly suggest that the capacity to control does exist and is being applied. Some present uncertainty about the outcome of this application surely does not justify the cessation of the inquiry. And even if only the potential for wise control is now present, the realization of that potential can be stimulated and encouraged only with the advance of the inquiry in question. Professor Livermore, although reflectively, gives up hope for Dame Reason; I judge that one is ill advised to join him in despair.

(i)   His anxieties—and indeed most arguments of this variety—stem largely from the "elemental" nature of the capabilities in view. But there is confusion hidden in the slippage between fears arising from the alleged probabilities of disaster (probabilities not ever established) and fears arising from the allegedly special, extraordinary properties of the knowledge to be discovered. The implicit suggestion that the knowledge sought is too godlike for human fraility gives seeming (but unjustifiable) plausibility to the claim that its acquisition will bring catastrophe. Once it is clearly seen that knowledge is not to be feared, that it may prove valuable everywhere, risky anywhere, and intrinsically improper nowhere, it will also be seen that the specialness of the discoveries in view, though in some ways real, ceases to serve as any ground for prohibition. Research in genetics, as in every science, moves ever onward; inquiry into the controlled manipulation of genetic macromolecules is a natural and inevitable phase of that advance. Fears that human beings are incapable of dealing with the products of their own intelligence are as much and as little justified on every other research continuum as on this one.

A differing effort to provide a tenable variant of this principle, 2b, is made by Sinsheimer. He writes (1):

> [One] may extend inquiry into the ends of inquiry and question whether, in particular instances, we want to know the answer in every case. Given the nature of man and of human society, are the secondary consequences of such knowledge, on balance, likely to be beneficial? Here it may be that the highest wisdom is to recognize that we are not wise enough to know what we do not want to know, and

thus to leave the ends of inquiry unrestrained. Indeed, I expect there are only a few instances where prudence would be in order. But the set may not be null.

What does this say? Sinsheimer is guarded, tentative, unsure. From his questions and modalities emerge at last his suggestion of four actual spheres in which—because the secondary consequences of knowledge are not likely, on balance, to prove beneficial—he seriously believes that inquiry might appropriately be restricted.

(i)   "Should we attempt to contact presumed 'extraterrestrial intelligences'?" Sinsheimer believes that "the impact upon the human spirit" if it should develop that there are vastly superior forms of life, and the impact of that knowledge upon science itself, would be "devastating" (1).

(ii) "Research upon improved, easier, simpler, cheaper methods of isotope separation?" Sinsheimer doubts whether such research is in man's best interest, yielding "slightly cheaper power, far easier bombs" (1).

(iii) "Research upon a simple means for the predetermination of the sex of children?" Sinsheimer appears to believe that the resultant potential for "a major imbalance in the human sex ratio" shows advances here to be undesirable (1).

(iv) "Indiscriminate research upon the aging process?" Sinsheimer appears to believe that the stated goal accompanying legislation to advance such research, "keeping our people as young as possible, as long as possible," is not, on balance, desirable (1).

It is hard to know what to make of these suggestions. They are very cautiously put, half in interrogative form. But (whatever Sinsheimer believes or may find desirable), if such speculations are treated as arguments for the principle of restriction here involved, they fail utterly. Two observations will suffice.

First. The most that such apprehensions could establish—supposing that everyone shared them—is that it might be that certain inquiries will not prove beneficial on balance. Of course that may be. What follows? Precisely that argument has been presented (with greater force than in most of Professor Sinsheimer's instances) against every scientific advance: against Galileo, against Darwin, against Freud. Such obstinacy (it was urged that good men not even look through Galileo's "infernal glass") is now considered indefensible. It is not an iota more defensible now than it was then. It was entirely correct for the opponents of these seminal thinkers to insist that inquiry of the kinds that they opposed might not prove beneficial in the end. What does that tell about the principle invoked? If, now, the same principle is used not merely to discourage but to prohibit research—in recombinant DNA, for example—one will operate under strictures of essentially the same character as those that persuaded so many rational men to condemn the teaching of the Copernican hypothesis.

Second. The illustrations given by Professor Sinsheimer of the applicability of his principle, his pleas for "prudence," where ignorance seems to him more desirable than knowledge, are self-convicting. If principle (2*b*) when applied means—as it appears to him—that researches into the aging process, into nuclear power, into extraterrestrial contact and so on are to be blocked or restricted because of what may transpire if they are successful, the upshot of the argument is revealed. He helps one to see the extreme consequences forced upon everyone, unacceptable to most, if the principle of restriction that he has put forward is taken seriously.

Finally, respecting 2*b*, one variety of great injury that some foresee (unlike catastrophic decisions having unintended impact) deserves remark. It is the gradual deterioration of human culture resulting from ever more extended subjection to technical control. As technology comes to pervade culture (some contend), human values must retreat—even wither. So we can defend our most humane interests only by disarming the technologists. Well, the probability of this feared outcome is very difficult to estimate. Its likelihood, in my judgment, though not trivial, is not great. To prohibit research on such grounds would be intolerably repressive. On this interpretation of disaster, too, the principle would cut against technological advance in every sphere, not that of molecular genetics alone.

For principle 2*c* [that research be prohibited when the probability of its very injurious consequences is high or very high] the case is different. Were we to believe that in a given case, the prohibition of that specific inquiry would be, I judge, at least arguable. In such circumstances—the only persuasive candidate I can think of is research into nuclear explosives—the alleged probability would have to be explored, documented, established as fully as resources would then permit. Even here potentiality for benefit would also need to be weighed. Recognizing that rational men may ultimately differ in the resolution of such cases, one must allow that for some research ventures the probability of disastrous use of the products might be so high as to justify prohibition.

Again, two observations respecting principles 2*c*. First. Our rational commitment to freedom of inquiry is such that, in judging any claim of highly probable disaster, the burden of proof clearly rests upon those who would prohibit on that basis. They must present a convincing account of what concrete disasters are envisaged, what the methods are for determining the probability of such outcomes, and how those methods establish the high probability of the catastrophe pictured. The burden here is not light, nor ought it to be. Scientific inquiry should not be blocked simply upon the presentation by critics of a parade of imagined horribles of unspecified nature and doubtful likelihood.

Second, principle 2c does not, in any case, apply to research with recombinant DNA. There is some probability (it may be supposed) that, after years of further development, the products of such research might be put to malevolent use—as instruments of war (although better, more convenient killers are already at hand) or in the realization of some (now farfetched but then realizable) brave new world. There is some probability of that, one must grant. But that probability, on the best evidence now available, is slight; at its gravest interpretation, which very few would accept, that probability is no more than moderate. A high probability it is not. Hence in this sphere principle 2c does not apply.

I conclude that principle 2 is generally inapplicable in most of its forms, and that in none of its forms may it properly serve to prohibit any research in the biochemistry of genetics now contemplated.

3. "Research should not be permitted when it aims specifically at the development or perfection of instruments for killing or injuring human beings."

This is a reasonable principle for both persons and institutions; it is now accepted and applied by some universities. There are times, circumstances and some institutions to which it might not properly apply.

Although this principle may serve as major premise in a practical syllogism forbidding some kinds of research in some contexts, it cannot serve in a syllogism forbidding recombinant DNA research in any context at present. Research now contemplated with recombinant DNA does not faintly resemble the "munitions" development that would be the target of such a principle. If such an aim were proposed, or even seriously entertained, this principle might, indeed, be called into play, Under present and foreseeable circumstances, however, principle (3) simply has no bearing on the problem at hand.

I conclude that no tenable principle for the prohibition of research based upon its product can now serve to restrict research into recombinant DNA.

Principles Based upon the Process of Research (#4–#6)

4. "Research should not be permitted when it is not conducted openly, for all to examine the ongoing process and results."

This attractive principle is of a kind very different from the others so far reviewed. Rightly understood, it presents not a limitation upon research but the statement of an ideal—one that we all properly share and promote. The realization of this ideal cannot and should not be taken as an inviolable condition for the conduct of the inquiry. It simply is not that. National security, proprietary interests (of firms or individuals) justly acquired, or other special concerns may render full openness an impractical ideal in many circumstances, certainly not justifying the cessation of all research whose conduct falls short of that ideal.

Openness—both in the research process and for the results of research—is an ideal widely and genuinely honored. Some universities do not permit in their precincts (or are inhospitable toward) research so classified as to restrict access to its results. But complete openness of the research process is very deliberately not applied as a necessary condition by universities, governmental institutions or private enterprises. If it were to be so applied, a great deal of research progress would have to be discontinued. Much research that is planned would not (under such restriction) go forward. One would not seriously wish to insist upon this discontinuation and blockage. I conclude that the publicity principle is not even remotely acceptable as a basis for prohibition. Since it cannot serve for prohibition generally, it cannot serve so for research in recombinant DNA unless this inquiry can be shown to be specially prone to the evils of secrecy. That cannot be shown. There is no reason to believe that review of the research process in this field, and timely publication of its results, will fail to meet normal standards of the scientific community.

In fact, the circumstances surrounding DNA research lead to the very opposite conclusion. The nature, location and conditions of recombinant DNA research have been subject to a publicity surpassing that in any other comparable scientific sphere. Largely as a result of initiatives taken by the scientific community itself, public scrutiny in this area has been intense, and the researchers' standards of openness have been, and are sure to continue to be, much higher than normal. Therefore, even if the demand for openness in the process were an appropriate ground for restricting some inquiries in some contexts it could not serve to restrict, in this context, inquiry using recombinant DNA.

5. "Research should not be permitted when its continuation is unfair, either to the subjects of the experimentation or to those otherwise involved in the research."

This principle is sound; research whose process does substantial injustice ought not to be pursued, no matter its kind. Often disregarded in the past, this principle is now generally accepted, and is applied concretely to research activities in which unfairness can become a problem. To this end, all experiments proposing to put human subjects at risk—in any way, and to any degree, even the slightest—must be screened by a specially organized human-subjects review committee

(HSRC), one kind of institutional review board. No such research may go forward without the explicit approval of an HSRC—and this restriction applies fully to any work with DNA that proposes to put human subjects at risk in any way.

Restrictions for the protection of human subjects being already in force, they need no special restatement for a specific sphere of inquiry. Experiments with recombinant DNA involving human beings have not yet been proposed. If they become a real possibility, and are proposed, the task of the HSRC screening that enterprise may be a delicate one. One will surely agree, in any case, that fairness (noncoercion, full information and so on) toward proposed subjects must be a condition for the continuation of that investigation.

Fairness to other researchers—respect for the present state of the work of others, full information to all investigators involved about what is in view and what is at risk—is also a demand reasonably made. But it, like openness, is a principle for the guidance of conduct, not the restriction of it. It cannot serve for the prohibition of research of a given kind.

Principle 5, I conclude, although important and sound, has no special application to research in recombinant DNA. So far as it does have institutional applicability, its application must (and is) general, screening out research protocols in every sphere that fail to measure up to the requirements of justice. But measuring up to this standard is not a function of the subject matter of the research.

6. "Research should not be permitted when the conduct of such research (as distinct from its product) presents risks (either of misfire or of accident) so great as clearly to outweigh the benefits reasonably anticipated." A misfire might be, for example, the creation, through DNA recombination, of an organism with unintended pathogenic capacity against which ordinary antibiotics proved ineffective. An accident might be, for example, the undetected escape, from a laboratory thought to be sealed, of a micro-organism giving rise to contagion.

The principle also is sound and applicable. Bringing it to bear upon DNA research, however, is complicated, and the outcome of that application is uncertain. Just here lie the major technical problems that have been and remain the focus of much scientific debate. The problem of containing the recombined DNA either by physical retention within the laboratory or by weakening the host organism so as to render it not viable outside the laboratory, has been the major topic in controversy over what is unreasonably risky and what is not. Only in the light of the present state of effectiveness of containment measures can risks be rationally estimated. Hence the emergence of guidelines (laid down by the National Institutes of Health) for containment, and for permissible risks given known levels of effective containment. Hence, too, the need to reassess what is reasonably safe to do in the light of existing technical capacities to contain—especially since the capacity to contain biologically, by "disarming the bug," is being steadily improved. Recombined DNA molecules do create special dangers, which do, rightly, require special attention to the conditions and precautions under which specific research activities are carried on.

Still, the principle accepted here, that risks must be minimized, and never allowed to exceed the reasonably anticipated benefits, is one of general application. It applies to all research in medicine, in physics, in biology, in aeronautics, and so on. It has a bearing upon DNA research, to be sure—but only to the extent that the risks encountered in a specific experimental project within that domain appear to equal or to outweigh the anticipated benefits of that project.

Critical here are the risks to persons other than those involved in the research (principle 6*b*): the people in the street who may be endangered by accident or misfire. Risks to experimental subjects (principle 6a1) must be screened by human-subjects review committees [described above under principle 5] designed precisely for that purpose. Risks to researchers themselves or to others formally involved in the research project (principle 6*a*2) may be grouped for present purposes with risks to outsiders. It is for this combined group that the key question arises.

That key question, now heatedly argued, is this: Are the risks of recombined DNA, whether of misfire or of accident, of such enormity and such probability as to justify prohibition of further research in that sphere?

The first thing to notice about this question is that although here framed in the singular, it must in fact be asked about a host of very different research proposals. Estimating risk-benefit balance for some proposed investigation is often a vexing task. But it should be emphasized that the decisions called for in this family of cases are not, in principle, different from those we are commonly obliged to make when data are incomplete, the time-frame long, the object risky but promising. In this family of cases, as elsewhere, we will do the best we can.

Some contend that in this sphere of research, unlike others, the general sum of anticipated risks, taking into consideration both their degree of seriousness and their probability, outweighs the general sum of anticipated benefits, their value and likelihood similarly weighed. Therefore (they conclude) the proper estimate of risk-benefit balance calls for cessation now of all further research in this sphere.

This argument is unsound. Consider:

First, the key premise is false; it makes a stronger claim than the evidence supports. Granting that the present state of knowledge is short, all indications are that, if one weighed, as on a balance scale, the expected goods in view, multiplied by the likelihood of their probability, the result would be very different from that supposed by this argument. Everyone will allow that there are dangers, some not yet fully known, but the most careful and sophisticated discussions of these dangers, taking severity and probability into account, do not begin to show that they outweigh, or even approach, the sum of advantages likely to accrue from such research over the long term.*

The fact remains that some future proposals for investigation in molecular genetics, because of the special risks entailed in the process (and with full consideration of the nature and probability of the benefits in view) may, after thoughtful deliberation, then be rejected by the research institution. Clearly, such rejections, if they transpire, will require the continuing activity of an institutional review board, on the model of a human-subjects review committee, but with a differing focus. Rejection on the grounds of excessive risk is a step that must not be taken lightly, but it probably will be taken in some cases.

Since general prohibition is not in order, and approval for individual proposals must be given on a case-by-case basis, it is appropriate that such continuing review bodies set the conditions for the permissible pursuit of the risky inquiry proposed. Here lies the operative force of the general conviction that, whatever the level of risks found tolerable, there must be a commensurate level of precaution. Only through a process of ongoing review will it be possible to adjust restriction to enterprise rationally. Only through such deliberation can improvements in the effectiveness of containment (both physical and biologic) be weighed, as well as any special likelihood of misfire that the specific nature of the investigation at hand may present.

Finally, it is the singularity of this sphere of research, the uniqueness of its risks and promises, that is so specially provocative. That singularity must be taken account of, but should not be overplayed. Special attentions are rightly given, on a continuing basis, to proposals

---

*A number of bodies, with members both in the sciences and in the humanities, have deliberated long and carefully upon the likely balance of risk and benefit, over long term and short, of recombinant DNA research. Perhaps the two most thorough and probing studies are the Report of the Working Party on the Experimental Manipulation of the Genetic Composition of Micro-Organisms. Presented to Parliament by the Secretary of State for Education and Science, London, January 1975 (this document is widely known as the Ashby Report, after the Chairman of the Working Party, Lord Eric Ashby), and the Report of the University Committee to Recommend Policy for the Molecular Genetics and Oncology Program, presented to the vice-president for research at the University of Michigan, Ann Arbor, March 1976 (attached to this document, widely known as the Report of Committee B, is a dissent and a reply to the dissent). Subsequently appeared a critique of the Report of Committee B, a response to that critique by Committee B, and a separate endorsement of the Committee's original report, all presented to the Regents of the University of Michigan.

for experimentation with recombinant DNA, all agree. But it has been my aim to show that the form of the questions to be answered is not essentially different in this domain of scientific research from that in any other. There is a temptation to treat the recombination of DNA as fundamentally different just because of the character of the knowledge aimed at. But if I am right about the first set of principles entertained (#1–#3) pertaining to the product of research, this temptation should not be yielded to.

In summary, for the general prohibition of scientific research in any sphere, the only arguments that might suffice require premises vastly stronger than any now available or likely soon to be available . There is no valid practical syllogism, having true premises, whose conclusion is that research into recombinant DNA should be stopped. Of proposed arguments to this end it may be fairly said either that the major premise (the principle of prohibition) is false, or when true principles are provided, that the minor premise (specifying recombinant DNA research as defective in the ways indicated) is very far from established.

REFERENCES

1. Sinsheimer R: Inquiry into inquiry. Hastings Cent. Rep. 6(4):18, 1976.
2. Livermore S: Statement of dissent, Report of the University Committee to Recommend Policy for the Molecular Genetics and Oncology Program, Office of the Vice President for Research, University of Michigan, Ann Arbor, March 1976. Appendix B1, pp 46-47.

Biocatastrophe and the Law: Legal Aspects of Recombinant DNA Research

Roger B. Dworkin

Research involving recombinant DNA molecules, like other risky human behavior, poses two rather different kinds of legal issues. First, what should the law's response be after something goes wrong? Second, and much more important, how can we prevent things from going wrong in the first place, and how great a set of risks are we willing to run in exchange for what and how certain a group of benefits? In essence the questions are compensation and liability on the one hand and conduct control on the other.

The difference, of course, is not entirely clear-cut. After-the-fact compensation systems often have conduct control as *one* of their goals, as people believe that one is less likely to behave in an undesired way if doing so will expose him to potential liability than if it won't. However, the effect of such compensation systems on conduct is highly questionable, and surely compensation after-the-fact is an indirect, clumsy, and inefficient way to achieve conduct control if that is one's primary goal. On the other hand, whether adequate compensation to injured people is available may influence one's decisions about whether and how strictly to attempt to control conduct to prevent injuries from occurring.

COMPENSATION AND LIABILITY

Recombinant DNA research poses liability and compensation issues that differ from conventional ones only in magnitude. Standard theories can easily be applied to assign liability for injuries caused by accidents arising out of recombinant DNA work. The size of the potentially injured populace is so great, however, that adequate compensation becomes a major problem, perhaps so major that it transforms a difference in size into a difference in kind.

Workmen's Compensation

The easiest problems to deal with are those of workers who are injured or contract a disease from employment that exposes them to DNA recombinants. Typically, workers injured or made ill on the job can obtain compensation from their employers (or their employers' insurance companies) merely by showing that the injury or disease arose out of the employment. The employee need not show that anyone did anything wrong; he must only show that the injury came from the job.

Workmen's Compensation awards are usually low and are determined by formulae that consider the age and occupation of the employee and the nature of the injury. Often Workmen's Compensation is the only liability system open to injured employees, the view being that the virtual certainty of receiving some payment under Workmen's Compensation makes up for the lost chance to gamble on a higher tort law recovery. Thus, for example, laboratory workers adversely affected by working with DNA recombinants will have to be satisfied with their standard Workmen's Compensation benefits, and at the injured-employee level recombinant DNA research poses no special legal problems.

Tort Liability

Much more sensational and difficult for assessing potential liability is the area of common law liability to persons other than employees affected on the job who suffer injury from recombinant DNA research, as, for example, by acquiring an antibiotic-resistant infection from an escaping organism. Here compensation is provided, if at all, by the law of torts (1).

Tort law compensates people for their injuries by imposing liability on those who caused the injuries either intentionally or negligently, and occasionally despite the absence of either intentional wrongdoing or negligence. Most sensible concern with recombinant DNA research is with accidental injury, and the present discussion will be limited to issues that arise in litigation about negligence and about so-called "strict liability" or liability without fault.

In order to recover for injuries allegedly caused by someone's negligence, a claimant (plaintiff) must prove that the person being sued (defendant) acted negligently, that his negligence caused actual injury to the plaintiff, that the defendant owed plaintiff a duty not to act negligently toward him, and that his negligence was what lawyers call the "proximate cause" of the injury. Even if all these conditions are met, a plaintiff may still be denied compensation or have his recovery reduced (depending on the jurisdiction) if his own negligence contributed to his injury or if he "assumed the risk" of being injured. Thus the elements and defenses of a negligence cause of action are:

Elements	Defenses
1. Negligence	1. Contributory negligence (full defense) or comparative negligence (partial defense)
2. Causation	2. Assumption of risk
3. Injury	
4. Duty	
5. Proximate cause	

Negligence means unreasonably dangerous conduct and includes dangerous acts of both omission and commission. Obviously what is "unreasonably" dangerous is a matter of judgment that will vary from case to case and time to time. Recently, the concept has shown considerable slippage as courts have increasingly imposed liability on defendants, allocating losses in what they believe to be economically sound ways while continuing to speak the language of negligence and fault.

In deciding whether conduct is unreasonably dangerous, courts and juries may consider a number of factors, including the likelihood of the conduct's injuring anyone, the severity of the injury risked if it occurs, the burden on the defendant of asking him not to engage in the conduct or to do it in a different way, the social utility of the defendant's conduct, and ostensibly unrelated questions such as efficient loss allocation or who can best bear or spread the cost of injuries. These factors, of course, are not susceptible to quantitative analysis. In fact, they are rather unscientifically applied by lay people—the judge and jury—working in a complex relationship with each other.

In weighing the factors, compliance by the defendant with the law, with the custom of others engaged in similar work, or with official or semiofficial codes of safe practice will make a finding of negligence more difficult to support than it would be in the absence of such compliance, but will not make such a finding impossible. For example, in a famous case about the tugboat industry (2) Judge Learned Hand, one of our most distinguished jurists, said that liability could be imposed even if the tug were equipped in the customary way, noting that at times an entire craft or industry may lag behind the level of safe practice that society may reasonably demand. More recently, the Supreme Court of Washington held that the failure of an ophthalmologist to perform a glaucoma test on an asymptomatic patient under forty years of age is negligence as a matter of law, even though the chances of such a patient's having glaucoma are one in twenty-five thousand and ophthalmologists do not usually perform the test on such patients (3).

While compliance with custom, law, and safety codes will not exempt defendants from liability, deviation from law almost surely will result in a finding of negligence, and deviation from custom or safety codes will have the same effect unless the defendant can show his practice was safer than the customary one. Thus, in the recombinant DNA context, compliance with NIH Guidelines will not necessarily preclude a finding of negligence, but deviation from the Guidelines except by exceeding their demands will almost certainly constitute negligence.

Two kinds of negligence issues may arise in connection with recombinant DNA work. An injured person may claim that doing recombinant DNA work is negligent in itself or that a particular practice or the failure to take a specific precaution was negligence. In either case lay decision makers will weigh the risks and benefits of the defendant's work. Expert testimony about recombinant DNA research and safety techniques will probably be required, although as in medical and other technical areas where proof is more accessible to defendants than plaintiffs, the courts may use a procedural device called *res ipsa liquitur* to force the defendant to prove his lack of negligence instead of expecting the plaintiff to prove negligence. Of course, nobody can predict in advance how a particular recombinant DNA negligence suit will turn out.

If the trier finds negligence, liability seems very likely. The plaintiff may have difficulty proving the defendant's negligence *caused* his harm, but if he can surmount that obstacle, the other requirements of a negligence suit should pose little problem. Courts are likely to hold that researchers owe a *duty* of care to everyone who may foreseeably be injured by their work; and if the injury that occurs is the kind whose risk resulted in finding the practice negligent, they will say the negligence *proximately* caused the harm unless the court is unwilling for policy reasons to expose defendants to the extent of liability such a finding would cause. The concepts of duty and of proximate cause (which has nothing to do with causation and little to do with proximity) are devices for limiting the liability of negligent defendants (4). Only the excessiveness of possible liability seems likely to move a court to use either device in the recombinant DNA context. Similarly the doctrines of contributory (or comparative) negligence and assumption of risk will be of no help to defendants except in the rare case where the plaintiff himself negligently contributed to his own harm or voluntarily chose to run a known risk.

Recovery may be made even easier for a plaintiff by a court adopting the view that recombinant DNA research is an abnormally dangerous activity. People who engage in abnormally dangerous activities must compensate those they injure regardless of how carefully they performed (5). They are said to be "strictly liable" for injuries arising out of the abnormally dangerous activity.

Abnormal dangerousness, like negligence, is determined by considering a number of factors, including (i) the existence of a high degree of risk of harm; (ii) the great gravity of the harm if it occurs; (iii) the impossibility of eliminating the risk by the exercise of reasonable care; (iv) the fact that the activity creating the risk is not a matter of common usage; (v) the inappropriateness of the activity to the place where it is carried on; and (vi) the value of the activity to the community (6). Here all would agree that very grave harm may occur, that merely reasonable care cannot eliminate the risk, and that recombinant DNA research is not a matter of common usage. People will differ over the degree of risk and the value of the research as well as over the propriety of doing the work in any particular place. Some will think that any properly equipped university microbiology laboratory is appropriate; others will insist that only areas remote from population centers are appropriate for recombinant DNA research. The resolution of this question and those of risk and value will determine whether to impose liability without fault for injuries caused by recombinant DNA research. In resolving doubts courts will probably be influenced by the modern notion of "enterprise liability,"—the view that those who are engaged in an enterprise should bear its costs, including the costs of injuries it causes. This view predisposes courts who hold it to find activities abnormally dangerous in order to shift a loss from an injured individual to an enterprise. Strict liability for recombinant DNA–caused injuries seems likely. Any doubt will be removed if Congress passes legislation pending as this chapter goes to press that would explicitly create a federal strict liability cause of action (7).

Under either negligence or a strict liability approach, the question of *who* is to be liable will arise. Obviously under a negligence scheme an individual whose negligence caused an injury will be liable for it. In addition the principal investigator will almost surely be personally liable regardless of whether liability is based on negligence or is "strict." This is true despite any lack of personal negligence by the principal investigator. The same principle, that the superior answers for the torts of his employees, will make the employer of the principal investigator liable as well. If the employer is a state agency, such as a state university, statutory provisions in some states will preclude or limit the amount of liability. Typically the employer's liability is in addition to, not instead of, the individual's liability, although, of course, an injured person may only obtain one full recovery. Other potential defendants, although their liability is much more speculative, include members of the institutional biohazards committee, funding sources, and conceivably even the federal government.

Finding a party on whom to impose liability is a small problem. Finding a party who can bear the liability may be well nigh impossible. If a recombinant DNA experiment makes a few people sick or kills them, normal resources of such organizations as universities and drug companies can pay the damage awards. But if a doomsday scenario occurs and so many people are adversely affected that damages run into many millions or even billions of dollars, nobody can cover the losses. What are we to do to prevent the theoretical availability of recovery from becoming a sham?

One possibility would be to join all plaintiffs in one gigantic lawsuit and have the defendant(s) pay off claims at a rate of a few cents on the dollar as in a bankruptcy case. Such an approach would obviously be unsatisfactory. Injuries might arise and become manifest at different times, thus making the chances of joining all parties small and creating the question of how to compensate latecomers. Worse, paying off at the rate of a few cents on the dollar will *both* leave victims inadequately compensated *and* destroy defendants (pharmaceutical companies, universities, etc), thereby doing great social damage for grossly inadequate social gain.

A slightly more attractive alternative would be to put the defendant into a receivership to permit it to continue to generate funds to pay claims. Unfortunately, this approach is not feasible either. First, under a receivership claimants would be compensated in small amounts over long periods, which may be acceptable if the imagined catastrophe is the creation of a chronic disabling condition, but which is inadequate if the injury is acute. Injured people need compensation as soon as possible after being injured in order to pay bills and get started down the road to rehabilitation and resumption of a normal life.

Even if the delay is tolerable, however, other factors make the receivership approach unworkable. In the most famous case where the approach was used (8), the defendant was a circus and the injured people members of the audience. While large, the number of people in a circus audience is knowable. The numbers who may be injured by recombinant DNA may be orders of magnitude larger, thereby making administration of the receivership extremely difficult. More importantly, circuses are profit-making businesses in a low-competition industry. Probable recombinant DNA defendants are nonprofit institutions and highly competitive pharmaceutical firms. One can hardly imagine a university surviving if it had to increase tuition enough to pay thousands of injury claims; and can anyone employed by a public university believe that any state legislature will add a recombinant DNA receivership fund to the university's budget? In the drug company context, competition will prevent the receivership from working. A company in a competitive industry cannot raise prices on all its products enough to pay injury claims that its competitors do not face. Market factors and similarity of exposure to risk are presently adequate to permit companies to function. A change as major as a recombinant DNA receivership will prevent those normal forces from working, with the result that the receivership approach will only delay, not avoid, the evils of the bankruptcy approach: destruction of the company and inadequacy of compensation of the injured.

Other compensation possibilities might be creation in advance of a special fund to compensate people injured by recombinant DNA (9) and/or the payment of compensation by the government. A fund is not practical because no one can predict in advance either the likelihood or the extent of the risk; thus predictions about the proper level of the fund and contributions to it and about its adequacy if needed are impossible. The same factors make private insurance schemes unworkable. A government compensation plan may be feasible if one assumes the availability to the government of enough money to pay claims and the government's ability to recoup its losses through increased taxes after-the-fact. Such an approach could be structured either to render standard tort and Workmen's Compensation doctrines irrelevant or to leave them available for manageable situations with the government fund serving as a backup.

Compensation by the government would place the costs of recombinant DNA injuries on the public, which is the appropriate loss-bearing pool, because the public is the supposed beneficiary of recombinant DNA work. The fact that the most plausible compensation scheme would require the public to pay for injuries caused by recombinant DNA technology raises dramatically the fundamental issue: Is the public willing to run the risks the technology poses? The inadequacies of the compensation portion of the law demonstrate the critical need to deal wisely with the conduct-control aspects of the problem.

CONDUCT CONTROL

Conduct control may be attempted in a variety of ways, each of which will reflect both substantive and institutional choices. Substantive choices include decisions about whether some or all recombinant DNA procedures ought to be prohibited, discouraged, treated neutrally, encouraged, or required; which activities to place into each category; and in the middle ranges of the continuum how vigorous the encouragement or discouragement should be. Other substantive choices include decisions about the degree of safety to require (risk to tolerate); uses of recombinant DNA procedures to be prohibited, required, etc. (e.g., production of insulin, "gene therapy"); and who is to receive the benefits of the technique (only "the public," private developers, etc.).

Institutional choices concern how and by whom substantive decisions are to be made and pursued. They include choices about level of institution (national, state, local); type of institution (court, legislature, administrative agency); specific institution (NIH, EPA, etc.); and style of response (criminal sanctions, funding with regulations, etc.) to use to formulate and accomplish substantive goals. Although institutional issues are much less appealing to the nonlawyer than substantive ones, they are almost always more important. In a general sense the implications of institutional choices affect our entire social life, and even at highly specific levels they often affect substantive issues in unexpected ways. The nonlawyer's failure to understand this fundamental feature of legal reality contributes significantly to the frustration lay people often feel in dealing with the law and their inability to understand why what looked like sweet victory has turned sour in their mouths.

The most interesting legal aspect of the recombinant DNA controversy is that many of our options, both substantive and institutional, have been precluded by the rush of events. While a chapter of this sort is hardly the place to examine the factors in scientific growth and institutional deficiencies that make that true, we would be well advised to remember that the recombinant DNA experience may teach important general lessons about the relationship of science, law, and society.

For those opposed to recombinant DNA research, the most emotionally appealing response might be to urge that it be prohibited and that criminal sanctions be imposed on people who engage in it. That would be utterly absurd. The technique already exists, and it is used around the world; prohibiting it here will serve only to provide us with its risks while denying us its benefits. Moreover, progressive criminal law scholarship has demonstrated beyond cavil that the criminal sanction is overused in our society and that the costs of this overuse are enormous. Recombinant DNA research is an example of an area in which criminalization would cause more trouble than it could cure. To borrow (and oversimplify) from the classic work of the late Herbert L. Packer (10), criminalizing recombinant DNA research would violate several limiting principles of the criminal law:

1. Only conduct that is viewed without significant social dissent as immoral should be criminalized. This principle is designed to avoid both the risks of imposition upon minorities (or even apathetic majorities) and the spectre of unenforceability, as in the cases of prohibition of alcohol and marijuana use. Criminalizing recombinant DNA research would violate this fundamental principle.
2. Criminalization must serve the goal of crime prevention. This requires, among other things, establishing a high risk of being caught and punished and avoiding the criminalization of conduct to which people feel deeply committed. The privacy of the laboratory, the inaccessibility of sophisticated science to the layman, and the scientist's commitment to pursue knowledge and truth reduce the deterrability of research.
3. Conduct that is far from the harm feared should not be criminalized. Doing so raises too many doubts about culpability and may prevent socially useful activity. Different people fear different harms in the recombinant DNA context. The principle here speaks to those who raise the spectre of human genetic engineering. If human genetic engineering is the harm feared, prohibit it. Do not throw out the baby with the bath by prohibiting something as far removed from human genetic engineering as recombinant DNA research, thereby losing all the potential benefits of the research.
4. Conduct should not be criminalized unless one seriously wants the law enforced and violators imprisoned and wants the criminal process to devote time and manpower to enforcement. Surely, no sensible person wants to imprison our finest scientific minds or to divert police, prosecutorial, and judicial time away from serious crimes such as murder, robbery and rape to the enforcement of laws against recombinant DNA research.
5. Laws should not be created that cannot be enforced, that invite unequal enforcement, or that require litigable enforcement techniques. Research prohibitions cannot be effectively enforced for reasons mentioned under item 2 above. Any enforcement efforts will have to involve intrusions into laboratories, thereby inviting litigation about the Fourth Amendment prohibition against unreasonable searches and seizures; and because enforcement will be so difficult, it can only be spotty (unequal) at best.

Thus the "easy" solution of criminalization is not useful. Neither is mere reliance on case-by-case adjudication in courts. Such adjudication is too slow to provide effective social protection; too tied to the facts of specific cases to provide predictability and clear rules of conduct; and disabled from acting in a vigorous prospective way because it always operates after-the-fact, deciding cases that have already arisen out of conduct the law did not operate to control.

On the other hand, strict regulation by the legislature is unlikely to prove satisfactory for a large number of reasons, of which I shall mention only a few. If the common law is too flexible and tied to facts, legislation is not enough so. Legislation is inherently inflexible and unable to respond to specific cases. In an area such as recombinant DNA research, where countless different experiments and experimental conditions exist all with different risks and benefits, some degree of flexibility is necessary to prevent absurdly inappropriate contols. Related to its inflexibility is the inability of legislation to keep pace with science. Little reason exists to pass legislation today that will be outmoded almost before it can be applied. A current example of legislative inability to keep pace with science is the eager rush of "progressive" legislators to enact "brain death" as the legal definition of death, while many neurologists now regard it as an outmoded concept and think neocortical death is the currently best "definition" of death. We know too little and are learning too much about recombinant DNA to pass comprehensive legislation about it now.

Not only can legislation not keep pace with science, but legislatures are ill equipped to obtain, understand, and know how to use even the best current scientific information. In the 1960s state legislatures throughout the United States enacted legislation requiring newborn infants to be screened for phenylketonuria (PKU), a serious disabling genetic disease. In the late 1960s and early 1970s many legislatures also required screening for sickle cell anemia or sickle cell trait. All this legislation was well meaning and designed to alleviate suffering, but it was enacted in ignorance and was premature. The legislation did more harm than good, so that by 1975 a distinguished committee of the National Academy of Sciences called for an end to mandatory genetic screening (11). The lesson of the genetic screening experience is that modesty in the face of science is the wisest legislative policy.

A final factor suggesting caution about seeking a legislative response to any problem is that by design legislatures are amenable to considerations unrelated to devising the best possible solution to the problem at hand. In matters as important and complex as the control of an emerging scientific breakthrough, something is to be said for maximizing efforts to keep strictly to the point.

All the above criticisms of legislative response apply to federal, state, and local legislation. Criticisms multiply as the jurisdiction of the legislature decreases. State legislatures have even less access to expert information and ability to evaluate it than Congress does, and local governmental units have even less. Moreover, research technology does not recognize local boundaries. Flexibility of response is desirable; a crazy-quilt pattern of different responses is not. If control is important, it should be at least national in scope.

Even if local legislation is desirable, it may well not be feasible. Laws vary from state to state, but in many jurisdictions substantial doubt exists about whether local governments can exercise control over state agencies, including state universities. In addition, Congress may if it wishes preempt state control efforts because of the impact of recombinant DNA research on interstate commerce.

Much more promising than legislative control is response by administrative agency. Administrative agencies are designed to be able to make optimal use of experts in difficult, technical areas and to use informal procedures to produce flexible, easily changed, realistic controls. Of course the reality often falls far short of the theory. Because the principal actual response to control of recombinant DNA technology so far has been administrative, outlining that response briefly and using it to illustrate the benefits and shortcomings of an administrative response may be useful.

The primary legal control mechanism in the area of recombinant DNA research is a set of Guidelines issued by the National Institutes of Health (NIH) (12).* The Guidelines govern all recombinant DNA research funded by the NIH. They grew out of and are in part based on a series of recommendations adopted by a conference of distinguished scientists interested in recombinant DNA work. In preparing the Guidelines the NIH obtained extensive expert scientific advice as well as some lay advice.

The Guidelines, which were issued in June 1976, are to be revised to take into account new data and accumulated experience in the near future. Therefore, a detailed account of their provisions would be of little value. Simply stated, the Guidelines prohibit the five kinds of recombinant DNA experiments perceived to be the most dangerous and experiments involving more than ten liters of culture with recombinant DNA's known to make harmful products; they permit all other recombinant DNA experiments subject to certain safeguards. The safeguards consist of physical and biological containment requirements for each kind of experiment to be performed. Physical containment ranges from P1 (standard microbiological practices) for experiments of minimal risk to P4 (extraordinary engineering, equipment, and practice require-

*The NIH Guidelines appear in Appendix II.

ments) for high-risk experiments. As a practical matter, only a small number of P4 facilities will become available. Universities interested in recombinant DNA research typically seek to construct a P3 facility.

Biological containment involves the use of organisms that have been designed to have decreased ability to survive and be disseminated outside the laboratory. The Guidelines provide three levels of biological containment, EK1 through EK3.

Under the Guidelines each kind of recombinant DNA experiment must be done under prescribed conditions of both physical and biological containment. Thus, for example, a given experiment may be performed only with P3 physical containment plus an EK2 host-vector, while another experiment might be performed with P2 physical containment and an EK2 host-vector, etc.

In addition to establishing containment requirements, the NIH Guidelines impose responsibilities for their enforcement and for the proper conduct of recombinant DNA work. Primary responsibility rests with the principal investigator. However, this responsibility is shared by the investigator's institution. The institution must establish an institutional biohazards committee, which among other responsibilities must certify its approval of recombinant DNA experiments at the time of application to NIH for funding and annually after that. NIH study sections, which consider the scientific merit of research proposals, are also required to review the research for biohazards and compliance with the Guidelines. An NIH Recombinant DNA Molecule Program Advisory Committee is to resolve questions presented to it by a study section or the NIH staff. And the staff itself is to assure compliance with the Guidelines and proper review. Thus four levels of review beyond the principal investigator are required.

Commentators differ over the adequacy (or excessiveness) of the NIH Guidelines. To the extent their disagreement concerns the proper categorizing of particular experiments, a lawyer's observations are hardly the most useful. However, some things can be said here about the Guidelines as a method of control and about the process that produced them.

The Guidelines demonstrate both the strengths and the weaknesses of an administrative response. They reflect the expertise brought to bear in their drafting; they are flexible and capable of application to widely differing situations; and they anticipate continuous review and early revision. By being tied to funding, moreover, the Guidelines represent a positive rather than a negative legal response. That is, they provide an incentive to comply rather than a sanction of noncompliance. This is not only an attractive symbolic stance for the law to take, but it also avoids much of the paraphernalia of more negative responses—trials, involvement of counsel, delay, etc.

On the other hand, being tied to funding also points up one of the Guidelines' most significant failings. They control only those who need NIH funding. Other federal agencies are following the Guidelines in their grant programs, but drug companies and others who seek no federal funds are left completely unregulated. This selective regulation phenomenon illustrates a typical shortcoming of administrative control. An agency's jurisdiction is limited by the statutory authority it is granted. Several statutes, such as the Toxic Substances Control Act (13) and the Occupational Safety and Health Act (14), which were not designed with recombinant DNA in mind, may be susceptible to constructions that would provide substantial control over even nonfederally funded recombinant DNA work, but no specific legislation putting control of recombinant DNA programs under one agency yet exists. Current legislative proposals would correct this situation (15). To the extent that they do no more than provide a budget and grant an agency authority to exercise control, they would avoid the problems of comprehensive legislation that tried to control without significant administrative involvement (16).

The second serious drawback of the NIH Guidelines is that the manner of their preparation invites charges of self-serving. A recurring problem in administrative lawmaking is that reliance on experts to make or advise on regulations usually means that the people to be regulated (being the only experts) control the terms of their own regulation. This often leads to agencies being "captured" by the very groups they are to control. Even if an agency is not in fact captured, critics will be able plausibly to argue that it has been. Thus, in the recombinant DNA context, researchers themselves prepared the guidelines on which the official Guidelines were based and played a central role in preparing the official document. Critics consider this an impropriety. I consider it only an unfortunate inevitability, but it does illustrate one problem with control by experts. Even a generous view of the motivation of experts cannot avoid two realities: (1) Control by experts is control by people with a similarity of world view, which view may be very different from that of most other people. (2) Even if no impropriety is involved, control by experts creates the appearance of impropriety and thereby undermines public respect and confidence in the law. It specifically lends credence to the charge that the most basic issue—here, whether to permit any recombinant DNA research at all—received inadequate consideration. In my view the charges in this instance are incorrect and the behavior of the interested scientists, who, after all, first brought the recombinant DNA issue to public attention, deserves praise. However, the procedure followed does lend plausibility to the criticism.

The final failing of the Guidelines is a failing of our entire legal system. The Guidelines came too late. They were prepared *after* recombinant DNA methodology was widely practiced and widely publicized. One reason that outright prohibition received little attention is that it makes no sense to prohibit a technique that is being done around the world and that can be performed by bright high school students. The public did not learn of recombinant DNA technology until it was too late to stop it. Yet the recombinant DNA episode represents the high water mark of scientific responsibility and early public involvement in science policy formulation. For the present we should probably simply provide authorization to extend controls such as those in the NIH Guidelines to everyone doing recombinant DNA research and then turn our attention to exploring new institutional approaches to permit us to deal more adequately with future scientific-social developments.

NOTES

1. See generally, F. Harper & F. James, *The Law of Torts* (1956 with 1968 Supplement); W. Prosser, *Torts* (4th ed. 1971).
2. *The T. J. Hooper*, 60 F.2d 737 (2d Circ. 1932).
3. *Helling v. Carey*, 83 Wash. 2d 514, 519 P.2d 981 (1974).
4. See generally, L. Green, *Rationale of Proximate Cause* (1927).
5. Restatement (Second) of Torts § 519 (Tent. Draft No. 10, 1964).
6. *Id.* § 520.
7. S. 621, 95th Cong., 1st Sess. § 7 (1977).
8. *Jacobs Administrator v. Ringling Brothers-Barnum and Bailey Combined Shows, Inc.*, No. 71519 (Conn. Superior Court 1944). For a discussion of this case and an analysis of the receivership device in mass tort litigation, see Note, *The Equity Receivership in Mass Tort*, 60 YALE L. J. 1417 (1951).
9. For an example of a fund suggestion in a similar area, see Estep, *Radiation Injuries and Statistics: The Need for a New Approach to Injury Litigation*, 59 MICH. L. REV. 259 (1960). The Estep contingent injury fund is designed and more useful for resolving problems of attributing an injury to a specific cause than for handling problems of the enormity of damages.
10. H. Packer, *The Limits of the Criminal Sanction* (1968).
11. Committee for the Study of Inborn Errors of Metabolism, Division of Medical Sciences, National Research Council, *Genetic Screening: Programs, Principles, and Research* (National Academy of Sciences 1975). For a particularly useful account of the PKU screening experience, see Swayze, *Phenylketonuria: A Case Study in Biomedical Legislation*, 48 J. URB. L. 883 (1971).
12. National Institutes of Health, *Guidelines for Research Involving Recombinant DNA Molecules* (1976).
13. 15 U.S.C. §§ 2601 *et seq.* Mr. Thomas O. McGarity of the Office of General Counsel, United States Environmental Protection Agency, first called my attention to the relevance of the Toxic Substances Control Act to recombinant DNA issues. See Memorandum, *Request for a Search of E.P.A. Legal Authorities Concerning Recombinant DNA Activities.* from Thomas O. McGarity to

Dr. D. S. Barth, United States Environmental Protection Agency, Washington, D.C. (Nov. 23, 1976).

14. 29 U.S.C. §§ 651 *et seq.*
15. S. 621, 95th Con., 1st Sess. (1977).
16. S. 621, if enacted, would go somewhat further than the recommendation here.

Why We Avoid the Key Questions: How Shifts in the Funding of Scientific Inquiry Affect Decision Making about Science

Max Heirich

The controversy over recombinant DNA research brings into the open some unresolved issues in science—value questions that grow in importance as the character of scientific inquiry continues to change. This particular controversy is only one case of selective attention to such issues in science. It points up how arrangements for funding research have reshaped universities and scientific activity in the last few decades, and it provides a context for identifying other issues that need attention but that are not addressed effectively in the scientific community. Changes in arrangements for decision making about scientific research may be necessary in the years ahead.

We again seem poised on the edge of a scientific breakthrough that will change conditions for human life. A generation ago the breakthroughs came in nuclear physics; now they center in molecular biology.

Several people have influenced the ideas presented in this paper. My greatest debt is to Susan Wright, whose tenacity in keeping the recombinant DNA issue alive on the University of Michigan campus led to a much deeper consideration of issues at several points; she really is responsible for the development of this book. I am sorry that her own work is not included here. The argument underlying *this* paper emerged gradually, helped greatly by conversations with Les Howard, Laurie Lehne, Dorothy Mack, Michael Moch, and Marc Ross. Reactions to an earlier version of this manuscript, given at a faculty seminar on Science and Human Values (at the University of Michigan), helped sharpen the presentation here: I particularly wish to thank Carl Cohen, Billy Frye, Gordon Kane, and Eric Rabkin for their comments at that time.

Few doubt that recombinant DNA technology may lead to discoveries that could release potentials for good and ill at least comparable to those engendered by the earlier discoveries in physics (1). This time such potentials could be released not only by *applications* of basic research findings but also by the very process of inquiry itself. The scientists most directly involved have argued vigorously about how to minimize inadvertent disruption of the environment with their research. Perhaps because of past experience with breakthroughs in nuclear physics, this debate among microbiologists has mobilized publics both inside and outside the scientific community who hope to affect current decisions.

In contrast to the setting for earlier breakthroughs in physics, today's social context for carrying on scientific work is more conducive to public controversy. I believe that changes in the arrangements used to carry on research are producing subtle shifts in the nature of scientific inquiry and that these shifts will create demands for new forms of decision making in the sciences.

Certain changes in the social context of scientific activity seem particularly relevant. First, there has been a major change over the past 30 years in who grants funds, approves projects, and defines scientific problems and research priorities. Second, many of the traditional boundaries between "basic" and "applied" science have dissolved, so that larger issues become germane to so-called "pure science." As this has occurred, the range of social and political locations from which decision makers come and, consequently, the range of perspectives brought to bear for any given decision have narrowed. Over the past 25 years nonspecialists have gradually been shut out from these processes. Third, in response to these two developments new interest groups, including scientists themselves and various publics, have sought to influence policy decisions concerned with so-called "pure science" and not simply decisions about application of scientific principles. People are beginning to notice that scientific research produces winners and losers outside the scientific community itself, and they argue that such consequences should affect decisions whether to pursue particular scientific questions. At present there is no satisfactory way to resolve these conflicts.

In the case of recombinant DNA, conflicts of interest between the research needs of scientists and the concerns of the general public have been examined by several officially appointed groups in the United States and Great Britain. In each case the *ad hoc* groups have tried to make a "technology assessment," the equivalent of cost/benefit analysis, of the impact the new development is likely to have. While these assessment efforts have been going on, members of the scientific community and the public have raised at least eight important issues. Only one of the eight has been given sustained attention, however, by any of the assessing groups—a curious record, given the kinds of issues and the eminence of some of the scientists who have raised them.

That curious reaction depends at least partly on the research biases of current organizational funding arrangements. Hopefully this discussion can lead to greater clarity in the debate—not only about changes in arrangements for decision making needed in the years ahead, but also about the kinds of people and organizations needed within those arrangements in order to redirect attention to issues now neglected.

THE CHANGING PATTERN OF FUNDING IN SCIENCE

Within the United States over the past four decades, the federal government has assumed the dominant role in funding scientific research. At both a local and a national level this change has increased the autonomy of star researchers, decreased the role that nonspecialists play in the determination of science policy, and as a consequence narrowed the focus of attention concerning what values should be a part of scientific decisions.

Before World War II the bulk of federal research funds—grants made primarily through the Department of Agriculture and the Public Health Service and frequently given to land-grant colleges and universities—went for what was clearly "applied research" (2). Funds for basic research came largely through internal grants made from unrestricted funds held by a university or through foundation grants. Such endowment income and foundation money tended to go to a relatively small number of prestigious universities (3). As a result, the presidents and deans of these universities played an important role in the development of scientific inquiry on their campuses. Because of their contacts with foundations and wealthy donors, they acted in a patron-client relation to many of their faculty. Moreover, to the extent that local funding was the basis for their experiments, individual scientists often had to justify their explorations to colleagues, administrative superiors, and local sources of private wealth in terms of issues that transcended their specialty.

College presidents, in particular, played a major role in the shaping of scientific research by channeling funds to which they had special access. Their impact on policy more generally became strikingly apparent in the late 1930s, when three college presidents and the head of a private laboratory decided that there should be a federal coordinating agency to mobilize science activity for participation in the coming war (4). When Congress showed no interest in war mobilization, but rather sought to maintain U.S. neutrality, the Rockefeller Foundation funded the venture, with the approval of President Roosevelt. Thus, when the United States entered World War II in December of 1942, a central coordinating office for science policy was ready and waiting, and one of the presidents, Vannevar Bush of Carnegie Institute, became the wartime science czar for the United States.

During World War II the relation of the federal government to scientific research changed drastically. Bush, through the Office of Scientific Research and Development, became the major funder of scientific activity in the country. With central coordination of funds, an emotionally united and patriotic scientific community, and a keen administrator, scientific research in the United States made quantum advances both in "basic" research and "applied" areas. Among the most spectacular breakthroughs were those in atomic research and in medicine, including the development of antibiotic drugs (5).

Bush demonstrated the effectiveness that came with federal subsidy of scientific research. In the process he engendered considerable ill will from rival university presidents and from congressmen whose state universities did not get a proportionate share of research funds being distributed across the country. (Bush insisted that wartime needs required that research grants and contracts go to "proven producers"—which meant that the relatively small circle of elite schools got the bulk of federal research money.) To quiet his opposition, Bush promised to step down as science czar as soon as the war ended and to propose a plan for federal funding of the sciences that would prevent anyone else from assuming the degree of power over the direction of scientific inquiry that he was exercising at the time.

Bush kept both promises. The result has been a continuing and growing federal support base for the sciences, funneled through several federal agencies, most notably the National Science Foundation (NSF), the National Institutes of Health (NIH), the National Institute of Mental Health (NIMH), the National Aeronautics and Space Administration (NASA), the Department of Health, Education and Welfare (HEW), the Atomic Energy Commission (AEC), and the Pentagon. No federal bureaucrat controls the administration of science monies beyond his special-interest agency; and, except for the Pentagon, agency regulations specify the mechanisms by which decisions will be made, decision rules designed to preclude undue influences by the administrator of the agency. Review panels of research scientists from around the country judge the feasibility and importance of grant proposals. Funding, thus, has shifted from local to national sources, moving increasingly away from patron-client access routes that include nonscientists and toward access to funds through what is called "peer-review."

Review panelists often are chosen by the funding agency on the recommendation of their scientific peers. The most important committees come from nominations by the National Academy of Science, an organization of scientists who select new members by internal vote on the basis of the eminence of their reputation. For review panels that are more specialized, nominations often are solicited from present members (6).

The arrangements developed to prevent "czarism" in the allocation of science funds, in consequence, have given the "stars" within various areas of science the major role in determining the direction that scientific research will take.

With the shift from private to federal funding of science, the role of university presidents became much less central. The demise of many presidents during the student unrest occasioned by the Vietnam War further dissolved informal network ties to private donors, and inflation has reduced the flexibility previously available through endowments. Consequently, scientists became even more their own funding judges and regulators (7).

The main check on these developments was the finite total sum made available for science, as recommended by the executive office and funded by Congress. During the period of the cold war, Congress was in little mood to limit funds for scientific research. In the 18 years from 1956 to 1973, for example, federal funds for research in the physical sciences increased sixfold and support for social science research (in 1956 one-twentieth of the amount available to the physical scientists) increased 14-fold (8). Pressure from Congress usually took the form of an organized lobby to appropriate more funds for applied research in such areas as the study of cancer; such pressure tended to increase the pot of money available, rather than to limit it. The National Academy of Sciences, its Committee for Science and Public Policy (COSPUP), and the Office of Science and Technology have played a crucial lobbying role at the White House and in Congress in terms of the annual budget (9). For the most part, scientists have become their own regulators. It has been much like having gas producers decide appropriate prices and regulations for gas production.

Implications for the University

As a consequence of such shifts in funding, many of the larger, more prestigious universities have become dependent upon funds generated not by the university president and the "development team" or by state legislators but by their individual scientific researchers. The University of Michigan's budget illustrates a process that occurs much more gener-

ally. Approximately 26 percent of its budget comes from the state legislature. Another 25 percent comes from charges for medical services.
Student-generated income (from tuition, fees, room and board) amounts
to about 19 percent. Another 13 percent comes from endowment, gifts,
bequests, and income from such things as parking fees, the University of
Michigan Press, and sales for athletic events. The remaining 17 percent
comes from federal grants and contracts.

Federal grants are particularly useful to administrators of a university because of formulas for dividing the money. In the typical federal
research grant, a considerable portion of the money goes directly to the
university administration in the form of unrestricted funds for "indirect
cost recovery." (In 1978 most federal grants give 74 percent of salary
and wage budget figures to the university administration for this purpose; in the typical contract indirect cost recovery amounted to about 43
percent of the *total* grant moneys. Many private foundations, in contrast, limit such "overhead" figures to 10 to 20 percent of the total grant.)
Thus it should be clear that the shift toward federal funding of research
has provided an important degree of flexibility to university budget
planning (10).

Particularly during the 1970s, a period of spiraling inflation costs and
uncertain return from investment income, "indirect cost recovery" funds
from the federal government became an important cushion, providing
flexibility and discretionary income for carrying on the larger educational tasks of the university. Thus, the research faculty now generated
funds for the university at least on a par with those coming from other
sources. Since such money depends on the fund-raising ability of star
researchers, whose reputation among their colleagues nationally is critical for the flow of support, the former patron-client relation of university
administrators and faculty has been neatly reversed.

As this system has evolved, scientists increasingly have become entrepreneurs, oriented to a national funding market controlled by their
peers across the nation. Their attention frequently is turned outward
from their campus, rather than inward to other colleagues and administrators, and their relevant peer group consists of fellow specialists across
the country.

Implications for the Nature of Scientific Inquiry

At the same time that funding changes have made scientific researchers
dependent upon fellow specialists rather than upon other intellectuals
or on administrators, there also have been subtle pressures to change
the nature of the research question pursued. These pressures have aris-
en, not from discussion within the various scientific specialities, but

from two aspects of the new federal funding arrangements. A national budgeting practice known as "serial-funding" awards grants for limited periods, with renewal depending upon the kinds of results obtained in the interim. Frequently the renewal grants are awarded on a competitive basis, which puts the individual researcher (or research team) under considerable pressure to "produce" within a fairly brief period. Such limited time spans for results affect the definition of a pursuable topic in science and of what a payoff is.

The new arrangements, of course, do not *require* a redefinition of research topics. Many large-scale projects can be broken into smaller units, with payoffs being defined in terms of segments useful for the larger puzzle being pursued. Since peer-review is involved, someone with a reputation for exciting work often can persuade fellow scientists that the smaller-scale pieces are important for their contribution to a larger picture. Serial funding, thus, does not in itself force a shift of focus. Rather it subtly changes the emphasis of research problem formulation. One is more attracted to certain kinds of problems—those that have a shorter time-scale for investigation or that can be broken down into easily understood, simpler units. Serial funding also means that a fair amount of pressure develops to produce results within a time period that is "competitive" with that used by other scientists working on problems of comparable scope (11).

A second influence toward redefinition of scientific research strategies has come from congressional and federal agency interest in funding science to "solve practical problems."

During World War II Vannevar Bush encouraged a melding of "basic" and "applied" research questions in many fields of inquiry. During the cold war of the 1950s and the hotter war in southeast Asia of the 1960s, many federal agencies continued to offer research funds for "mission-oriented" projects (12). Universities that had operated war-related research facilities for the government during World War II often continued to seek contracts for the laboratories that had been set up, and they encouraged scientists to undertake applied research in conjunction with their pursuit of more "basic" questions in their disciplines (13).

Moreover, a well-organized and well-financed cancer research lobby in Washington regularly succeeded in enlarging the federal research budget to include more funds for work on cancer (14). Biological researchers quickly discovered that a number of "basic" scientific questions could be funded more easily if the work were done using carcinogenic agents, so that the project could qualify for special allocated research funds (15). In short, many scientists found it both interesting and financially advantageous to pursue their investigations of "basic" research questions within a potentially applied context, or else to move back and forth quite regularly between "basic" and "applied" foci for their investigations.

It soon became apparent in a number of fields that the old distinction between "applied" and "basic" research—which seeks organizing principles independent of possible applications of findings to solve problems of particular interest to some public—was becoming artificial. Applied research often gave rise to heuristic breakthroughs that added to the fund of more basic knowledge. And the "turnaround time" from abstract principles to applied settings also began to be quite short (16).

In the classic depiction of science, "basic research" is value-free (except for espousing the value of adding to knowledge) while "applied research" is value-engaged (it consciously intervenes in the environment to further specific ends, which may or may not be of equal benefit to various interest groups). As researchers used applied settings to further basic research questions, and sought early applications of more general findings, the old distinctions between "basic" and "applied" science became blurred at many points.

With the development of recombinant DNA technology, some observers saw the distinction between "basic" research and technologically oriented "applied" research breaking down at a still more fundamental level. The new research method uses restriction enzymes to break apart very large DNA molecules that contain many genes, creating smaller DNA molecules that contain only a few genes each. These small DNA molecules then can be rejoined to one another to give new arrangements of genes; or, if the restriction enzyme DNA fragments come from different organisms, the process can produce entirely novel combinations of genes. Then this newly invented sequence can be introduced into a living cell, where it may cause the cell to exhibit new characteristics (e.g., to produce new enzymes, to have altered interactions with other cells, or to grow more rapidly or more slowly). Some observers see this as fundamentally akin to applied science or technology: it involves using basic scientific principles to invent a new substance that is introduced into the environment for the benefit of a particular group of interested parties. In this case the interested parties are scientists intent on acquiring "basic" scientific knowledge; but they are changing the environment in a way that would not occur in nature, much as technology does. [The intervention is far more severe than in previous genetic research, which could only study changes consistent with a general evolutionary program already in effect. Now researchers can create new DNA combinations at will (17).] This amounts to value-free applied science—a deliberate application of scientific principles to alter the natural environment, but for the benefit of scientific curiosity rather than to improve the situation of some segment of the public.

Consequently, the range of issues appropriate to judging the desirability of *applications* of scientific knowledge begin to apply to the process of "basic" scientific inquiry as well.

Together, these trends have encouraged rapid exploration of puzzles within clearly articulated scientific paradigms, so that the pace of scientific progress has been impressive. The impetus toward short-term payoffs has increased pressure toward dramatic intervention as part of the research design. And the blurring of distinctions between "basic" and "applied" research has confronted the scientific community with a range of ethical and value issues that previously had seemed to apply to technology, but not to "basic" scientific inquiry.

As basic research becomes increasingly "technological" in its environmental implications—as the distinction between basic and applied science becomes blurred—the larger intellectual community will be forced increasingly to face the value implications of that work. Other kinds of people are likely to insist that their interests be represented, and to raise issues, be they realistic or imaginary, where they fear that the scientific community is likely to put them in jeopardy. Consequently, the ability of the intellectual community to develop sophistication and skill in technological assessment of trends in science is likely to become not only a moral good, but a practical necessity.

It has become difficult, however, for intellectual communities to confront value questions concerning the implications of applied scientific activity, even when these questions are presented to them in extreme and dramatic form.

Implications for Intellectual Inquiry in Other Fields

As the funding patterns just described became firmly established, they accelerated changes that had begun earlier in the nature of intellectual work, especially at many of the universities where the star researchers were located. The new funding formulas have helped accelerate a move away from value issues within the social sciences and the humanities, so that challenges to value assumptions of research scientists were less likely to emerge even if interaction with a wider range of intellectuals was easily available.

Elsewhere (18) I have chronicled the strategy used by one of the last great entrepreneurial university presidents, Clark Kerr, chancellor of the University of California at Berkeley and later president of the entire University of California system for a ten-year period during which the shift to federal funding became institutionalized. Berkeley reallocated internal moneys to capture a larger stable of star researchers in their bid to compete with Harvard. To do this they systematically rebuilt a number of departments—including statistics, philosophy, linguistics, and the social sciences—recruiting scholars who would orient them either toward problems methodologically useful to the "hard sciences" or toward *quantifiable* research questions that would allow successful competition for research moneys becoming increasingly available under the new system. And through a variety of organizational innovations they shifted the educational focus increasingly toward graduate teaching. It proved to be a mutually beneficial arrangement for researchers and administrators: as the stable of stars enlarged, Berkeley increased its share of federal moneys in many discipline areas. The "overhead" from these ventures allowed continued acceleration of these trends. The Berkeley example became known across the country and began to be imitated so far as was practical in many other universities.

For at least a half century earlier a trend toward reshaping many intellectual disciplines in the model of the "hard sciences" had been visible. This involved recasting questions toward analysis of formal properties, and turning away from questions that could not be made "operational" in terms of objective, quantifiable measures (19). As faculty attention shifted increasingly from teaching to research, and toward concern with the more narrow research questions that underlie much graduate teaching (in contrast with the larger issues that frequently organize undergraduate education), processes already underway gained further momentum. More and more areas of intellectual inquiry shifted their problems toward quantifiable, "operational" questions that tried explicitly to be "value-free." Fewer intellectuals were devoting themselves to values issues, or to the formulation of criteria in terms of which decisions might be made where values were in conflict.

Counter Movements

These trends have not existed in isolation, of course. As universities became increasingly national in their orientation—depending upon federal funding, relating research increasingly to national political priorities—countermovements began to emerge. They could be seen at work on campuses, among dissident members of the national associations of the various disciplines, and in the community at large. The antiwar movement of the 1960s, which swept most university campuses, fanned many issues that had been smouldering since 1945, with the release of the first atomic bombs and the emergence of the scientific community's continued involvement in the international arms race during the cold war of the 1950s and the undeclared war in southeast Asia.

During the first two decades of federal funding for science, following the end of World War II, countermovements seemed limited primarily to physicists (uneasy about the applications of nuclear research), to persons (academic and otherwise) concerned about the spiraling arms race, and to small groups of political radicals who saw science and the universities more generally as handmaidens of a capitalist class. Although such organizations as the Society for Social Responsibility in Science and the Federation of American Scientists emerged to provide a common voice for scientists concerned about the value implications of trends in scientific research, they remained a minority voice—and one with little access to decision-making structures within the scientific community itself. Indeed, the National Academy of Sciences, which was benefiting so handsomely from the new institutionalization patterns, did not favor appointments of "boat rockers" to committees determining priorities either for directions of research or for allocation of funds (20).

Containment of dissent became much more difficult, however, during the 1960s. The rapid expansion of graduate programs across the nation, made possible by the federal funding patterns already described, intersected with the large-scale antiwar movement on American university campuses. Student radicals began protesting the involvement of their universities in war mobilization and began to cast a spotlight on specific types of applied research that were occurring on their own campuses. Soon national meetings of the various disciplines found themselves confronted with political resolutions about the war and challenges to kinds of research that members of the association were undertaking. Objectors lacked access to key committee spots, so that they rarely won their campaigns within the associations. But the encounters sensitized many scientists (and especially those of the emerging younger generation) to ethical issues involved in the new patterns for carrying on scientific activity (21).

The emergence of an Ecology Movement in the late 1960s widened the range of challenges being put to the scientific community about the consequences of its interventions into the environment. Disillusionment with the war in southeast Asia already had taken its toll in terms of public confidence in the government, in universities, and in the scientific establishment (22). Now groups of scientists outside the "establishment" began mobilizing to fight projects that they feared might endanger the public safety. The development of atomic power stations, environmental battles over oil development and distribution, and air and water pollution issues began to engage increasing numbers of scientists and made clear the limits in science of value neutrality where applications of findings are involved (23). Scholars became aware of second and third-order consequences from previously acclaimed scientific interventions in the environment. Observers began to relate the population crisis in many parts of the world to the introduction of pesticides and "miracle drugs" 20 years before. They noted that the "green revolution" intended to solve the food crisis created by this population growth was creating its own forms of misery, forcing many small-scale peasant farmers off the land in areas where cities could not provide them a means of livelihood. Reassessment of the consequences of applied science continued in many areas, and scientists themselves began to take a more searching look at their own hopes to apply still more technology to solve the problems that emerged. Wide segments of the public, including numbers of intellectuals who previously had not questioned the value or direction of scientific inquiry, began to voice uneasiness that their interests, or the interests of humanity more generally, might not be furthered by unrestricted growth of scientific inquiry (24).

Three kinds of responses have grown out of this intersection of developmental trends in science and counterresponses to them. At a public level has come the mobilization of citizens' groups fighting harmful technological applications. At a policymaking level has come a renewed evaluation of the costs and gains involved in proposed areas of scientific research and development. And among scientists themselves has come a growing sensitivity to some of the value issues involved in their research, so that they have become both instigators and defendants in some of the battles that have emerged.

The recombinant DNA technology debate has brought all these components into focus in a way that may sharpen our understanding not only of the dynamics involved, but also the limits that present decision-making mechanisms impose on the kinds of issues that can be confronted.

THE RECOMBINANT DNA CONTROVERSY

The controversy about recombinant DNA technology began among a small circle of microbiologists, grew to include an international set of researchers, and then began to spill over into other segments of the public—politicians, university scholars from other disciplines, environmental advocates, the business community, and governmental units. In the process of its expansion it laid bare some of the important structural constraints on decision making in science.

I will recount the developments only briefly, since they have been well documented elsewhere (25). A small group of fellow scientists asked Paul Berg, a Stanford University scientist active in recombinant

DNA research, about safety conditions in his laboratory: he was planning to use *E. coli* bacteria, which normally inhabit the human gut, as hosts for experimental combinations of DNA fragments. Berg was asked how confident he was that the new creatures would not create a virulent disease reaction in humans, or that the creatures could be confined to the laboratory. (The concern was that an Andromeda Strain, as described in Michael Crichton's novel, might now be made real in a series of laboratories around the world.)

Berg saw the dangers that would ensue if pathogenic agents were created in such experiments and escaped. He called together a group of prestigious researchers, who announced that until safety standards could be worked out, they would observe a self-imposed moratorium on using the technique with tumor viruses and with certain other potentially harmful sets of genes. The National Academy of Sciences subsequently asked Berg to help organize an international conference to consider safety standards for laboratories engaged in recombinant DNA research. Implementation of the decision reached in these discussions fell largely to the National Institutes of Health, which fund most of the work in recombinant DNA research going forward in the United States. Meanwhile, politicians in the United States and Great Britain, university administrators, and other public figures began to conduct hearings and to set up commissions to advise policymakers about the promises and perils inherent in research of this kind.

As the debate was formulated by the microbiologists, few questioned whether new organisms which might produce cancer should be created for research purposes, or whether some random recombinant DNA techniques were an appropriate way to conduct scientific inquiry in these areas. Rather, they asked what safety standards were necessary to protect the population while work of this kind went forward (26). This is an understandable formulation of the problem from the standpoint of microbiologists, but it is not necessarily an appropriate limitation of issues from the standpoint of the general public.

Avoiding Control by Vested Interests

It seemed reasonable to bring in judgments from persons with a wider range of backgrounds. The British Parliament commissioned a group of distinguished scientists from other fields to assess costs and benefits from this method of research (the Ashby Report), the University of Michigan administrators used a cross-disciplinary group for the same purpose (Report of Committee B), the Cambridge, Massachusetts, City Council set up a citizens review board, and the director of the U.S.

National Institutes of Health (the chief funding source for research in this area) consulted his own citizens advisory committee. The World Health Organization's Advisory Committee on Medical Research also issued a public statement. Furthermore, U.S. Senator Kennedy, the New York attorney general's office, and the California legislature have held hearings concerning issues involved in this new area of scientific development.

But the problem is not so easily solved. Expanding the range of personnel considering these issues guarantees that the researchers who have vested interests will not set policy regulating pursuit of their own interests. But it in no sense guarantees that different questions will be addressed than those formulated by the interest group itself.

Each of the ad hoc assessment groups represents an attempt to break out of the narrow interests that have been involved in most decision making in science in the past few decades. The range of role-perspectives drawn into these assessment groups varies noticeably. Yet the tenor of their reports and conclusions has been rather similar. This would be comforting, were not the various reports also similar in another respect: each focuses upon the questions raised by the microbiologists themselves and gives only passing attention to a wider range of issues raised by other scientists, environmental advocates, or other segments of the public.

Factors that Influence Decision Outcomes

Decision theorists note that two ingredients are critical for predicting which decision will be reached when a group is weighing alternative options. One concerns the particular goals a set of decision makers share. This might be restated as their *definition* about what is at issue (i.e., what are these people trying to accomplish?). A second key factor concerns the range of alternative choices that become visible to the decision makers. Once these two factors are in place, it becomes fairly easy to predict which choices will be made, under conditions for rational problem solving. Each visible alternative choice can be ranked in terms of its likelihood of success in reaching the goal selected and the probable cost if it is used. The option offering the best chance of reaching the goal with the least cost to the participants should be the outcome chosen (27).

It should be clear that the definition of the problem to be addressed becomes critical for what else happens. Unless participants in the decision-making group represent interests that see a different issue at stake, the initial definers of the problem to be solved determine what options will appear relevant or irrelevant for consideration. Thus the group that defines the initial problem to be addressed strongly influences the outcome, whether or not this group participates directly in the actual decision making.

The microbiologists who first questioned the appropriateness of using recombinant DNA technology for certain kinds of cancer-related research began with the assumption that their paradigm puzzle should be solved, and within the shortest possible time. They further assumed that it should be used to study the genetics and molecular biology of cancer-producing viruses. These unexamined assumptions grew directly from their intellectual commitments and from the realities of competitive funding arrangements. Within this context, they debated the costs and gains involved in various laboratory safety procedures that should minimize the risk of accidental escape of organisms from the laboratory. Given their commitment to rapid and widespread research and their concern about cancerous organisms or other pathogens escaping, they limited the discussion to containment procedures for clearly dangerous carriers.

Such a restricted focus for the question, however understandable in terms of the immediate perspectives of the microbiologists themselves, ignores some critical elements that should be part of any technology assessment. To the extent the microbiologists' questions dominated the thinking of the various ad hoc groups asked to give advice on this matter, it matters little whether specialists or a wider group examined the range of options available.

Technology Assessment

All the assessment groups were given similar charges: they were asked to judge whether gains possible from using recombinant DNA technology outweighed the risks that might occur if laboratories working with pathogenic organisms proceeded with this kind of research. Microbiologists who use this technique were excluded from most assessment groups.

Joseph Coates, in an article on "Technological Assessment at NSF," suggests ten elements or modules that are common to all major comprehensive technology assessments. They include the following:

1. A statement of the problem to be considered—usually with a broader restatement or recasting of the problem after analysis is underway.
2. Definition of the system (technology) and of specific alternatives that could accomplish the same objectives (micro alternatives).

3. Identification of potential impacts—a creative enterprise requiring imagination and speculation.
4. Evaluation of potential impacts—a mixed effort of firm-handed analysis and informal judgment necessarily conducted on "semisolid" footing.
5. Definition of the relevant decision-making apparatus—a step which is often neglected.
6. Laying out options for the decision maker. Since traditional categories may now be inadequate, new inventions and imaginative development of options are usually appropriate and often needed.
7. Identification of parties of interest, potential "winners" and "losers," including both overt and latent interests.
8. Definition of "macroalternatives"—not alternative technologies as considered in module 2, but broader system alternatives such as energy conservation or solar energy generation rather than the Alaskan Pipeline; this step provides a standard to challenge conclusions drawn from modules 1 through 7.
9. Identification of exogenous variables—events that may disturb the system (natural catastrophe, war, embargoes, depressions, changing birthrates, etc.)
10. Conclusions—and possible recommendations (28).

The reports of these various advisory groups pay a great deal of attention to the first four elements in this list—so long as the problem is stated in terms akin to those used by the microbiologists originally. Rather little attention is given to steps 5 through 9, which assume that the application of science normally has "winners" and "losers," that there may be a variety of alternatives for dealing with the problems needing solution, and that nonnormal occurrences that could disturb the system should be taken into account as part of the scenario (29). These common omissions are especially fascinating in view of the efforts of a minority to raise a series of issues that would have made such considerations highly relevant.

Issues in Dispute

During the assessment proceedings eight issues concerning the use of recombinant DNA technology have been raised, often fairly eloquently, and usually by nationally known scientists. Only one issue, however, has been addressed with any degree of seriousness. This selective attention deserves comment.

The questions raised have to do with safety, with political decision making, and with larger ethical issues.

Safety

1. The first area of controversy concerns what might be called operational safety issues. Microbiologists themselves raised the question of appropriate safety standards to be used in individual laboratories where work with pathological organisms would go forward. This issue has received prolonged and sustained attention from a wide segment of the scientific community; the issues involved have been sufficiently complex to preclude participation by many beyond the circle of scientific specialists, laboratory safety officers, and experts in statistical probability theory. But short-term safety issues *have* received considerable attention (30).

2. A second issue concerns longer-term laboratory safety questions that arise because of the proliferation of research. (How many experiments should go on at one time, and where, to minimize risks of accidents over time? Where will the cohorts of graduate students trained in the new techniques find jobs? How will this affect the number of experiments occurring, the rate of application of findings, the ability to control application of safety standards? (31). These questions view research activity, not from the standpoint of the individual laboratory, but as a *system* of activity that will generate its own accidents, misapplications, and problems over time, unless central coordination and planning is achieved early. This issue was raised, but never addressed directly.

3. A third safety issue has been all but ignored. How can the public be protected from unwise *applications* of new genetic principles? Some observers fear widespread environmental disruption could occur if the research is successful. Here are some examples of what they fear: new organisms developed to clean up oil spills might evade efforts to contain them, once the spills were under control. What would they do to other oil pools? Insulin-producing organisms that could produce cheap quantities of medicine also could wreak havoc if they escaped because they would send a normal population into insulin shock. Again, the ability to specify the sex of an unborn child, though a blessing to individual sets of parents, could seriously disrupt the ratio of men to women within a generation. They urge that the pace of discoveries be slowed down until the public can be protected from inappropriate application of such principles (32). Thus far, no assessment commission has seriously addressed this problem.

Political Decision Issues. A second series of issues has been raised with considerable rancor, frequently outside scientific and academic halls, although a few scientists also have been involved. Those who raise these issues recognize that the application of science regularly produces winners and losers—those who gain from the application and those who pay heavy costs. This political aspect of scientific inquiry—an embarrassment to the value-free intent of paradigm pursuit—has tended to be ignored within scientific discussions and among academics more generally. Thus it has been raised rather stridently, by representatives of interest groups who distrust the impact that current directions in scientific research are likely to have on their own self-interest.

4. One set of questions concerns appropriate locations for research laboratories. The Cambridge, Massachusetts, city council, for example, demanded assurance that its citizens would not be endangered by laboratory experiments at Harvard or MIT. And some scientists have urged that all work be conducted in a limited number of national laboratories (33).
5. Environmental lobbies and other citizen groups have demanded that "the public" and not simply the interests of researchers be represented in science policy decisions (34).
6. More radical groups ask why major funds should go toward additional research that could endanger the public instead of toward already-known solutions to public health problems. For example, why not remove carcinogenic agents in the natural environment, rather than discover ways to outwit their effects once they have entered the human body (35)?

These questions have roused considerable heat. Few members of the intellectual community, however, have been willing to deal with the issues underlying them. For they threaten to take away the autonomy that scientists and intellectuals more generally have established for themselves. Many intellectuals see no substitute for self-regulation other than regulation through political means. They question the freedom of politicians to resist special-interest pressures, and so they fear the nature of any regulation that might occur.

Larger Ethical Issues. Beyond the question of who decides winners and losers from scientific inquiry, and how conflicts of interest are to be resolved, a more disturbing set of questions has been raised by a number of thoughtful scientists and scholars. They sound like contemporary reflections of two classic themes from literature, which might be described as the problem of the Sorcerer's Apprentice and the Faust dilemma.

7. Is it appropriate, some ask, to use nature's method for preserving species integrity to violate that integrity—just to see what will happen? At this stage of scientific discovery we do not know why species integrity exists as an organizing principle in nature or what the consequences of its violation are likely to be for the evolutionary process. The question is absolutely fundamental to the technology being used. It has been stated eloquently by some of the most respected scientists in the nation (36). Yet it has been almost totally ignored.*

8. Others have asked whether we have any business exploring an area that will unlock major new power for human use before we show evidence that we have either the wisdom or else safeguarding mechanisms that will prevent disastrous misapplications of the principles (37). All participants agree that the research issues concern an area at least as important for application as was the development of atomic energy theory a generation ago, and fraught with equal potential for good use or bad. And all agree that the new technique is sufficiently simple and inexpensive to become available soon even to high school students. Yet intellectuals stand committed to the belief that the pursuit of knowledge is good and that the problems arising from its use remain someone else's specialty (38).

All of these questions seem basic and important. Yet only the first, which concerns safety standards for individual laboratories, has attracted serious and sustained attention. For this I think there are at least two reasons. First, it is the only question that does not challenge, at some level, the present arrangements for carrying on scientific inquiry. It is also more easily quantifiable and researchable than many of the other questions, but this is not sufficient reason to explain its selection for attention to the neglect of others.

The question about proliferation of laboratories and research personnel, for example, would seem a direct and logical extension of this question, yet scientists have moved vigorously to eliminate consideration of this issue as part of the safety question. (They have attempted to make them serial considerations, without acknowledging that decisions requested for the first issue could preclude solutions to the second part of the series.)

The question of winners and losers also seems amenable to assessment technologies common to science. But people have preferred to

---

*Editors' note: See chapters by David A. Jackson, Rolf Freter, and especially the chapter by Bernard D. Davis for discussions of this point.

assume that there would be no losers, rather than to examine the question. Thus, while other issues *could* be pursued with some of the techniques currently at hand, these issues begin to challenge the arrangements by which the enterprise currently goes forward.

The issues concerning proliferation of research, for example, and concerning control of application of discoveries can be dealt with only if someone begins to assert control over the direction and pace of scientific inquiry. As we have seen, these issues threaten the autonomy now enjoyed by scientific communities, and they also would threaten the resource base that has become central to many universities across the nation. Small wonder that neither scientists nor the academic community more generally have been willing to devote sustained attention to these problems. They have been raised and discussed but then dismissed before anyone attempted the kind of careful assessment that has gone into questions of individual laboratory safety. This refusal to think about science from a systems perspective is extremely shortsighted: it will be very difficult to establish controls later after a certain research momentum has developed. But it seems clear that the scientific and academic communities are not prepared to face such issues directly.

The issues about who *decides* policy when interests are not in common, and what priorities should take precedence, are not welcomed by those who benefit most from the present decision arrangements. They have only been pursued outside the circles of scientists and academe.

The last two issues are perhaps the most fundamental and the most difficult to address. One could imagine that answers to the earlier questions might be formulated by national policymakers beyond the scientific community itself. Questions of rational system planning and of the involvement of relevant publics in the decision-making process are, after all, the kinds of questions that policymakers are accustomed to addressing. But the larger ethical questions, concerning appropriate strategy for scientific inquiry and the exploring of areas where findings could be misused, remain central to scientists themselves, rather than to other groups. They are much easier to evade than to face. Thus these questions have initially been overlooked, or dismissed with a brief argument, then resisted because addressing them would slow down the pace of research.

Many thoughtful critics have asked, why the hurry? The problem, as I see it, lies less with the race for eminence (for the Nobel prize, as some cynics have suggested) than with the dependence of individual scientists, and of the universities themselves, on federal moneys allocated with the expectation of rapid breakthroughs. To raise these questions not only puts individual careers in limbo, it also threatens the economic base upon which academia now depends. And in the absence of other mechanisms, it threatens to throw the scientific enterprise into the laps of vested political interests, because the questions strike so fundamentally at present arrangements for decision making in science.

UNRESOLVED ISSUES IN SCIENCE MORE GENERALLY

The issues have been formulated above in terms of the immediate controversy concerning recombinant DNA research. Stated more generally, however, they extend to much larger areas of scientific inquiry. In question are socially desirable limits to the amount of scientific research, effective controls on the application of scientific findings that could damage population groups, the rights of various publics to take part in decisions that may affect their futures, and priorities for use of public funds: what kinds of questions deserve massive outlays of public funds, under what circumstances?

It should not surprise us that scientists do not find these questions attractive—for any answer to them is likely to challenge their present autonomy and privilege. So long as groups of scientists frame the questions that are addressed by policy-recommending groups, issues such as these are not likely to get sustained attention. Yet the recombinant DNA controversy makes it clear that the nature of scientific inquiry is changing. Questions such as these can be ignored only at real cost to all concerned, including the scientific community as well as the larger public.

The larger ethical issues also extend beyond the question of DNA research. They concern possible limits that scientists should impose *on themselves*, to avoid disruptive environmental consequences either from the pursuit of knowledge or from its application in new settings. They pose haunting questions that need answers from the collectivity of scientists. Because they challenge assumptions about progress through knowledge, assumptions that have guided scientific inquiry since its inception, they become especially difficult for intellectuals to pursue. Yet they cannot be ignored.

In the absence of alternative ways to make decisions, science finds itself in a dilemma. The present trend toward the merging of basic and applied science means that value issues no longer can be avoided; science begins to intervene too directly in the environment not to be considered a vested interest, to which others will respond. Yet, for the reasons just outlined, the scientific and intellectual communities are ill-equipped even to recognize, much less to respond effectively, to the more fundamental value issues being generated by their own work.

We need some new procedures for making decisions about science.

We need them quickly, before it is too late to deal with some of the issues now being ignored. No new procedure will have much impact, however, unless it affects how funds are allocated for scientific research. As I see it, this would involve at least three changes in the way decisions now are made. First, some new device would have to be created for making decisions about the *pace* and *direction* of scientific inquiry. Second, some procedure would have to be developed for ensuring that an enlarged set of criteria are an explicit part of the decision process. Third, the range of parties involved in making such decisions would have to be expanded in a way that prevented the scientific community, the affluent special interests that would benefit from their work, or narrow political interests from defining the range of questions to be addressed. A social invention is called for, and soon. Who will make the first proposal?

NOTES

1. The National Council of the Federation of American Scientists describes molecular biology as having reached a stage "brilliant with both danger and promise." Among the dangers they foresee is "a potential biological proliferation threat that might rival that of nuclear weapons." They also see a chance for "a new attack on many cellular diseases," leading ultimately to genetic engineering that may hold forth "the risky potential of improving man himself." [*F.A.S. Public Interest Report*, **20**(4) (April 1976)].
2. See James L. Penick, Jr. et al., eds., *The Politics of American Science: 1939 to the Present*, p. 7.
3. See Penick, *Politics of American Science*, pp. 8–9.
4. The men involved were James Conant, president of Harvard; Karl Compton, president of MIT; Vannevar Bush, who had just moved from a post as vice-president of MIT to become president of the Carnegie Institute; and Frank Jewell, head of the Bell Telephone Laboratory and president of the National Academy of Sciences. In the early days of the New Deal, Compton, as president of the Science Advisory Board, had attempted unsuccessfully to develop a comprehensive coordinating organization of scientists. In the late 1930s the four men again suggested that a national agency was needed to redirect scientific inquiry—this time into areas needed for military defense. For a fuller account, see Penick, *Politics of American Science*, pp. 8–11.
5. See Penick, *Politics of American Science*, pp. 9–34.
6. See Organization of Economic Cooperation and Development, *Reviews of National Science Policy: United States*, pp. 163–173.
7. Joseph Ben-David, The Scientist's Role in Society, pp. 166–168.
8. In fiscal year 1956 the federal government spent $0.6 billion for research in the physical sciences. By 1973 that figure had grown to $3.8 billion. Social science research expenditure amounted to $30 million in 1956 and had grown to $412 million by 1973. See Michael Useem, "Government Patronage of Science and

Art in America," in Richard A. Peterson, ed., *The Production of Culture*, 123–139.

9. See Stuart S. Blume, *Toward a Political Sociology of Science*, pp. 187–214.

10. The indirect-cost recovery formula varies by type of research and circumstance. In the health sciences indirect-cost recovery figures occasionally are higher. When work is done off campus, the figure is lower. Source: Alvin Zander, Associate-Vice President for Research, and James Brinkerhoff, Vice-President and Chief Financial Officer, University of Michigan.

11. It would be interesting to see how both volume and proportion of research proposal topics has changed since these funding practices went into effect. Such a compilation, however, would require a substantive knowledge of subfields within science, which puts the task beyond the scope of most social science research teams, including the present author.

12. In 1969 the federal budget for research and development was $16,891,000,000, of which $1,350,000,000 went directly to universities. The breakdown by governmental programs was as follows:

Department	$ Million
AEC	1504
Agriculture	276
Commerce	81
Defense	8194
HEW	1211
Interior	202
NASA	4495
NSF	257
Others	207

See Martin Brown, ed., *The Social Responsibility of the Scientist*, pp. 1–2.

13. Between 1954 and 1964, for example, the federal government increased its expenditures for research on university and college campuses from $380 million to $1590 million (approximately 18 percent of the 1964 funds went for applied research and another 3 percent for development). At the same time, it increased its expenditures for university-managed contract research centers from a 1954 figure of $140 million to a 1964 total of $630 million (of which 32 percent went for applied research and another 38 percent for development). Source: Organization for Economic Cooperation and Development, *Reviews of National Science Policy: United States*, pp. 192 and 193, based on reports from the National Science Foundation.

14. The American Cancer Society assumed new political clout in 1945, when an advertising executive, Albert Lasker, and the chairman of the board for a pharmaceutical company, Elmer Bobst, became active in its leadership. Later Mary Lasker continued the effective lobbying work that had begun earlier. A portion of this story is told in William Gilman's *Science U.S.A.*, pp. 198–201.

15. As one Nobel prize winner wryly, but anonymously, commented, "The scientist who finds a cure for cancer will not be popular among his fellow

scientists" (quoted by David M. Rorvik, "Do the French Have a Cure for Cancer?" *Esquire* **48:** July 1975, pp. 110 ff.)

16. See Spencer Klaw, *The New Brahmins: Scientific Life in America*, pp. 171–172.
17. See, for example, Liebe F. Cavalieri, "New Strains of Life—or Death," *The New York Times Magazine*, August 22, 1976, pp. 8 ff.
18. Max Heirich, *The Spiral of Conflict: Berkeley 1964–65*, pp. 51–58.
19. See, for example, Herbert Marcuse's discussion of shifts in the focus of philosophy, in *One-Dimensional Man*, pp. 170–199.
20. Blume, *Toward a Political Sociology of Science*, pp. 193–214.
21. Charles Schwartz, for example, describes attempts to bring such issues to the attention of physicists, in Martin Brown, ed., *The Social Responsibility of the Scientist*, pp. 19–34.
22. National surveys show a steady erosion of public confidence in the government, and in other major institutions. See, for example, A. H. Miller, "Political Issues and Trust in Government," *American Political Science Review*, **68** (September 1974), pp. 951–972, or J. S. House and W. M. Mason, "Political Alienation in America, 1952–1968," *American Sociological Review*, **40** (April 1975), pp. 123–147.
23. See Joel Primack and Frank von Hippel, *Advice and Dissent: Scientists in the Political Arena*, pp. 128–235.
24. Cf. Barry Commoner, *Science and Survival*, M. Taghi Varvar and John P. Milton (eds.) *The Careless Technology: Ecology and International Development*, statements from Science for the People, or Robert Sinsheimer, "Inquiry into Inquiry," *Hastings Center Report*, August 1976, p. 18.
25. In addition to accounts that appear in this book, interested readers might want to read the description that appears in *Recombinant DNA Research*, Vol. I, prepared by the Public Health Service and the National Institutes of Health, or the *New York Times Magazine* article by Liebe F. Cavalieri, "New Strains of Life—or Death," August 22, 1976, pp. 8 ff.
26. For documents from the Asilomar conference on Recombinant DNA Molecules, see *Recombinant DNA Research*, Vol. 1, pp. 44–68.
27. These arguments were first formulated by John von Neumann and Oskar Morgenstern in *Theory of Games and Economic Behavior*, and have been developed since then by scholars with a number of interests. See, for example, Robert Axelrod, *Conflict of Interest*, Anatol Rapoport, *Fights, Games and Debates* and *Two–Person Game Theory*, or Thomas C. Schelling, *The Strategy of Conflict*.
28. Joseph F. Coates, "Technology Assessment at NSF" (National Science Foundation), Chap. 2, pp. 11–17 in Sherry R. Arnstein and Alexander N. Christakis, eds., *Perspectives on Technology Assessment*.
29. It would be an oversimplification to assert that none of these topics was discussed by evaluating committees, or that they are not mentioned in their reports. Rather they were not the subject of sustained examination, measurement, and continuing attention. Where they are discussed, they tend to be dealt with in a paragraph or less, based on arguments for which substantiating data do not seem to have been sought or reported.
30. See, for example, the Report of Committee A, University of Michigan, or the extensive set of reports and commissioned studies carried out for the National

Institutes of Health committees and included in their volumes of documents relating to "NIH Guidelines for Research Involving Recombinant DNA Molecules."

31. See the letter to Dr. Donald S. Fredrickson, National Institutes of Health, dated February 5, 1976, from Professor Robert Sinsheimer, chairman of the Division of Biology at the California Institute of Technology, reprinted in *Recombinant DNA Research*, Vol. 1, pp. 436–438, or an article prepared for the University of Michigan Board of Regents by Max Heirich, "Proliferation and Options for New Actions" (available upon request to the author).

32. See *The New York Times'* coverage of the debate on the following dates: June 3, 1976, 3:1; September 23, 1976, 17:1, October 10, 1976, VI, 8; October 13, 1976, 42:1; October 28, 1976, 21:1; and October 31, 1976, IV 9:1.

33. The Harvard–Cambridge City Council controversy was covered by *The New York Times* in the following issues: June 17, 1976, 22:4; July 8, 1976, 12:6; August 22, 1976, VI, 8; August 29, 1976, 36:6 and IV 6:3; and September 1, 1976, 32:4. See also Barbara J. Culliton, "Recombinant DNA: Cambridge City Council Votes Moratorium," *Science* **193** (July 23, 1976), pp. 300–301.

34. See, for example, letters from the New York Public Interest Research Group, Inc. (reprinted on p. 539 of *Recombinant DNA Research*, Vol. 1), from *Friends of the Earth* (pp. 542–544 of that volume), or statements made to the University of Michigan Board of Regents by the Ann Arbor Ecology Center.

35. Jonathan King, of MIT, spoke at the University of Michigan Forum on Recombinant DNA Research, March 3–4, 1976, as a representative of Science for the People. He developed this theme at some length.

36. The strongest statements have been made by Erwin Chargaff and Robert Sinsheimer. See, for example, Sinsheimer's letter of February 12, 1976, reprinted in *Recombinant DNA Research*, Vol. 1, pp. 443–445.

37. See Appendix B1, Statement of Dissent by Professor Shaw Livermore, Jr., to the Report of Committee B, University of Michigan (pp. 46–47), March 1976, or Robert Sinsheimer, "Inquiry into Inquiry," *Hastings Center Report*, August 1976.

38. See, for example, the statement by Professor Fred Neidhardt, Chairman of the Department of Microbiology and of Committee A, the University of Michigan, to the meeting of the Board of Regents, May 24, 1976.

REFERENCES

Ann Arbor Ecology Center, Statement to the Board of Regents, University of Michigan, May 24, 1976.

Arnstein, Sherry R., and Alexander N. Christakis, eds., *Perspectives on Technology Assessment*, Jerusalem, The Academy of Contemporary Problems, Science and Technology Publishers, 1975.

Axelrod, Robert, *Conflict of Interest, A Theory of Divergent Goals with Applications to Politics*, Chicago: Markham, 1970.

Ben-David, Joseph, *The Scientist's Role in Society*, Englewood Cliffs, N.J.: Prentice-Hall, 1972.

Blume, Stuart S., *Toward a Political Sociology of Science*, New York: The Free Press, 1974.

Brown, Martin, ed., *The Social Responsibility of the Scientist*, New York: The Free Press, 1971.

Cavalieri, Liebe F., "New Strains of Life—or Death," *The New York Times Magazine*, August 22, 1976, p. 8 ff.

Coates, Joseph, "Technology Assessment at NSF," in Arnstein, ed., *Perspectives on Technology Assessment*.

Commoner, Barry, *Science and Survival*, New York: The Viking Press, 1963, 1966.

Culliton, Barbara J., "Recombinant DNA: Cambridge City Council Votes Moratorium," *Science* **193** (July 23, 1976), pp. 300–301.

Farvar, M. Taghi, and John P. Milton, eds., *The Careless Technology: Ecology and International Development*, Garden City, N.Y.: Natural History Press.

Federation of American Scientists, *Public Interest Report* **29** (4) (April 1976).

Friends of the Earth, Letter to Dr. Donald S. Frederickson, National Institutes of Health, published in U.S. Department of Health, Education and Welfare, Public Health Service, National Institutes of Health, *Recombinant DNA Research*, Vol. 1: Documents Relating to "NIH Guidelines for Research Involving Recombinant DNA Molecules," February 1975–June 1976; August 1976.

Gilman, William, *Science U.S.A.*, New York: Viking Press, 1965.

Heirich, Max, "Proliferation and Options for New Actions," Statement to University of Michigan Board of Regents, May 1976 (mimeo).

———, *The Spiral of Conflict: Berkeley 1964*, New York: Columbia University Press, 1971.

House, J. S., and W. M. Mason, "Political Alienation in America, 1952–1968," *American Sociological Review*, **40** (April 1975), pp. 123–147.

King, Jonathan, Response to David Baltimore, Forum on Recombinant DNA Research, University of Michigan, March 3, 1976.

Klaw, Spencer, *The New Brahmins: Scientific Life in America*, New York: William Morrow and Company, 1968.

Marcuse, Herbert, *One-Dimensional Man: Studies in the Ideology of Advanced Industrial Society*, Boston: Beacon Press, 1964.

Miller, A. H., "Political Issues and Trust in Government," *American Political Science Review*, **68** (September 1974), pp. 951–972.

Neidhardt, Fred, Remarks to Board of Regents, University of Michigan, May 24, 1976.

New York Public Interest Research Group, Inc., Letter to Dr. Donald S. Frederickson, Director, National Institutes of Health, *Recombinant DNA Research*, Vol. 1, p. 539.

New York Times, articles appearing in 1976 issues for June 3, June 17, July 8, August 22, August 29, September 1, September 23, October 10, October 13, October 28, and October 31, 1976.

Organization for Economic Co-operation and Development, *Reviews of National Science Policy: United States*, 1968.

Orleans, Harold, ed., *Science Policy and the University,* Washington, D.C.: The Brookings Institute, 1968.

Penick, James L., Jr., Carroll W. Pursell, Jr., Morgan B. Sherwood, and Donald C. Swain, eds., *The Politics of American Science: 1939 to the Present,* Cambridge, Mass.: The MIT Press, 1965, 1972.

Petersen, Richard, ed., *The Production of Culture,* Beverly Hills/London: Sage Publications, 1976.

Primack, Joel, and Frank von Hippel, *Advice and Dissent: Scientists in the Political Arena,* New York: Basic Books, Inc., 1974.

Rapoport, Anatol, *Fights, Games and Debates,* Ann Arbor: The University of Michigan Press, 1960.

————, *Two-Person Game Theory, The Essential Ideas,* Ann Arbor: The University of Michigan Press, 1973.

Report of Committee A, The University of Michigan, March 1976 (mimeo).

Report of Committee B, The University of Michigan, March 1976 (mimeo).

Rorvik, David M., "Do the French Have a Cure for Cancer?" *Esquire,* **48** (July 1975), p. 110 ff.

Schelling, Thomas C., *The Strategy of Conflict,* Cambridge, Mass.: The Harvard University Press, 1963.

Schwartz, Charles, "A Physicist on Professional Organization," in Martin Brown, ed., *The Social Responsibility of the Scientist.*

Sinsheimer, Robert, "Inquiry into Inquiry," *Hastings Center Report,* August 1976, p. 18.

————, Letters to Dr. Donald S. Fredrickson, Director, National Institutes of Health, dated February 5 and February 12, 1976, in *Recombinant DNA Research,* Vol. I, pp. 436–438, 443–445.

United States Department of Health, Education and Welfare, Public Health Service, National Institutes of Health, *Recombinant DNA Research,* Vol. I: *Documents Relating to "NIH Guidelines for Research Involving Recombinant DNA Molecules," February 1975–June 1976,* Washington: U.S. Government Printing Office, August 1976.

Useem, Michael, "Government Patronage of Science and Art in America," in Richard Petersen, ed., *The Production of Culture,* pp. 123–142.

Von Neumann, John, and Oskar Morgenstern, *Theory of Games and Economic Behavior,* Princeton, N.J.: Princeton University Press, 1953.

Who Decides Who Decides: Some Dilemmas and Other Hopes

Donald N. Michael

After carefully considering the situation, one part of a community seeks to undertake a scientific activity, in which it has deeply vested interests, that will put another part of the community at a problematic risk for what the proponents of the activity believe to be socially worthy reasons. Who, on what basis, decides whether the action is permissible? And who decides who decides when circumstances are sufficiently unconventional to raise questions about the procedures and legitimacy of conventional decision-making structures and personalities? Indeed, who decides that circumstances are sufficiently unconventional to merit *new* decision-making procedures and participants?

We are faced here with what seems like an infinite regress, the result of many circumstances but especially the result of the dissolution of an accepted set of values about the good and the right and the processes for establishing and maintaining them—including norms regarding who decides who decides about what. This dissolution is importantly but not exclusively a result of science and its powerful technologies and of the influence of scientists themselves, some of whose words have helped define and extend the conflicting and transforming normative issues burdening this society, this world. Because of the pervasive and ambivalent role science and technology play in our lives, an important consequence of this dissolution of shared norms and values, and of the decision-making procedures that represent and reinforce them, has been to focus a variety of disparate concerns on the conduct and consequences of scientific research.

In this chapter I shall use the question of who should decide whether to undertake recombinant DNA research in a publicly supported university to illuminate some aspects of the increasingly pressing problem facing this society: what persons and procedures should determine whether to undertake publicly supported esoteric science that is potentially hazardous? The recombinant DNA issue is a prototype of things to come, especially in research conducted in the biological and possibly in the social sciences. The University of Michigan, as a publicly supported institution, has an obligation to serve the public interest: it is a prime example of that large variety of organizations whose very existence depends on direct or indirect support from funds produced through taxes. Hopefully, we can understand better the nature of the general problem we face by relating its abstract aspects to a real-life example. This I shall try to do by alternating between abstract exploration of the problem and attention to aspects of the University of Michigan experience that give substance to it (1).

A conventional response to the questions raised at the beginning of this chapter would be that the decision, being based on esoteric knowledge and intentions and being undertaken at least in part for the public good, should be decided by scientists involved and by the relevant administrators (in this case, those of the University), probably with occasions provided for comment and suggestion from the community at large. But when all is said and done, the decision should be made by the conventional decision-making structure which, it is presumed, has the best interests of all parties at heart. This is especially so in cases of scientific research because disinterested good will can be expected to prevail and new knowledge can be expected to advance human welfare. Furthermore, all that could reasonably be done would be to minimize the risks, but risk is part of life and part of the cost of gaining new knowledge from which humankind ultimately will benefit.

An alternative response, the one that undergirds the questions that give cogency to the chapter, and one that seems to be subscribed to by growing numbers of lay citizens, would argue that: whatever level of risk is involved, those who might be victims should it become fact should have a formal part in deciding whether and under what circumstances to accept the risks.

The argument continues: given the nature of recombinant DNA research, in principle, all citizens of the world should have a part, since the consequences of accidental leaks of material from laboratories conducting that research could well be worldwide in scope. At a minimum, according to this argument, the community immediately surrounding the research environment (in this case Ann Arbor, Michigan) should be directly involved in deciding whether the University should undertake such research, certainly when the research is supported by public funds and conducted in a publicly funded organization.

To better appreciate the argument for this position and the dilemmas and difficulties that arise in attempting to transform the general logical or ethical case into operational terms it is useful to be more specific about the nature of the risks themselves in the recombinant DNA case.

Two types of risks have been delineated: process risks and product risks. Process risks pertain to the consequences of *accidental* dissemination of research substances into the environment outside the laboratory. Product risks pertain to the consequences of *deliberate* dissemination into the environment beyond the laboratory of the products finally produced from the research effort. The arguments for and against recombinant DNA research revolve around both of these risks. In the case of product risks the arguments have to do with whether the hoped-for but undemonstrated benefits of such research will outweigh the feared and unknown costs. The costs are unknown in part because of our ignorance concerning interactions of these new, "chimeric," life forms with natural life forms and in part because such research carries the potential for irreversibly changing human life itself (2). This enormously complex issue generates problems that go far beyond the costs and benefits of more conventional technologies, but it is not the topic of this chapter except to observe that beliefs about the long-run balance of product costs and benefits probably influence feelings about the degree to which process risks should be accepted in the short run.

Arguments about process risks have to do with how perfectly the laboratories and their biological contents can, in practice, be insulated from the community that surrounds them and whether, if such substances were leaked into the larger environment, they could be expected to have deleterious impact. The esoteric issues of biology and of probability calculations involved in such an assessment are not the topic of this chapter, either, beyond the observation that it is generally conceded that (i) extant probability calculations assume ideal performances by all researchers and other professional and nonprofessional staff associated with the laboratories, whereas such perfection cannot be expected to prevail given human fallibility and the unintentional inevitability that, sooner or later, some emotionally disturbed person(s) will be involved in these activities; (ii) we really don't know what the consequences would be of accidental leakage into the environment because we don't know what products would be leaked and we don't know what knowledge and ignorance about the environment would be involved in coping with that leak (3); and (iii) as can now be calculated, if one disregards human fallibility, it looks as if the odds of accidental dissemination from *a* specific installation are very, very small. In sum, there are legitimate questions about just how small those odds would be in "real life" and there are very serious unanswered questions about the consequences of those low-probability events if they should occur. The consequences could be enormous, and herein lies a major area of concern for both scientists and nonscientists alike.

It remains to be observed that even if the likelihood of leakage is small, history amply evidences that "rare" accidents do, in fact, happen (4). For all these reasons some in the University and some in the larger Ann Arbor community saw a compelling need to face the question of whether or not to undertake the research.

Given the problematic nature of the risks associated with esoteric research, very difficult operational issues will attend efforts to evolve and implement new decision-making procedures that include the larger community in decisions about undertaking the research. To further appreciate the nature of the task it will be useful to review some of the sources of discontent with and repudiation of conventional decision-making procedures, based on expertise and duly constituted authority, and of the conventional overriding priority assigned to freedom of inquiry. The four sources examined here are not the only ones, but they illustrate especially well the extraordinariness and the complexity of the decision-making task with which scientific endeavors such as recombinant DNA research now burden our changing society.

In the first place there is growing subscription to the ideological and psychological virtues of direct or at least less indirect citizen participation in decision making. The ideological argument asserts that such participation is a right of any person or group that might suffer the consequences of unilateral decisions made by a formal organization. Psychologically, it is argued that decisions can be improved and consensus and a sense of community enhanced if the recipient publics participate in the formulation of the questions and the design of solutions to them. In this way both the problem and the solution become "theirs": they understand the tasks and problems involved in defining and implementing the decision and thereby they experience a deeper sense of responsibility toward and commitment to the decision. In addition, and of central importance here, participation provides the occasion for recognizing, for discovering, ethical issues. By itself participation does not resolve them. But it does provide the occasion for creating, for *learning* new ethical norms. And the situation we face is in every respect one we shall have to learn about and learn what to do about—as I shall emphasize throughout this chapter.

A second source of pressure for new decision-making practices is a growing challenge to the autonomy generally accorded to scientific research and to the very high priority assigned to it in this society. The challenge revolves especially but not exclusively around questions of autonomy and priority when the research is supported by public funds. Significantly, these challenges come not only from lay opinion leaders but also from vocal and accomplished scientists. Increasingly, questions are asked about *what* research the public should pay for (i.e., what research contributes to the public weal) and under what circumstances scientific research, its methods, and resultant technologies are appropriate for seeking answers to or dealing with the problems and possibilities of the human condition. There seems to be a substantial, perhaps growing, antitechnology undercurrent that, while by no means exclusively correlated with emphasis on citizen participation, may often be found in close ideological association (5).

A third challenge to the conventional decision-making processes expresses itself in pervasive questioning of the legitimacy of existing organizations—that is, the validity of their entitlement to make decisions affecting those outside the organization and of the processes by which they do so. The questioning includes reexamination of conventional definitions of what constitutes competency to make such decisions. Throughout society there is much distrust of large organizations. Since scientists are mainly associated with large organizations, this contributes to rejection of the image of scientists as motivated exclusively by a disinterested devotion to truth. Rejection of this image is strengthened by growing recognition that intense competition among scientists, plus heavy dependency on public funding of scientific research by the institutions supporting the scientists, result in deeply vested mutual interests—interests reflected in decisions that often are not the same as those which would seem right to the publics who pay the bills and sustain the risks.

The fourth factor, which exacerbates all the others and especially challenges conventional authority and decision-making processes, is a widening recognition that science is not ethically neutral and, thereby, that decisions regarding science and technology cannot be made exclusively in terms of scientific and technical arguments even though these must be critical contributions to the decisions (6). Inevitably, the scientific and technical facts and data are incomplete, especially in new areas (such as recombinant DNA research). What is more, the available facts and data result from earlier decisions about what merited most attention and what could be learned with available time and money. As such, the available facts and data are expressions of the value judgments (or biases) of those who collected and those who funded the collection of the data. Such judgments necessarily go beyond purely logical, technical, issues into realms of political feasibility, esthetic norms, rightness, and goodness.

All these considerations were part of the local and national dialogue that informed the University of Michigan experience. My informal canvassing of the motives and expectations of those more or less directly involved indicated that the Forums, the most explicit and dramatic invitation for the University and Ann Arbor community engagement, were seen variously as mere ritual, or as building a new consciousness about the relationship of publicly supported institutional research and the surrounding community, or as a "laboratory" for developing new means for University-community decision making. Only a minority argued that Ann Arbor citizens should have a formal part in decision making about whether to undertake moderate-risk recombinant DNA research; the "duly constituted authority" position was the dominant one within the University. However, the University administration and the scientist proponents for the research recognized that an informed community was needed and that the University might benefit from community advice: the administration had established Committee B much earlier, and the administration and the Senate Assembly financially supported the Forums. And Committee B's report urged community representation in the governance of recombinant DNA research: citizen members on the research monitoring committee and on a proposed "oversight committee." Membership on these follow-on committees, rather than a part in deciding whether there should be research at all, was, then, the University's not unconventional response to this unconventional problem. This expressed, surely, the conventional reluctance of those with the power to make decisions to diffuse their prerogative. But another consideration preoccupied many and will continue to do so as the decision challenge reoccurs: protection of the freedom of inquiry.

It is a basic belief of most University faculty members and, indeed, of educated people everywhere in the West that freedom of inquiry must not be constrained in any arbitrary manner, especially not by persons outside the community of peers associated with the inquiry. It is, however, a belief that is increasingly challenged, both by some who well understand its centrality for conventional definitions of an open society and by others more preoccupied with other priorities (7). Research in the biological and social sciences has added new intensity to the challenge.

From this perspective, if the University were to forfeit, through citizen involvement, its exclusive right (within NIH regulations) to determine whether recombinant DNA research should be undertaken, it would very likely be establishing a precedent with regard to freedom of inquiry not only in this area but in any other area of the natural or social sciences where members of the community could argue that they were being put at physical or emotional risk by the research process itself or its possible products. Given changes in attitudes toward science, participation, and decision making, reviewed earlier, such a precedent would profoundly disrupt the elaborate and subtle mechanisms that motivate and guide science and systematic inquiry in general. Consequences would be as unpredictable and possibly as societally catastrophic as those feared from the DNA research itself. However, some would argue (myself included) that the very fact of growing challenges to the ethic of freedom of inquiry and to its maintenance through "duly constituted authority" makes it all the more necessary to begin now to discover new ways that might reconcile the demands for participation by those putatively at risk with demands for protection of freedom of inquiry.

Both demands carry very heavy costs as well as very great benefits: it was the recognition of these and the need, therefore, for a new overarching ethic that endowed the recombinant DNA research decision-making issue with both symbolic potency and unique potential for initiating learning about what such an ethic might be and how it might express itself in decision making about such activities. It is going to take time and much experience to learn what values and techniques work, and the hour is already late. What problems then, need new solutions so that we may agree and act upon "who decides who decides"?

The first question we could ask is: "What is the appropriate geographic and temporal scale from which to draw the decision makers"? With chimeric biological materials it is impossible to anticipate how widespread will be the consequences for natural life forms (8). Therefore the appropriate decision-maker pool would seem to include the whole world as well as future generations, since everyone, especially future generations, may be the deliberate or inadvertent beneficiaries and/or casualties of this research. But there is no such world-level decision-making capability; the initial examination of the risks in recombinant DNA research, undertaken by involved scientists during a self-imposed moratorium, is as close to world-scale participation as we've gotten (9).

Lacking world-scale, or even a world-regional-scale, decision-making capability, we are thrown back on the nation as more appropriate than the immediate environs around the research laboratory for making decisions that have such profound consequences over space and time. The NIH Guideline deliberations were an exceptional and on the whole admirable experiment in this direction, though they lacked sophisticated studies delineating the long-term social cost and benefits of the research, in part because we know too little to do very much in this direction. Moreover, the question of who would be entitled to participate in decisions about process risk exposure in the proximate area where the research would be done was left unexamined. Instead the main emphasis was on how to balance the need to minimize risks for those outside the laboratory against the risk that if the constraints were too stringent some scientists would disregard them (10). But the fundamental flaw in the NIH approach was that it reinforced the usual mode of operation wherein geographically *separate* institutions compete for funds and for the prestige won through successful research. This mode inevitably puts a premium on getting there "first with the most," and it focuses concern at the local level over whether to incur the associated risks (11). (Even though a local accident might result in worldwide consequences, the "acceptability" of the risk probably depends on one's perceived geographic proximity to the source.) At the same time and place, those seeking to do the research are acutely motivated by recognition that they are in competition with scientists in other locales who may not be delayed by local demands for community involvement. Therefore, localization of risk, on the one hand, and pressures to get on with the research, on the other, can be expected to be a likely setting in which new forms of decision making will need to be created and implemented. That context is assumed in what follows.

Under these circumstances who, then, should be involved in decision making? How are they to be involved? And how are they to become involved? Criteria for choosing revolve around questions of (i) the "right" to involvement by virtue of some special capabilities or competences; and (ii) expediency—the consequences of recognizing or ignoring claims. Here, one's role and, engendered and sustained by that role, one's expectancies about self and others critically influence the preferences and prejudices one brings to this task of choosing who should participate.

Probably the first criterion applied would be entitlement to participation by virtue of competence. This criterion is clear enough when the issue is technical or scientific competence. However, what would constitute "competent" community participation? Usually it is assumed that whoever represents the community should be competent in the esoteric subject matter itself. Others, however, fear that persons from the community, competent in the scientific and technical issues, are likely to be scientists or engineers who, thereby, are likely to weigh their judgments by the same criteria as the scientists more directly involved. From this perspective there are other more relevant forms of competence, such as the ability to sense and express the fears or hopes and the confusions of lay persons, all of which are held to be data equally as cogent for decision making as technical facts. Yet it is necessary that community representatives (or otherwise participatory community members) understand the scientific-technical issues enough to appreciate the technical bases for the arguments for and against the research. How to provide both kinds of competence is a central and unresolved problem, though the growing capability of consumer information and action groups suggests the challenge is not insuperable (12).

Another kind of competence belongs to those with formal organizational responsibility and associated skills to contribute to decisions and implement them. (Such legal or operational competences are represented at the University of Michigan by the Regents, the laboratory directors, the researchers, certain Deans, the Vice President for Research, Committee B, and so on.) What are the corresponding competences and responsibilities in the community? Members of the Community Council? The Mayor? The leaders of the various socially active religious groups? Unofficial but influential groups like the Ann Arbor Citizens Council or the League of Women Voters? Different groups will themselves have differing views as to what constitutes competence and appropriate responsibility for participation.

Finally, there are the competences needed to represent future generations. How are these to be defined and who is to judge?

A second general category of claims on participation in decision making pertains to "turf" protection. Whether or not the research is done will affect the status of persons, the dominance disciplines, the comparative power of administrators, research directors, deans, and so on. For example, in the University of Michigan case, in many eyes important contributions to the University's prestige and, therefore, to its future overall research budget (its "turf" vis-à-vis other universities) would depend on vigorous involvement in recombinant DNA research. (Others argued that the University would gain prestige by leading the way in rejecting the research.) If the community were to be involved in the decisions, analogous concerns with "turf" protection would arise there too.

Related claims on participation would be in terms of risk to personal reputation and income (including consulting fees) if the research were not done and from physical exposure to these synthetic biological entities if it were done. And if accidental leaks from the laboratory cause damage, who would be at risk financially if the University were sued by all those allegedly harmed?

All claims on the right to participation will also be influenced by the focus accorded to the decisions to be made. That is, what is the purpose of the decision? What is its scope? How inclusive is it to be? Is the decision chiefly scientific, political, or ethical, or is it an operational matter? Is it to be advisory or binding? Note, too, that at least the initial focus for decision making itself depends on the perspectives and interests of those who decide who is to decide.

These examples of claims to entitlement in the decision-making process are not exhaustive: their purpose is to emphasize that claims will be a function of the role of those who put them forward and that different values and norms will be involved, including many that extend well beyond issues of scientific and technical competence.

I turn now to the question of what steps must be carried through for there to be effective community involvement in the decision process. What must happen preceding the decision making in order that it can be effectuated? The preliminary steps are conventionally accomplished all the time and for this reason are, for the most part, unremarked on; the procedures of due process, the functions of duly constituted authority, and the linkages out into community membership are such that these steps get done more or less routinely. In the present situation, however, claims on participation are problematic, raising profound questions for the conduct of free inquiry; and those outside the conventional decision-making network who claim a right to participate do so on the basis of values and norms not necessarily compatible with those characterizing the conventional system. So, the almost unnoticed steps, the comparatively routine steps, taken in a conventional decision-making system now become serious questions of procedures and tactics. Getting from "here to there" will require new social inventions and probably new norms to legitimate those inventions.

These, then, seem to be steps necessary to set in motion a process prerequisite for inventing new decision processes.

Step 1: The community must organize itself to make its claim for participation. Not only must community interest be generated and focused, but the question must be answered, "Who constitutes 'community'?" Crudely, how many people in the community, or which groups in the community, need to be engaged so that (a) they can claim successfully to represent the community interest in whether research should be undertaken, and (b) the organization doing the research can accept them as representing that interest? Putting it another way, is it possible to deal with this situation through some institutionalized community process rather than the makeshift approaches that would tend to be used if the community were to hostilely confront the institution (13)? Can we reach out tentatively with the intention of learning how to do these things deliberately and in good spirits instead of waiting until anger, confusion, and multiple extraneous interests collude to force a messy and uninformative confrontation?

Step 2: Assume that, one way or another, the community has

created some kind of representative entity to engage the organization in discussion of community participation in the decision-making process. Who is to be approached for this purpose? That is, how does the community inform the organization of its intention to participate? Clearly, it must reach a person or persons who take seriously the community interest, at least as a matter of public relations and hopefully out of recognition of the institution's ethical obligation to the community. What is more, the person or persons approached must have enough "clout" within the organization to converge and hold its members' attention to the question of community participation in decisions that have been exclusively the prerogative of the organization.

At the University of Michigan, individuals, ad hoc groups (especially the one formed around Professor Susan Wright's memoranda to Committee B), and the University Values Program, all espousing the need to face the question of community participation, but not representing Ann Arbor as such, sought out members of the Board of Regents, the Vice President for Research, members of Committees B and A, the Senate Assembly, and the Senate Advisory Committee on University Affairs (SACUA). The result was University-wide moral and financial support for the Forums (including invitations to outside experts to participate). The Forums and further conversations and memoranda also contributed to the extraordinary attention the Regents devoted to the issue.

These activities were influenced by some *in* the University who put much effort into alerting others in and out of the University to the need for a public airing (14). Their varying interpretations of the situation converged in a belief that these activities offered a real potential for new processes of University-community interaction. But how all this might have gone if it had been evident that the Ann Arbor community was going to insist on an active role in the initial decision is problematic. That never happened, nor was it expected to when various groups in the University agreed to support the Forums. However, it may well be that some justified complacency allowed this institution to be more innovative than it would have been if the community had been more assertive. Thus, in the absence of crisis many in the University entered the issue in ways that yielded learning that might be useful in less tranquil circumstances, should they eventuate.

Step 3: An answer is required to the question: how does the community participate in the organization's procedures by which it decides whether and how the community is to be involved in the later, risk-relevant decision process? Intensification of a recombinant DNA-type issue could make this a very real question indeed. The very social forces that produce the demand for a broader decision base also produce the demand that decisions about the "if" and "when" of that wider decision base themselves involve the participation or concurrence of the potentially wider base. The end of this seemingly infinite regress would appear to lie within the organization, since it is being petitioned by outsiders and since it has the organization and traditions for making decisions about the extent to which it is willing to alter its decision process. These decisions depend on *beliefs* about who could act, in what kind of decision, conducted according to what procedures. They depend, too, on the *process* by which the organization would arrive at a decision that participation is permissible. And this process depends on the operative definitions of competence and "turf" described earlier.

However, not all the options lie within the organization. The community could seek legal redress, in which case the decision about who has a right to be involved might be made outside the organization. In the Ann Arbor-University situation the Regents played a more active role than usual in deliberating about the proposed moderate-risk recombinant DNA research. On the basis of their legal obligations to protect the general interests of the people of the State vis-à-vis those of the University, the Regents might have sought to involve persons from Ann Arbor in the decision. Under circumstances of greater urgency, even though the University might steadfastly protest interference with freedom of inquiry, it is conceivable, even though unlikely, that the Regents or judicial authority might require the University to include Ann Arbor citizens in the decision making and might specify the criteria for selection as well as the decision-making procedures.

Collaboration and inventiveness in seeking answers in Step 3 depend on the extent to which trust can be built up between the interested parties and, through trust, appropriate norms evolved. Denial of organizational legitimacy and insistence on fuller participation are in part the result of acute distrust of the conventional decision-making processes in organizations. Trust and shared norms probably can be reestablished only under circumstances that encourage and reward experiments with—and acknowledgment of the need for—new decision-making methods and norms explicitly designed to make decisions about who is to make risk-relevant decisions. The shared experience of *learning together* how to do these things seems prerequisite for creating decision-process norms commensurate with the enormity of decisions affecting the impact of esoteric and powerful science on an increasingly complex and vulnerable world.

Step 4: Assuming that a decision is made to involve citizens in subsequent decisions, we next confront the question: how can the decision-making process be operated to give the citizen members a truly potent role in influencing outcomes? It has taken many years to invent and refine decision-making processes in more conventional areas of an open society, and the same development can be expected here. There is no reason to expect that new procedures will initially work well, even if it were clear what "well" means—which it certainly isn't. However, whatever else is to be sought, a primary condition to be met is insuring that learning will occur that leads to improvements. In Roland Warren's words, "We need to find ways of channeling change which will assure that you and I will reach the optimum agreement possible, but that our remaining disagreement will neither immobilize us nor result in our destroying each other and those around us" (15).

How are decisions to be arrived at? By consensus? By vote? By referendum? By what proportion? How is information to be presented and evaluated? By advocacy? By collaborative synthesis? Such questions bear not only on decision-making procedures but on the proportional composition of decision-making groups. Whether decision-making entities in fact set policy and operations or whether they merely make recommendations to other entities that make the decisions would be additional considerations. Anticipation of how these considerations will be dealt with will influence decisions and actions associated with Steps 1 through 3.

At this stage various proposals come into play for combining technical and social considerations in decision-making for public policy— such proposals as the science court (16), judgment analysis (17), and judicial evaluation (18). Intriguing and hopeful—and controversial—as these are, they do not of themselves diminish the new and difficult tasks of getting to the stage where they can be tried out: *their use implicitly assumes that decisions have already been made about who will make the decisions facilitated by these procedures.* However, explicit intentions to experiment with such procedures might somewhat simplify questions regarding appropriate competencies and the focus of the decision-making task. At least such intentions could color expectations about what is to be done and how, and this, in turn, could contribute to the building of trust and shared norms.

Community members may well find themselves in a minority status in the decision-making entity if, owing to their role, they take a different perspective from those who represent organizational and scientific interests. Sometimes other members of the deciding entity may find themselves in the minority. Either way, but especially because the community members may have different interests, minority positions must have access to special resources in order to (i) make the best case they can *as they develop* their position(s), and (ii) disseminate their views to potentially supportive constituencies in and outside the organization proposing the research. These resources will be especially necessary if a "minority" position is being espoused regarding an upcoming decision choice. By the very nature of the issues, perspectives that lead to rejecting or questioning the conventional wisdom of the "experts" about the costs and benefits of proposed research or the appropriate context for evaluating them are likely to be minority positions. Yet in areas as novel and momentous as those involving powerful new scientific knowledge and technique the minority position may well be precisely the one that most merits intensive and early amplification and attention if wise decisions are to be made later.

Minority positions, then, must be able to command:

1. Access to sufficient information and to skilled resources to develop that information into the best case they can make. If the position is held by community representatives, their technical understanding may need augmentation, and they should have the funds necessary for access to supplemental information sources. Funds may be required for staffing of alternative technology assessments and/or social or environmental impact studies.
2. Sufficient "presence." A "devil's advocate" will not be enough. S/he is good for the conscience but usually insufficient for effective influence (19).
3. Sufficient resources and public access to disseminate their position broadly so that others who might find it attractive will learn about it. Typically, minority positions lack both dissemination capabilities and legitimacy. Therefore, part of the task of a decision-making entity ultimately responsible to the public interest should be to ensure that its minority positions are supplied with both.

Conclusion

One fact is clear in all the swirling ambiguity of positions and counter-positions about the state of society and what needs to be done about it: we are too ignorant of our own condition and its potentialities and problems to engineer our way into the future either materially or socially—we cannot get there the way we got to the moon. Instead, we must learn to create a new set of norms, values, and supporting behaviors that will allow us to *continue* to be a *learning society*, learning where we think we are, where we think we want to go, whether we are getting there, and whether we still want to. Rapidly changing circumstances permit no other mode of rational conduct.

Making decisions in areas of changing values, risks, and ambiguities requires profound—perhaps radical—changes in the norms by which decision-making entities in research-oriented organizations operate and in the ends for which they operate. In essence, these entities have to learn how to design themselves as effective learning systems, aimed at improving the effectiveness of community and organization participation in decisions about esoteric scientific activities that involve potential community risks as well as potential benefits.

More specifically, we may list three requirements for decision-making entities, seeking to learn their way through newly emerging issues wherein the public interest seems to confront freedom of inquiry:

1. A shared learning relationship instead of an adversarial stance. A zero-sum approach, an assumption that there is one right answer and that only one side can win, can lead only to disaster.
2. An openness to continuous reexamination of the norms and values by and for which they operate—and especially of the means for estimating and evaluating social costs and benefits. Alternate scenarios will need to be explicated, so that the community and the research organization will have the broadest possible perspective for decision making in these ambiguous and ambivalent areas.
3. The devotion of effort, money, and study to learning *how* to learn in these situations, how to integrate persons, ideas, and actions based on new normative modes. Whatever the decision-making entities decide and however they do it, it will unavoidably occur by way of experiment, by research and development on the norms and process of decision making.

Surely this sounds utopian, yet, as Bertrand Russell observed years ago, a utopian perspective is the only practical one in the kind of world exemplified by the recombinant DNA issue. We must make the same kind of intense commitment of imagination, experiment, and time to learning how to conduct decision-making processes involving potentially risky esoteric research as we do to learning about natural processes in the physical universe. If we do, then we can hope that, even though a particular mode of participation or outcome may not satisfy everyone, the norms developed in *arriving* at it will be rewarding enough to provide a sustaining sense of community while other processes evolve.

For some of us, the University of Michigan experience was a beginning of the kind of learning that could move toward realization of that hope.

NOTES*

1. It is right that at the outset I give my personal position on the topic of this chapter. In November of 1975 Professor Susan Wright shared with me and a few others a memorandum she was addressing to Committee B requesting more attention to certain aspects of the risks associated with recombinant DNA research. (Until then I was unaware of Committee B or of the question of recombinant DNA research at the University of Michigan.) I immediately became involved in efforts to bring the community into the picture through participation in a small ad hoc group inspired by Professor Wright's concerns; as a member of the group guiding the University Values Program, and, later, as a member of the committee designated to plan the Forums.

   I became involved because of my concern with the issue *per se* and because the recombinant DNA research issue was an invaluable occasion for the University and the community to begin to learn how to deal with such issues. My personal, cautious, inclination is toward community involvement in the basic decisions. Cautious, because I also acknowledge the dilemmas and difficulties described in this chapter.

   It remains for me to acknowledge that we who ponder on and seek to act regarding the place of science in society are caught in a maze of distorting mirrors that reflect the currents and conflicts in our culture and its many subcultures and, therefore, in ourselves. We too are mirrors caught up in the maze and contributing to the maze. No matter how much we act with good will and seek to be unbiased we are, ineluctably, mirrors.

2. See Leon R. Kass. "The new biology: What price relieving man's estate?" *Science*, November 19, 1971, **174,** 779–788.

3. "Environmental science, today, is unable to match the needs of society for definitive information, predictive capability, and the analysis of environmental systems as systems. Because existing data and current theoretical models are inadequate, environmental science remains unable in virtually all areas of application to offer more than qualitative interpretations or suggestions of environmental change that may occur in response to specific actions." National Science Board/National Science Foundation. *Environmental Science*, 1971, p. viii.

4. Recall that two large commercial aircraft collided over the Grand Canyon. An Air Force bomber hit the top floors of the Empire State Building. The ocean liner, *Andrea Doria,* sank after a collision with another ocean liner in clear, calm weather in mid-ocean. The oil tanker, *Torrey Canyon,* went aground on well-known shoals, spilling oil all over the southeast English coast. Three astronauts burned to death in a routine test on the launching pad. The unsinkable *Titanic* sank on its maiden voyage.

   Specifically with regard to recombinant DNA research see Nicholas Wade. "Dicing with nature: Three narrow escapes." *Science*, January 28, 1977, **195,** 378.

*A few particularly apposite references have been included in these notes that were published after the completion of the manuscript (December 1976) and before the book went to press.

5. See Todd R. La Porte and Daniel Metlay. "Technology observed: Attitudes of a wary public." *Science,* April 11, 1975, **188,** 121–127.
6. See William Bevan. "The sound of the wind that's blowing." *American Psychologist,* July 1976, **31** (7), 481–491. Also see Clifford Grobstein, "Recombinant DNA research: Beyond the NIH guidelines." *Science,* December 10, 1976, **194,** 1133–1135.
7. See the sophisticated statements pro and con limiting freedom of inquiry found in Hans Jonas. "Freedom of scientific inquiry and the public interest," pp. 15–19; and R. L. Sinsheimer and G. Piel, "Inquiring into inquiry: Two opposing views," 18–19. *The Hastings Center Report:* Institute of Society, Ethics and the Life Sciences, August 1976, **6**(4). For other straws in the wind see B. Culliton. "Kennedy hearings: Year long probe of biomedical research begins." *Science,* July 2, 1976, **193,** 32–36; and T. Seay. "Stoned in Peoria." *APA Monitor,* June 1976, 11–12. The latter article is about Congress' refusal to fund research already approved by the National Institute on Drug Abuse.
8. Notes 2 and 3 are also relevant here.
9. This extraordinary and laudable social invention itself evidences the changing norms in science with regard to social responsibility. It certainly merits systematic study—which it hasn't gotten—for the deeper understanding it could provide about the social and psychological conflicts and clarities unfolding in today's science community.
10. That there was acknowledged concern about the possibility of arrogant disregard of "overly stringent" guidelines evidences another aspect of the normative and ethical disarray of this society—the same society that engendered the voluntary moratorium on recombinant DNA research. Susan Wright discusses some normative questions regarding how the NIH guidelines were decided and have been implemented in her "Doubts over genetic engineering controls." *New Scientist,* December 2, 1976, **72,** 520–521.
11. Apparently the anticipated risk was perceived as too small to justify the complexities and delays associated with serious examination of the possibility of restricting the chances of accidents to regional or national laboratories analogous to Brookhaven, Argonne, NIH inhouse research itself, or the great multinational research installation, CERN. Such facilities, if located well away from dense human habitat, would have eliminated the local issue of who is entitled to participate in decisions.

    Some of us—especially Professor Max Heirich—urged the Regents of the University of Michigan to seek to join with their counterparts at other involved universities to seek funds from the federal government for a jointly shared, isolated laboratory. Such an effort by the Regents would have been unprecedented and time consuming. But some of us urged that circumstances merited such a social invention from the group bridging the University to the larger community—the Regents being elected by the public at large. The Regents did not act on this recommendation.
12. Growing numbers of consumerist organizations are able to provide such information and knowledgable spokespersons. Scientists and engineers are prominent resources in most of these groups. Examples are Science for the People, Federation of American Scientists, Scientists' Institute for Public

Information, various offspring of Nader's activities, and ad hoc groups such as those arguing against nuclear reactors.

13. A dilemma: How to make the community aware that there is a risk and that the consequences may be grave indeed without inflating the issue to panic proportions. Panic would obviate deliberate and enlightened decision making and also destroy chances for emergence of an attitude that would make the eventual decision at least tolerable to most if not all parties. While not precisely this situation, the response by Cambridge, Massachusetts, to Harvard's research intentions is most informative. See Barbara Culliton. "Recombinant DNA: Cambridge City Council votes moratorium." *Science*, July 23, 1976, **193**, 300–301.

    A related difficulty merits comment. If the research organization is a university, the chances are (as at the University of Michigan and at Harvard) that some community interest will be stimulated by University personnel. While it needn't work this way, it is likely that signals of concern, especially the early ones, will be carried from the University to the community by University people. If community interest grows and if that interest is antagonistic to the conduct of the proposed research, the risk of polarization within the University itself will also grow. Polarization would destroy the openness necessary among University members if there is to be social learning and invention of the high order required to cope with such problems. See Nicholas Wade. "Gene-splicing: Critics of research get more brickbats than bouquets." *Science*, February 4, 1977, **195**, 466–469.

14. Committee B had been open to input from the community, but until the aforementioned groups became active some three months before the Forums, chiefly as a result of Professor Susan Wright's memoranda to Committee B, there had been little public or University-wide attention to the matter.

15. Roland Warren. *Love, truth, and social change.* Chicago: Rand McNally, 1971, p. 298.

16. See Task Force of the Presidential Advisory Group on Anticipated Advances in Science and Technology. "The science court experiment: An interim report." *Science*, August 20, 1976, **193**, 653–656; and P. M. Boffey. "Science court: High officials back test of controversial concept." *Science*, October 8, 1976, **194**, 167–169. For a knowledgeable and incisive criticism of the science court concept, see Leon Lipson. "Technical issues and the adversary process." *Science*, **194**, 890.

17. See R. L. Wolf, J. Potter, and B. Baxter. "The judicial approach to educational evaluation." April 1976. A transcript of an instructional tape on the Judicial Evaluation Model. The tape was presented at the Annual Meeting of the American Educational Research Association, San Francisco, April 1976. Information about this tape can be obtained from Dr. Robert L. Wolf, Education 325, Indiana University, Bloomington, Ind. 47401.

18. See K. R. Hammond and L. Adelman. "Science, values and human judgment." *Science*, October 22, 1976, **194**, 389–396.

19. For a most perceptive critique of the "devil's advocate" role and its limits see A. Hirschman. *Exit, voice and loyalty.* Cambridge, Mass.: Harvard University Press, 1970.

The Press As Guilty Bystander

Charles R. Eisendrath

I. DNA AND THE FIRST AMENDMENT

On June 11, 1807, shortly before retiring from the presidency, the nation's premier advocate and defender of the free press wrote an extraordinary letter to John Norvell, 17, of Danville, Kentucky. Young Norvell had asked the President about newspapering as a career. He received an eloquent reply. "I really look with commiseration," wrote Jefferson, "over the great body of my fellow citizens who, reading newspapers, live and die in the belief that they have known something of what has been passing in the world in their time" (1). It was the darker side of a man who on other occasions considered the press to be "the best instrument for enlightening the mind of man" (2), and who had grafted that conviction to the young constitutional tissue in the form of the First Amendment.

The theory behind guaranteed free speech is as elegantly simple as Jefferson's prose. A democratic society can scarcely be expected to opt wisely on its alternatives if it lacks information about the choices. This was the point that Jefferson successfully argued to the Federalists and other skeptics (3) of what was then the newly dubbed "Fourth Estate." Social learning is "of the essence" of liberty. It must overrule other considerations, such as the probability that the press will present information in ways that displease most of the people some of the time, and many of them, most.

What happens, however, if the press errs not in flawed conduct of its duties, but rather in omission? What if it decides to be free not only in the constitutional sense, but also to be sprung from any obligation to raise up before the public major entries for debate? Worse, what if the press fails to inform *itself*, and thus to recognize questions of fundamental importance to it and to every person and institution within range of its circulation or broadcast? These are the issues forced by the press's handling of DNA recombinant research at the University of Michigan.

In Jefferson's day, the only practicable tyranny over personal affairs was political, and it was political censorship against which he fought. Today, politics remains very much what it has always been—perhaps too much so. But something new has been added. Science and technology, fields as different from politics as is $E = mc^2$ from "I Like Ike," have participated in a new co-conspiracy for information blackout. It is inadvertent and often mindless, as are so many of the century's horrors. Science contributes by insistence on expressing its most important learning in absurd gobbledygook while simultaneously shouting that society must understand it. The press plays along by pretending to a ridiculous degree that science either does not exist, or isn't "news." The co-conspiracy has brought society censorship by ignorance.

Jefferson, particularly Jefferson the scientist, would have commiserated with his 4.8 million fellow countrymen living within a 50-mile radius of the Medical Science II building on the Ann Arbor campus of the University of Michigan in 1975. The world of molecular genetics had been in a state of uproar for nearly four years over whether DNA breakthroughs would reveal the "secret of life" or create "biological A-bombs," either of which, surely, would be newsworthy. In February, 140 specialists from 16 countries gathered at Asilomar, California. Among other things, they discussed whether and under what conditions they *dared* proceed with something so potentially dangerous (4). They produced news—the first voluntary research ban on safety grounds in the history of modern science—and even seemed to tailor it to the needs of Southeast Michigan journalists. University of Michigan professors were among those who pioneered DNA techniques and had joined the Asilomar discussions, neatly combining handy news sources with a hot local "angle." The professors, after all, might make Ann Arbor the site of a laboratory equivalent to the Mars landing, or of some ghastly biological bomb's accidental detonation.

Yet for what may prove to have been a period of critical importance in their lives, the people of Southeast Michigan learned nothing (5) from their nine daily newspapers, 74 weeklies (6), 45 radio stations, and eight television channels (7) about singular changes in things "passing in the world in their time." Not a single article or program concerning DNA work near their homes appeared for a full nine months after Asilomar. In the closing weeks of 1975, well after key campus decisions and opinions had been set, the *Ann Arbor News* carried two routine pieces on inside pages (8), and the University-sponsored WUOM-FM broadcast a brief announcement (9). Southeast Michigan's dominant publications printed nothing until the following year, nor did its wire services carry a line (10). With the single exception of WTVS, a Public Broadcasting Service affiliate, which aired a 30-minute documentary November 3, 1976, area television channels ignored the story. As of April 1977, the date of present writing, many broadcasters simply were unaware of it. Douglas McKnight, assistant editor of WXYZ-TV (an American Broadcasting Company channel) said in late March, "I personally have no knowledge of the research going on at the University of Michigan" (11).

Justification for the First Amendment rests on the press's ventilation of socially significant issues in time for public debate. The "guarantee" of free speech imposes the cost of providing intellectual exchange, which characteristically springs from "news events" or from the forces producing them. Adhering to the practice of reporting only "events," such as the recommendation by a faculty committee to proceed with DNA experiments, often implies a specious distinction, since coverage itself may lift an otherwise unknown happening to "public event" status. The First Amendment demands more—that the press exercise a discovery function.

Every four years, journalism meets this challenge more-or-less well by providing adequate knowledge of presidential candidates and their programs. In the Watergate scandals, American investigative reporting distinguished itself as never before. But elections and executive malfeasance grow in the well-furrowed fields of national politics, where reportorial produce is harvested, baled, and shipped to market on routinized schedules. Even an event of merely *potential* importance, such as whether Walter Mondale or Robert Dole should become vice-president, is reported by the short ton. It could easily be argued that in the case of either, victory would affect the lives of voters in Southeast Michigan only if their running mate died or were forced from office. Assuredly, such potentialities are real. But so are those involving DNA. Are discoveries that pertain to the secret of life, disease cures, self-fertilizing crops, uncontrollable epidemics, and genetic engineering for totalitarian regimes as *potentially* likely to influence lives as the question of which representative of two profoundly similar political parties wins election to a largely powerless office?

Politicization of science stories can gain them greater play. The SST project flew along in relatively smooth silence until grounded by the widely reported atmospherics of pollution hearings. Similarly, Mayor Alfred Velucci assured the DNA issue voluminous coverage in Cambridge, Massachusetts, by summoning Harvard and MIT professors to televised grillings in City Hall. The sessions were messy and acrimonious. Politicians who admittedly knew nothing about genetics unceremoniously demanded that specialists "spell it out" in plain English (12). The encounters did nothing to advance the theory of molecular biology, nor the practice of town-gown relations, but they unquestionably did accomplish something positive. They forced those who would potentially endanger the community to confront those charged with protecting it in the full glare of public and press attention.

No such thing happened in Ann Arbor. Not only is the town far more inclined to trust the University to do the "right thing" in technical matters, but its highest elected official took steps in May 1976 to prevent open debate. Critics of gene splicing had drafted a "Resolution Requesting an Environmental Impact Statement on Guidelines Governing DNA Research," because, they wrote in part, it "involves risks of serious, possibly irreversible damage to individuals and society." Mayor Albert Wheeler, himself a distinguished professor of microbiology at the University, polled City Council. Learning that the resolution would be defeated, he decided that opening the issue to debate by the unusually large crowd would not be "useful." He quietly tabled the motion "to avoid a lot of uninformed discussion" (13).

Wheeler's tabling of the DNA issue unquestionably contributed to the relatively high level of exchange on the subject in Ann Arbor. His action, however, also barred it access to the public forum best suited to debate by the community—and mostly likely to attract attention in the press. "Issues can bubble along below the surface without our covering them," notes *Detroit Free Press* City Editor John Oppedahl, "but when they reach the level of public officials, we are compelled to react" (14). In Ann Arbor, DNA never reached the point of "compulsion."

Analyzing the need for press coverage quickly becomes a question of defining "news," which is something like describing the taste of oysters to the uninitiated. The only route leads through analogies, all of them muddy, and the crucial decision usually has more to do with the credibility of the person doing the recommending than the quality of his imagery. News, or the taste of oysters, is what somebody says it is. To editors and reporters, "somebody" means themselves as second-guessed by each other. Can anything, then, be news? Only theoretically. In practice, nobody can afford to advise that oysters taste like *"oeufs à la neige."* There are limits.

Given the priorities of journalists concerned, the nature of the DNA story, the conduct of Ann Arbor officialdom, and the structure of the area press, it is relatively easy to pin the lack of coverage to specific reasons. Neither logic nor First Amendment theory, however, shines brightly through them.

1. Southeast Michigan, a major industrial and research center, lacks a single full-time science writer. The journalist closest to that description is Medical Writer Dolores Katz of the *Detroit Free Press*. Finding the DNA story "boring," she downplayed it accordingly (15). The *Ann Arbor News'* Larry Bush devotes "about 50 percent of my time that's left over from obits and such" to science reporting—but the *News* gave most of the DNA story to the man covering education (16).

2. Gene splicing is the most difficult kind of information to popularize. It is complicated. In Ann Arbor, public officials steered clear of it. It is so arcane that few scientists know much about it, and even among those who understand it, there is much disagreement as to what it means (17). It forces editors to invest expensive reportorial man-hours gathering information that they do not know how to evaluate. In some of this, DNA suffers by comparison to recent aerosol-can/ozone-layer controversies, in which the University also figured centrally. Unlike aerosols, DNA as a story is neither familiar nor easily pictorialized. As Benjamin Burns, assistant editor of the *Detroit News*, summarized (18), "Freon was easy. Most of our readers spray stuff on their hair or somewhere, but how many spray DNA onto their underarms in the morning?"

3. Ann Arbor suffers from an odd variant of a traditional journalistic foible known as "Afghanistanism," meaning that a man biting a dog in some faraway, exotic setting often draws more coverage than the same event on a sidewalk outside the city room window. Regional editors, however, do not apply pure Afghanistanism to Ann Arbor. Instead, they treat a town of major news potential located less than an hour's drive from most of them as if it really *were* somewhere near Kabul—remote, unfamiliar, uncoverable. No out-of-town newspaper, broadcaster, or wire service maintains staff representation in Ann Arbor, nor even sends reporters on a regular "beat" basis to keep track of its indisputably newsworthy resources.

Those three factors—dearth of specialized reporters, impediments of technical stories, and lack of Ann Arbor coverage—produce an unusual degree of dependence on the University Information Services

(UIS). There simply is no other conduit available when journalists require information. The regional press routinely relies on UIS's five public information specialists to predigest heavy material, and on its top editors to key them to upcoming "newspegs."

Highly controversial stories disrupt this relationship. Press-public relations interaction is an uneasy affair at best. In biological terms, it combines the coziness of symbiosis with the rapacity of predation. PR men generally are paid to put out material that makes their institutions look good, but score no points unless it is credible enough to be picked up. Reporters, on the other hand, try to profit from PR's free legwork without inadvertently toeing any institutional "line." The DNA story squeezed UIS between several interests and responsibilities.

First there were the scientists. Journalists know from experience that highly trained people confronting them for the first time frequently contract a virulent "childhood disease" sometimes known as "complexoria." The cause lies in an ill-understood phenomenon that makes scientists supply more details when fewer are needed (19). The symptoms are suicidal caution linked syndromatically to equally suicidal faith in the ability of verbal precision to conquer confusion, boredom, or both. If anyone gave you road directions the way complexoria victims brief reporters, you would spend so much time squinting in *every* direction at *each* intersection (noting curbing height, lunar phase, Latin names of nearby flora) that you would never arrive. If you did, it would be without knowing where, or why. Strangely, complexoria appears unrelated to native eloquence in the afflicted. Nor does it return after the "childhood" case is cured, generally by trauma.

One of the principal news sources of the DNA story was David Jackson, an associate professor of microbiology who had pioneered recombinant techniques, and who is blessed with ability to explain the intricacies of his expertise with remarkable succinctness. But in the fall of 1975, Jackson, who had never before encountered press demands for instant explanations (20), developed symptoms of the disease peculiar to these circumstances. He wanted the first UIS release to be scientifically perfect. Result: biologists and a PR man under UIS Director Joel Berger scrimmaged unhappily for almost two months, during which time no announcement went out to the press. What finally appeared February 25, 1976, was a 2000-word treatise that pleased nobody. Press usage, notes Berger, "was zilch" (21).

From the beginning, University administrators took a stance guaranteed to complicate the life of a thoroughly professional PR man like Berger. They were strongly predisposed to permit DNA research and equally determined to promote discussion of it along the way

(22)—an issue explosive with public-relations minefields. They made DNA part of the University Values Seminars in progress that year. When challenged by critics, they called in outside experts to inform the regents of potential hazards. The weekly *University Record* conscientiously reported DNA-related campus events. Charles Overberger, vice-president for research, even suggested that a national conference of science writers in Ann Arbor take up the question. (Remarkably, the chairman of the Council for the Advancement of Science Writing session turned down DNA as a "nonstory," according to Berger.) Caught between the scientists and administration attitudes that might well have produced adverse reaction, UIS took the safe course of doing nothing until February 25 (23).

When faculty "Committee B" three weeks later recommended that DNA experiments proceed, however, UIS geared up to overdrive. Berger immediately boarded a plane for New York to hand-carry UIS's release to science writers of the *New York Times, Time,* and *Newsweek* magazines and the *Wall Street Journal,* while an associate made similar rounds in Detroit (24). UIS's dramatic flourish drew little interest. Only the *Wall Street Journal* responded. It may be true, as one science writer told Berger on that trip, that "the Eastern Establishment Press takes more interest in private institutions like Harvard than in public ones, like Michigan" (25). Conversely, by saying very little until very late, UIS may have helped shield the University against controversy at the expense of later positive attention in the national press. *Time*'s cover story April 18, 1977, for example, mentioned DNA programs at every relevant US institution—except the University of Michigan, which had been first to raise the question in a community context.

Should UIS have done more? Should Berger have insisted that something go out before late February 1976? Should he at least have telephoned contacts in the press before the controversial aspects of DNA were decided in May of that year? Some of the most thoughtful regional journalists—several of whom admit to having inadequately reported DNA—think UIS is obligated to give the press an early leg-up on such stories (26).

"The University of Michigan matched the ineptitude of the press," said the *Detroit News*'s Burns. "It is a publicly supported institution of higher learning, after all. When questions come up about research affecting the public, it has the obligation to present this sort of information— even if it's embarrassing" (27).

The *Ann Arbor News,* with more experience dealing with UIS, offered similar views and an interpretation of why its staffers (who man typewriters four blocks from UIS's offices) received no contact or releases until *after* their DNA stories began. Managing Editor David Bishop sees it this way:

"Does the University have an obligation? I don't think it would be asking too much for them to give us a ring on something like this. But I also don't think it's realistic to expect them to. They won't hold a press conference on anything unless it will make them look good" (28).

Berger rightly counters that he can only help journalists, not order them (29). It is also true that scientific etiquette plays against him by tabooing the "grandstanding" of research findings—lobbing them into the press before they appear in scholarly journals, typically after months of delay. Should an enterprising reporter seek out a scientist before that time, UIS is unlikely to know about it (30). Berger's comments on such matters imply the need for greater trust by scientists in their own PR men, as well as a more affirmative approach to the press by the University.

"The timing of our releases," Berger explained, "was not a matter of covering up controversy at the University, but rather of not knowing the issue. We weren't much more aware of it than the press."

On the subject of calling a press conference to air both sides of the issue as it developed—potential benefits and possible risks—he added, "It would have been feasible, but of doubtful interest" (31).

Interest is often a function of familiarity, which must begin somewhere. In science, that point of origin is the laboratory. As developments there penetrate with deepening intimacy faraway and unrelated lives, the public's stake passes from finance to issues of safety. Scientific events capable of altering society assuredly impose new responsibilities on the public besides keeping track of tax expenditures for research. But so too they demand new responsiveness from scientists. The tradeoff long recognized in terms of financial accountability has come to the marketplace of ideas. It asks that science, as a major beneficiary of the First Amendment, meet its obligations in giving prompt public notice. The "right to know" today stands at the door of the laboratory as well as the political "smoke-filled room."

Scientists themselves, of course, generally lack the training and interest to operate as competent PR persons. Most of the responsibility for providing society with scientific information necessary for enlightened debate must come from the press. To do the job, publications and broadcasters need commitment far beyond what they brought to coverage of DNA in Southeast Michigan. But they also require the sort of briefings that were not forthcoming. Granted, universities are under no legal obligation to hold press conferences. The same forces that thrust new obligations on scientists, however, impose duties on their sponsors. There are at least two ways of looking at the problem. First, research institutions—particularly publicly funded ones—have grown so closely relevant to those living around them that, as ivory towers, they seem singularly stormable. Mayors, legislators, and governors aside, a second approach makes communication blockages unaffordable. To be faithful to their constituencies—and to free speech—research institutions must open themselves promptly to journalists on controversial developments such as gene splicing. Even if few reporters come to press conferences, briefings should be held for the same reason that professors lecture to light attendance on the first warm day of spring. Education for some is better than education for none. It is also the best means of assuring that, in Jefferson's phraseology, universities remove themselves from a process preventing the people from knowing "something of what has been passing in their time."

II. DNA AND THE PRESS

Recombinant DNA is a world-scale "story" as well as a phenomenon capable of tangibly affecting every living thing on earth. For purposes of this study, however, the scope was narrowed to the print and broadcast press of Southeast Michigan. It is there that gene splicing at the University may first be felt, and there most certainly that the community has the tightest claim to a "right to know" about the research.

"Southeast Michigan" more closely defines a set of problems than a geographic area. Its seven counties* share legislative recognition in the 113-member Southeast Michigan Council of Governments (SEMCOG), but in no strict sense do they occupy the entire southeastern quadrant of the state. Their real communality lies in attempts to cope with the urban blight caused by Detroit's spectacular decline as a manufacturing and residential city (32). SEMCOG has formally designated water quality, housing, and land use as susceptible to upgrading only in regional terms; transportation has been singled out for priority attention in the Southeast Michigan Transportation Authority (SEMTA). The seven counties contain 4.8 million people, which for comparative purposes equals the entire population of Norway, or half that of Michigan. Racial and economic problems center there. Because of them, the counties often find themselves facing the same kind of suspicion in the legislature in Lansing that New York City encounters from its famous "upstate" critics in Albany. During the recession of the mid-70's, Southeast Michigan was among the hardest-hit districts in the nation. Governor William

*Wayne, Washtenaw, Monroe, Livingston, Macomb, St. Clair, Oakland.

G. Milliken responded by naming a blue-ribbon commission to study prospects for renewal. Significantly, it was the only such body in the state.

Southeast Michigan is also the communications nexus. Corporate headquarters of Michigan's leading daily newspaper chains (Knight-Ridder, Booth) are located there, as well as the Observer & Eccentric string of twelve semiweeklies. The largest wire-service bureaus in Michigan operate from Detroit, as do the state's representatives of national general-interest publications (*The New York Times, Wall Street Journal, Time, Newsweek, U.S. News & World Report*). All three broadcast networks have radio/television affiliates in the seven-county market. What follows is an analysis of how these organizations covered the DNA story as it developed in Southeast Michigan, newsmen's explanations of their decisions, and their observations regarding handling of DNA information by the University.

Local Publications

> Hell, the leash law generated more interest.
>
> —David Bishop, *Ann Arbor News* (33).

If a pebble thrown into a pool represents a breaking news story, the concentric wavelets record dissemination, beginning with the local press. Its functions are discovery and relay, the latter aided by wire services, broadcasters, and the larger publications that conventionally monitor it for story ideas.

In Ann Arbor, the local print press is nearly synonymous with the *Ann Arbor News*,* with a circulation of 40,900 (34) and access to the considerable resources of the eight-member Booth chain.† The newspaper carried 31 news stories and three editorials on local DNA research in the study period beginning with the Asilomar conference in February 1975 and ending April 30, 1977, the date of present writing (35). In so doing, it quantitatively outperformed any paper in Southeast Michigan by a factor of nine. DNA being a "backyard story," this is natural enough. What is not clear is why its coverage began so late, why its editorials were so lightweight, and why nothing from its pages was picked up by regional publications, wire services, and broadcasters.

The reporters, editors, and editorial writers responsible for *The News'*

---

*The student-operated *Michigan Daily*, which ran a dozen DNA stories beginning December 12, 1975, has been exempted from this study (37).

†In October 1976 Booth Newspapers, Inc., was sold for $300 million to the Newhouse organization, the biggest newspaper sale in U.S. history (38).

DNA coverage cite four major reasons (36) for what they are quick to admit were shortcomings in their work, insofar as it failed to be of much aid to public discussion in 1975 and early 1976.

1. Lack of understanding of the issue.
2. No initiatives from UIS until sides had been taken and the campus debate was in full blossom.
3. Among the critics of DNA research, absence of a spokesperson whom reporters considered credible.
4. Scanty indications of interest in the subject from the community.

Arthur Gallagher, then editor of the *News*, put it simply. "Maybe it was just too complicated. It was one of those things that I didn't feel confident about; I didn't feel we should jump into a technical issue that quickly, nor try to tell the University what to do" (39).

Roy Reynolds, who wrote most of the DNA stories as higher education reporter, joined Managing Editor Bishop in faulting UIS for not having tipped the press at least by the fall of 1975 to something of extraordinary interest upcoming soon. But unlike Bishop, who felt that the University made contact only when it had good news to tell, Reynolds perceived the delay as indicating lack of trust between administrators and PR men.

"The University administration doesn't trust its own PR shop," said Reynolds. "Ever since the boycott of the blacks in 1970, when UIS put out the story before the administration was ready, we have noticed a hesitance in UIS. In covering DNA in 1975, I had the feeling that the administration just hadn't told UIS about it—and UIS got caught short" (40).*

Reynolds and Larry Bush, *The News'* part-time science editor, both noted a lack of authority behind critics of DNA research. This was particularly striking to Reynolds, whose coverage dealt with the "contra" side of the issue more than Bush's. "We couldn't help being skeptical of Wright (Susan Wright, the assistant professor of humanities who led the critics)," Reynolds observed. "She isn't a scientist, and she's young. If a few senior people had done what she was doing, it sure would have been easier to convince the city desk that we had a story" (41).

Even when *The News* swung into a rhythm of detailed coverage, feedback came in meager rations. To a considerable degree, this resulted from the town's "hands-off" stance toward the University, and the failure of DNA to be debated in City Council. But *The News* itself was also responsible. Its attention awoke late, and its three editorials were more

*For UIS's explanation, see part I of this chapter.

watered-down essays than fired-up position papers. The first (May 16, 1976) expended 600 words to say little more than its own seven-word headline: "U-M Faces No Ordinary Decision on Research." The second generally welcomed the appointment of a single representative of a town of 95,000 to a nine-member University committee overseeing DNA experiments. The third dressed down an official of the U.S. Office of Technological Assistance for suggesting that the University should have more seriously considered postponing DNA research. Snorted *The News:* "If established major research universities can't handle this research responsibility, no one can." Unsurprisingly, the letters-to-the-editor section drew a total of two entries (42).

Because of the blandness of DNA as a press issue, *The News* stories went no further than its own columns. Notes Bush, who as science editor recommends subjects in his specialty to the *Associated Press:* "AP is not very open to comprehensive news; they want spot stuff. I don't recall their picking up anything from us on DNA" (43).

For the last dozen years, successive new publications have sprung up in Ann Arbor claiming to cover what the "establishment" *News* omits. Currently, there are two. The *Ann Arbor Observer* (circulation 15,000) aims at "getting people involved in what's going on," according to editor-publisher Donald Hunt, but has not mentioned DNA in its columns (44). While the *Observer* excuses itself on grounds of recent birth (July 1976), the *Michigan Free Press's* position is somewhat different. An aggressively "radical" journal with pronounced political views, the *Free Press* would seem ideally suited to arguing a cause in the DNA dispute. Yet, oddly, the 20,000-circulation weekly ran stories on DNA in Cambridge without giving a line to the debate at its own doorstep. "We would have been too far behind the dailies," said editor/publisher George dePue (45).

Local Broadcasters.   Five radio stations serve Ann Arbor (46). Although researching of their programming is hampered by the industry habit of destroying records a few months after air-time, spokespersons for four of them indicated that they carried only minor pieces on DNA, and none before 1976 (47). The exception was University-operated WUOM-FM, a National Public Radio affiliate, which, through the efforts of News Director Frederick Hindley, came in somewhat lately, but nevertheless thoroughly.

In November 1975, Hindley had broadcast the University Regents' decision to renovate laboratories for DNA work. The following January, he produced a 15-minute program on the University's decision to call in outside experts to testify about the degree of risk involved. Hindley pledged on the air to keep his listeners abreast of the debate. He did. In addition to spot stories, WUOM "built" three in-depth features (110 minutes, 95 minutes, and 120 minutes, respectively) around the DNA hearings held in Ann Arbor March 3–4 (48).

Regional Publications

We did not distinguish ourselves.

—John Oppedahl, *Detroit Free Press* (49).

In most metropolitan areas, dominant dailies routinely scan papers from outlying cities of major news interest to "take the pulse." Not that this is purely an exercise in staying current. As inner cities declined, circulations of large publications followed readership to the suburbs, which have become deep troves of advertising dollars. In Detroit, for example, *The Free Press* and *News* vie statewide for readers as the home market empties.

Neither paper, however, conforms to the national custom of pulse-taking, at least concerning Ann Arbor. They do not monitor the *Ann Arbor News* (50). This is particularly striking because, in the case of *The Free Press*, Sunday circulation in Washtenaw county is more than two-thirds that of the local paper—27,400 against 40,900 (51). Inattention of the metropolitan press to Ann Arbor in general is reflected in the small image they gave readers of the DNA controversy. *The Free Press* did not cover it systematically. The *Detroit News* carried a three-part series in the spring of 1976, but mostly by accident. Stephen Cain, who wrote the articles, happens to live near Ann Arbor (52).

Free Press City Editor John Oppedahl had reservations about the DNA story. "It was important," he concedes, "but not as interesting to the people of, say, Roseville as PBB, MESC (Michigan Employment Security Commission)—or a three-legged duck" (53).

Nevertheless, neither Oppedahl nor other *Free Press* editors deny that the paper suffers from the lack of a science writer and "somebody to cover Ann Arbor regularly." Those holes seemed particularly glaring during 1975–1976 when, as Oppedahl recalls, "Half of our staff (of 45) was writing about scientific subjects—PBB, water quality and the like," many sources for which were in Ann Arbor (54).

Nobody in the newsroom recalls any personal contact from UIS on the DNA story, nor from anyone else involved in it—actions that Oppedahl claims would have "made all the difference in any decision to send a reporter up there right away" (55). In all, *The Free Press* ran five pieces on the University's DNA research. The first, February 17, 1976, omitted mention of the unusual nature of the program. The most comprehensive, April 26, 1976, came just a few weeks before the University was to make its final go-ahead decision. As the debate closed, medical writer Dolores Katz took up the subject of DNA; she wrote on May 16, 1976:

> Most scientists have already decided that we must risk the unknown and, they believe, remote danger for an unknown, but potentially enormous benefit. But the public has not yet voiced an opinion. Unfortunately, the decision will probably be made without that voice.

It was Katz' first mention of the "remote danger," but she was correct in her prediction. The decision to proceed did indeed arrive before the public's "voice." It might be argued that a principal cause was that Katz, who came closest to being a full-time science writer in Southeast Michigan, had not listened for the public's voice, reflected it, nor commented upon it. She states her reasoning forthrightly.

> In this business there are no issues that stand up all by themselves and demand discussion. The only criteria for coverage are things that I think would be interesting. I don't see how the public would benefit from a discussion of some people saying "great" about DNA and others saying "terrible." I found DNA boring. As for its being the "biggest science story since the A-bomb," I think that's dumb. Swine flu was going to be "the greatest scourge," remember? And then those cancer "cures". . . .

Given her disinterest, might not Katz have passed the DNA story to another reporter, after it became clear that scientists who were divided over the wisdom of conducting DNA experiments agreed wholeheartedly about the importance of the issue? Again, Katz's reply comes quickly (56):

> Not the way the *Free Press* is run, which is in anarchy. I have no single editor who would sit down with me and say "Okay, what are the big stories and what are you doing about them?" So it's easy for things like DNA to slip through the cracks.

Assistant Editor Benjamin Burns of *The Detroit News*, during a staccato question-and-answer session, summarized why he thinks readers were "badly served":

> Q: Whose fault was it?
> A: Editors'.
> Q: Why?
> A: Ignorance.

Q: Why?
A: Because nobody (reporters included) called them.

Burns prescribes as a remedy better editors ("Even when we train science writers, we don't educate the editors who control them and their stories") and briefings by scientists and their institutions. The worst reportorial problems result from disagreement among specialists, not the innate difficulty of mastering the material. Press conferences and backgrounding *before* the issue explodes, he argues, are the only practical means of educating the press to inform the public. As for the complexity of DNA stories, Burns quips: "Combining plants and animals? That's not hard to understand. In fact, it's already been done. Just notice the creepy prose style of a scientific journal, or for that matter, look at the weird judgment of some people in the news business" (57).

Stephen Cain, who occasionally reports scientific developments in the course of regular duties in *The News*, wrote by far the most probing pieces on the DNA controversy in Ann Arbor. His series took full advantage of the local and outside experts invited to testify before University committees, and it received top play by the paper. Cain considers his work, however, to have been "too little (three parts), too late (March 7–9)." In Cain's view,

> Newspapers should be part of emerging issues rather than a mirror to reflect them when they're finished growing. I performed the latter role, but not the former, partly because of my own perceptual lag, partly because of the nature of the paper.

Cain corroborates Burns' view of editorial awareness, which often determines *when* a story appears, as well as whether and how. When Cain first suggested a piece on DNA in January 1976, "My editors had never heard of it." Nor did they give him encouragement, let alone orders, to "get" the story. Cain ascribes this to the "provinciality" of *The News*, which, he says "had little interest in having its own people cover national stories." To reporters convinced that a story is particularly important, the situation calls for something to force their paper to run it quickly—a newspeg. "Emerging" issues, by definition unheralded by marked public interest, frequently lack them, and for that reason are delayed. "Without a peg," says Cain, "editors don't run your stuff, or bury it on the psoriasis page" (58).

The Ex-Urban Press

Publications circulating in the interconnected and often ill-defined markets around Detroit range from "giveaway" shoppers to conventional dailies of considerable size. Blanket interviews with their editors turned up no DNA stories before 1976. Only a handful have carried anything at all. The following quotations were selected from papers representing the largest of their kind—those most likely to have presented DNA to their readers as a matter of special interest.

Early in 1977, *The Oakland Press* in Pontiac ran its first DNA piece, an informative editorial. Ed Blunden, editorial page director, recognizes the lateness of coverage for the paper's 78,000 readers, explaining, simply, that other matters were more pressing, and that the story's complexities make it "a little rough" for the audience (59).

The Observer & Eccentric (O & E) organization excluded the DNA story on principle. Headquartered in Livonia, O & E runs a dozen twice-weekly papers averaging 45 pages in the northwestern reaches of Wayne and Oakland counties. Total circulation is 155,000; combined news staff is 61. Unlike many publications in outlying districts, O & E was well aware of the DNA controversy and of how the research could potentially affect its market area. The story was omitted, says Tim Richard, the group's chief editorialist, because (60):

> It isn't our job. We devote all our resources to covering the immediate circulation area, since nobody else does. Stories like DNA are for the local paper, the wires, the Detroit dailies and the national publications.

Wire Services

Aside from spot coverage in areas of traditional news interest, Southeast Michigan bureau editors of the Associated Press and United Press International see themselves as running essentially relay operations between local clients, and from them to broader markets. The discovery function of journalism is thus left largely in the hands of reporters in the immediate vicinity of breaking stories. This is particularly true in places such as Ann Arbor, where neither wire maintains a staff, and on subjects such as DNA, which neither has sufficient resources to cover without extensive backgrounding.

Those views lead UPI News Editor Paul Varian to consider the single story his staff produced (in December 1976, 22 months after the Asilomar conference) to have been "really in on the ground floor, before DNA mushroomed into a national issue." The "mushrooming" he cites, of course, was the attention given by the *national* press, which came months after Ann Arbor and Cambridge, having grappled with the implications of DNA for community welfare, had drawn their separate conclusions. UPI's lone science writer is based in Washington.* The Detroit Bureau's piece was written by William Silberg who, as Varian puts it, "is just out of college—Yale or someplace—where DNA was big." It required the unusual step of sending somebody to Ann Arbor.

"Obligation to help community discussion in Ann Arbor?" asks Varian, quizzically (62):

> I'd like to do that, but I don't feel obligated. The purpose of a wire service is to give newspapers and broadcast outlets what they need. If they want something special it's up to them to get it, or to ask us specifically to get it for them.

Michael Graczyk, news editor of AP's Detroit bureau, disagrees, but only on theoretical grounds. Wire services should have "closely monitored" the DNA story in Ann Arbor as an aid to public debate. "But we were just too thinly staffed," he adds. "There were too many other issues of competing importance—Project Seafarer in the Upper Peninsula . . . nuclear dumping in northern Lower Michigan. . . ." The bureau produced a single story, May 22, 1976. Even if UIS had peppered the bureau with DNA bulletins, AP's shorthandedness would have hindered delivering the information to the service's 32 outlets in Southeast Michigan.

"We've learned not to take UIS's stuff as the 'bottom line' on stories," says Graczyk. "The University, after all, operates with a lot of government revenue. Anything that makes it look bad jeopardizes that funding" (63).

Regional Broadcasters

Print journalists often think of broadcast counterparts as parasites who draw fat salaries by regurgitating a rich diet of basic research produced by wire services and newspapers. To a considerable extent, particularly on local levels, the impression is correct. Radio and television newsrooms typically operate with fewer reporters than comparable publications, relying heavily on printed pickups for story ideas and copy. DNA, however, was not among the morsels. Most major broadcasters serving Southeast Michigan said nothing about it. Detroit's private network television channels carried nothing at all (64).

Since broadcast newsmen explained their lack of coverage in terms very similar to those given by print counterparts to account for thin reporting, little would be added by quoting each briefly. Instead, the passages below are offered to suggest the main lines of response.

*The only UPI national DNA story picked up by Southeast Michigan's largest publications appeared February 16, 1976, in *The Detroit News*. It did not mention research in Ann Arbor (61).

Radio. David White, news director of WJR, which easily carries more in-depth news than its competitors (65):

> DNA just didn't come to our attention. I don't blame it entirely on the University Information people, although the story sounds a lot more important than the mountains of usual stuff we get from them about concerts, regents' meetings and sports. We're event-oriented people and although we may be ignorant on some issues, if somebody calls us and says "Hey, on Tuesday at Ferry Field there're going to be a bunch of scientists who can explain something important," we will be there.
>
> Ideally, we should have drafted somebody with the basic interest—interest is more important than knowledge—to go after DNA. But even then, we in broadcasting have problems making something like that comprehensible. Broadcasting isn't like print. It's sequential. If a listener really wants to hear about DNA, he's going to have to listen to the weather, too.
>
> I subscribe to the theory of the press helping the people learn about new developments for community discussion. That's what we're here for. But the situation is getting worse—both with us and with the University Information. I feel like I'm riding a train that keeps getting farther behind schedule. We omit stops along the way to try to catch up, but it doesn't work. All we do is cut service.

Television. George Noory, executive producer for WJBK, CBS's Detroit affiliate, listened closely March 22, 1977, as an interviewer described recombinant DNA techniques in the course of asking whether Noory's staff had aired any material. "Can you hang on a minute?" he asked. "I'm going to check with my science editor. This sounds like a fascinating subject." He continued (66):

> It's very definitely our responsibility to come into public discussion on something like this. But we depend on people to call us on things our reporters don't pick up in the normal course of spot news.
>
> Like when you mention DNA, I start thinking back to high school biology, and I don't find much there. We do need cues. There's no doubt about that. But hell, yes, give us a call and we'll cover it.

With few exceptions, news editors and qualified reporters with Southeast Michigan publications and broadcast organizations agreed that science stories of DNA magnitude require coverage as an aid to community debate. They also admit their variously qualified failure to provide it in their home markets. Lack of information caused by the University's lagging notice, and their own staffing weaknesses, were the explanations most often given. Additionally, each level of the press to some degree saw the story as somebody else's responsibility. Organizations closest to it tended to assume it to be a "national" affair. Regional publications and broadcasters left it to "the locals." Relay services waited for demand that never arose. Fortunately, these shortcomings submit relatively easily to improvement. But the timely dissemination of "hot" scientific information will not come without action based upon existing concern.

1. The press should be encouraged to evaluate science stories on the same scale used for politics. To the public, it matters little whether the environment is manipulated by chemicals or by legislation. The right to know springs from the *urgency* of knowledge, not its availability.

2. Science-sponsoring institutions can help the press fulfill its obligations by regarding research results likely to influence public welfare as major information events and behaving accordingly. Internally, *regularized* channels should speed the flow of information from laboratories to PR services as part of the responsible exercise of free speech. Scientists whose research draws major attention should be given instant reinforcement—extra telephone service, secretarial help, PR counseling—to avoid the loss of time and patience now associated with publicity. When dealing with the press, the University should increase efforts to speak "the language" of the trade. Unattended press releases generally travel directly from "in" baskets to wastebaskets.

NOTES

1. Mott, Frank L., *Jefferson and the Press*, Louisiana State University Press, Baton Rouge, La., 1943, p. 55.
2. *Ibid.*, p. 65.
3. *Ibid.*, p. 12.
4. Interview with David A. Jackson, March 1977, Ann Arbor, Mich.
5. Telephone survey and search of newspaper library files of 57 Southeast Michigan publications, radio stations, and television channels, spring 1977. Cited in subsequent notes as Survey, *op. cit.*
6. *The Michigan Newspaper Directory*, The Michigan Press Association, Lansing, Mich., pp. 67–84.
7. Taishoff, Saul, ed., *Broadcasting Yearbook 1977*, Broadcasting Publications Inc., 1977, pp. B109–111, C102–109 *passim*.
8. Survey, *op. cit.*
9. Interview with Frederic Hindley, April 1977, Ann Arbor, Mich.

10. Survey, *op. cit.*
11. Telephone interview with Douglas McKnight, March 1977.
12. Bennett, William, and Gurin, J., "Science that Frightens Scientists," *The Atlantic*, February 1977, **239**(2), 60.
13. Interview with Albert Wheeler, March 1977, Ann Arbor, Mich.
14. Interview with John Oppedahl, March 1977, Detroit, Mich.
15. Telephone interview with Dolores Katz, April 1977.
16. Survey, *op. cit.*
17. Jackson interview, *op. cit.*
18. Interview with Benjamin Burns, March 1977, Detroit, Mich.
19. Jackson interview, *op. cit.*
20. *Ibid.*
21. Interview with Joel Berger, March 1977, Ann Arbor, Mich.
22. Interview with Alvin Zander, Associate Vice President for Research, University of Michigan, December 1976, Ann Arbor, Mich.
23. Berger interview, *op. cit.*
24. *Ibid.*
25. *Ibid.*
26. Survey, *op. cit.*
27. Interview with Benjamin Burns, March 1977, Detroit, Mich.
28. Interview with David Bishop, March 1977, Ann Arbor, Mich.
29. Berger interview, *op. cit.*
30. Jackson interview, *op. cit.*
31. Berger interview, *op. cit.*
32. Interview with William Drake, member, State Advisory Committee on Transportation, March 1976, Ann Arbor, Mich.
33. Bishop interview, *op. cit.*
34. *The International Year Book,* Editor & Publisher Co., Inc., New York, 1976, p. 126.
35. Survey, *op. cit.*
36. Group interview with David Bishop, Larry Bush, and Roy Reynolds, March 1977, Ann Arbor, Mich.
37. Survey, *op. cit.*
38. Telephone interview with Richard P. Griffin, vice president for finance, Booth Newspapers Inc., April 1977.
39. Telephone interview with Arthur Gallagher, April 1977.
40. Interview with Roy Reynolds, March 1977, Ann Arbor, Mich.
41. *Ibid.*
42. Survey, *op. cit.*
43. Interview with Larry Bush, March 1977, Ann Arbor, Mich.
44. Telephone interview with Donald Hunt, April 1977.
45. Telephone interview with George dePue, April 1977.
46. *Broadcasting Yearbook 1977, op. cit.*, p. C102.
47. Survey, *op. cit.*
48. Hindley interview, *op. cit.*
49. Oppedahl interview, *op. cit.*
50. Survey, *op. cit.*

51. Telephone interview with Perry Whitney, Assistant Circulation Manager, *Detroit Free Press*, April 1977.
52. Interview with Stephen Cain, March 1977, Ann Arbor, Mich.
53. Oppedahl interview, *op. cit.*
54. *Ibid.*
55. *Ibid.*
56. Katz interview, *op. cit.*
57. Burns interview, *op. cit.*
58. Cain interview, *op. cit.*
59. Telephone interview with Ed Blunden, April 1977.
60. Telephone interview with Tim Richard, April 1977.
61. Survey, *op. cit.*
62. Telephone interview with Paul Varian, March 1977.
63. Telephone interview with Michael Graczyk, April 1977.
64. *Ibid.*
65. Telephone interview with David White, March 1977.
66. Telephone interview with George Noory, March 1977.

An Afterword

On the Dangers of Inquiry and the Burden of Proof

Carl Cohen

When it is argued that research with recombinant DNA is so dangerous that it should be prohibited or restricted, who bears the burden of proof? Must the researchers (or others) show that the inquiry is not dangerous before proceeding? Or must the critics (or others) show that it is dangerous before restrictions are imposed? I aim to answer these questions through an examination of the leading arguments in support of restriction.

These arguments commonly take the following form. Some general principle is relied upon, most often implicitly, as a major premise according to which research—if it can be shown to have certain very hazardous consequences or by-products—should be restricted or prohibited. [Examples: "Research should not be conducted if it is likely to lead to knowledge not wise to place in human hands," or, "When the risks of some research outweigh the benefits it may confer, that research should not be continued."] The restrictive principles called upon have varied greatly in thrust and stringency. I discuss them elsewhere [1].

To apply any one of the alternative general principles of restriction to research with recombinant DNA, it must be shown that these inquiries under attack do, in fact, present the great hazard specified in the major premise. To do this—that is, to establish the appropriate minor premise of the restrictive argument—leading critics of recombinant DNA research have repeatedly presented subsidiary arguments in two

Editors' note: This article is reprinted with the permission of the *Southern California Law Review,* and another version of this essay was printed therein.

patterns, marked both by the frequency with which they appear in the critical literature and by their colorful character. I call these patterns (I) *heavy question arguments,* and (II) *arguments by metaphor.* Both patterns exhibit serious flaws of argumentative structure; both—but especially the first—betray confusion about the placement of the burden of proof in matters such as these. That confusion, I will show, arises from two misunderstandings: a misunderstanding of the role of ignorance in drawing inferences, and a misunderstanding of the protections afforded by the principle of free inquiry.

I. HEAVY QUESTION ARGUMENTS

By "heavy question arguments" I refer to a somewhat heterogeneous class of arguments, all put in ponderous interrogative form, that serve to impede rather than to facilitate rational dialogue. As directed against research with recombinant DNA, these heavy questions are so framed as either to be unanswerable, yet effective in surrounding the research with the aroma of catastrophe, or to be literally answerable, but only upon the supposition of the truth of unfounded allegations about this research that are never plainly stated or defended. If arguments in this form had been proposed only by irrational persons, motivated perhaps by ideological enthusiasm or the thirst for personal advancement, they could be dismissed summarily. In fact, the most reputable of scientific critics of research with recombinant DNA have grounded their most fundamental reservations on heavy questions of precisely this kind.

Concrete illustrations, from leading essays on this vexed subject, are in order:

A. Foremost among the practitioners of the heavy question argument is Erwin Chargaff, Professor Emeritus of Biochemistry, Columbia University. In a now famous letter to *Science,* June 4, 1976, Professor Chargaff asks:

(1) *"If Dr. Frankenstein must go on producing his little biological monsters . . . [h]ow can we be sure what would happen once the little beasts escaped from the laboratory?"[2]

Further in the same letter he writes:

(2) "[T]he principal question to be answered is whether we have the

---

*Parenthesized numbers are mine for purposes of reference later in this essay. Bracketed numbers refer to notes at the end of the chapter.

right to put an additional fearful load on generations that are not yet born."

And again:

(3) "Is there anything more far-reaching than the creation of new forms of life?"

And yet again:

(4) "Are we wise in getting ready to mix up what nature has kept apart, namely the genomes of eukaryotic [higher] and prokaryotic [lower] cells?"

And still again:

(5) "Have we the right to counteract, irreversibly, the evolutionary wisdom of millions of years, in order to satisfy the ambition and the curiosity of a few scientists?"

These are exercises in rhetorical flourish, we may say, designed to call attention to what might otherwise be ignored. Perhaps. But the suggestion that without such rhetoric there would have been no attention to ethical aspects of this research is simply false; the moral issues had been carefully addressed long before the matter had become a public sensation [3]. Whatever their original purposes, these questions are designed so that they may be asked repeatedly, with mounting evocative consequences, no matter the care with which moral concerns had been previously addressed. They are devised as argumentative weapons shielded from rational response.

B. Professor Chargaff is master of the technique, but far from its only practitioner. Clifford Grobstein, Professor of Biology and Vice-Chancellor for University Relations, University of California, San Diego, presents (also in *Science,* December 10, 1976) a series of "ethical, social, and political issues" about which "substantial concern has been expressed" and which "remain to be evaluated." He continues:

(6) "Are there some kinds of knowledge, even though they offer health benefits, for which the price in other values is too high?"

(7) "Is it safe, in the present state of our society, to provide means to intervene in the very essence of human individuality, even to achieve humanitarian ends?"

(8) "Can genetic destiny, whether of human or other species, wisely be governed by human decision?"

(9) "Will genetic engineering widen or close the existing gap between knowledge-rich and knowledge-poor cultures and nations?"

(10) "Will it [genetic engineering] provide a new club in the hands of terrorists or dictatorial regimes?"

(11) "Will it [genetic engineering] render still more turbulent the currents of national and international power conflict?" [4]

C. Frances R. Simring, writing to *Science* as a representative of the Committee for Genetics, Friends of the Earth, applies the heavy question technique to the narrower issue of laboratory containment.

(12) "What scientist would claim that complete laboratory containment is possible and that accident due to human fallibility and technical failures will not inevitably occur?" [5]

D. Liebe F. Cavalieri, Professor of Biochemistry, Graduate School of Medicine, Cornell University aims his heavy question at any efforts to develop weakened strains of host bacteria that will be unable to survive outside the laboratory. He asks:

(13) "[H]ow can all the possible bacterial growth conditions outside the laboratory be simulated for the test?" [6]

E. Professor Robert L. Sinsheimer, Chancellor of the University of California at Santa Cruz, uses heavy questions in several different argumentive contexts [7]. In his essay, "Troubled Dawn for Genetic Engineering" [8], appear his heaviest queries:

(14) "How far will we want to develop genetic engineering?"

(15) "Do we want to assume the basic responsibility for life on this planet—to develop new living forms for our own purposes?"

(16) "Shall we take into our hands our own future evolution?"

Some two years later, at a Forum conducted by the National Academy of Sciences in Washington, D.C., Professor Sinsheimer sharpened and reapplied the heavy question method of argument [9].

(17) "Does anyone imagine that the roster of carcinogens or mutagens has been completed?"

(18) "Could their ingredients [i.e., those of slow viruses] lurk in these random bits of genome we now juggle?"

(19) "Are there really no evolutionary booby traps for unwary species?"

F. George Wald, Nobel laureate and Higgins Professor of Biology, Harvard University, helped to fuel the politicized disputes over DNA research in the City Council of Cambridge, Mass. Later at an Academy

Forum in Washington, Wald had some difficulty meeting the arguments of Paul Berg (Willson Professor of Biochemistry, Stanford University Medical Center) respecting the potential benefits of increased understanding of the regulatory mechanisms of animal genes. Wald responded with a heavy question:

(20) "I think the central problem before us in this direction is how many normal, healthy persons to put at risk in order to achieve the possibility, not at all clear, of eventually proceeding to cures." [10].

G. The reasons for the special attractiveness of heavy questions in this sphere I will turn to shortly. But the absence of real answers— answers that, as the questioners themselves sometimes admit, are impossible or nearly impossible to give—encourages heavy questions at the metalevel. Why don't they answer our questions? Might not the answers be fearsome? Jonathan King, Associate Professor of Biology, Massachusetts Institute of Technology, provides an example of this escalation.

(21) "Now I ask you, what is going to happen if by some chance, some small chance, the Walds, and the Hubbards [Ruth Hubbard, Professor of Biology, Harvard University], and the Chargaffs, and the Cavalieris are right; the experiment is done, and we get the answer—a disaster. Where will we be?" [11].

The list of illustrations need not be extended. What I shall say about this pattern of argument will refer exclusively to its use by major critics of recombinant DNA research, those cited above bearing fine scientific reputations. It is important to realize, however, that after persons of high repute have indulged publicly in this mode of discourse, an excited bandwagon quickly forms. The argument is taken up by serious but much less knowledgeable persons. And it is they—some representatives of public interest groups [12], some journalists [13], and some political figures [14], who are most effective in reaching large audiences. When it is also widely held that the issues of restriction or continuation of such research should be decided in the political forum, by an informed public, the instruments of public information come to be of crucial importance [15]. With the example of Professors Chargaff, Sinsheimer, and others before them, some political persons find the temptation of the heavy question too strong to resist.

Arguments by heavy question are very troubling. The instances cited here are at differing levels of sophistication, obviously; but they are registered seriously, with practical argumentative purpose, by able scientists whose lead has been widely followed. I emphasize that heavy questions are not peripheral to their positions, or merely incidental to them. To the contrary, such questions serve as the pillars upon which the attack has rested. Coming from reputable scholars, they have been quoted and requoted as putative profound critique of recombinant DNA research. Both the forms of these arguments and their substance, therefore, must be carefully looked to.

First, some observations about argumentative form.

1. Questions—and these questions very clearly—*do not assert any-thing*. This is an obvious but fundamental point. Whatever may be suggested by a question, or implied by the terms used in it, or inferred by the hearer from the attitude of the person doing the asking, the interroga-tive does not express any proposition whatever. It cannot, therefore, be true or false. Precisely for this reason, no responsibility need be taken by one who asks a question for any claims that it may suggest. If, in reply to the questioner, one points out that some suggestion implicit in his ques-tion is mistaken, the questioner may always (and sometimes honestly) reply: "But I never asserted that!" Quite so. The questioner asserts nothing; his utterance, therefore, cannot serve even as the premise of an argument.

2. Some of these interrogatives are textbook examples of the very common fallacy of the compound question [16]. What color is the presi-dent's beard? Interrogatives may be so framed—or loaded—as to permit response only upon the assumption of the truth of some related claims that are implied but not asserted[17]. It is difficult to grapple with some of these, because the related propositions whose truth they assume are not always easy to formulate. Where they can be formulated, they turn out to be either (a) very probably false, or (b) quite uncertain and highly debata-ble, or (c) entirely unverifiable. Such related propositions are often mix-ed, by this interrogative form, with matters that are capable of rational disposition. The resultant brew is pungent.

Look more closely at some of the illustrations above. Before answer-ing whether "we have the right to put an additional fearful load on future generations" (2, Chargaff), one is obliged to assume both that the inquiry in view will put a "load" on future generations, and that it will be a "fearful" one. In the sense that every acquisition of knowledge leaves its inheritors with the task of deciding how it is to be used, all science puts a load on future generations—although we may find the noun "load" to be loaded in the context. Whether, in this case, the "load" will be "fearful" is, in fact, precisely what is at issue—but as-sumed in a question whose covering suggestion, wholly unsupported, is that some people are doing what they have no *right* to do. How is one to deal with such argument? To reject one or more of the implicit as-sumptions one cannot simply respond to the questioner, who did not assert them, but must address some apparently neutral third party. And even to formulate that rejection appears to give substance to insinuations without their ever having been defended or even straightforwardly advanced.

Again, consider the logical difficulties in responding to "Have we the right to counteract, irreversibly, the evolutionary wisdom of millions of years, in order to satisfy the ambition and curiosity of a few scientists?" (5, Chargaff). Just to take the question apart requires great effort and more space than the answerer is normally given. Irreversibility of results is an important and an interesting matter. Defending some research, we might wish to deny that it has irreversible effects. Or we might wish to allow that its effects are indeed irreversible, but only in the same sense that the effects of any important experiments are irreversible. Experiments with many drugs, on humans, are not reversible; neither are experiments with a nuclear reactor. What is objected to as "irreversible" may in fact be troubling to the critic not because it is that, but because the effects have, he believes, an unmanageable scope. Whether research with recombinant DNA has effects with unmanageable scope is precisely part of what is at issue, and on that point its defenders give detailed and well-supported response. To frame the complaint in terms of an "irreversibility" that is undeniable because it is the logical consequence of experimentation itself, clouds an important substantive matter.

Before such matters of arguable substance can even be reached, however, this often-quoted question obliges the assumption by responder both that evolution over millions of years has been "profoundly wise," and that the research in question will "counteract" it. These are doubtful propositions, to say the least. Regarding the first of them, a leading microbiologist (and DNA researcher), Stanley Cohen, Professor of Medicine, Stanford University, replies for much of the scientific community:

> I would like to point out that this so-called evolutionary "wisdom" has given us bubonic plague, and smallpox, and yellow fever, and typhoid, and diabetes and cancer. The search for and the use of virtually all biological and medical knowledge represents a continual and intentional assault on what Dr. Chargaff considers to be evolutionary wisdom. Most post-Darwin biologists believe that there is no wisdom in evolution, only chance occurrences. Do we really desire to glorify chance evolutionary occurrences as "wisdom" and to accept without protest or countervening action the diseases and plagues that such "wisdom" has bestowed on mankind? I would suspect that most of us are not prepared to simply endure whatever nature may have in store. Thus, science continues to search for new ways to influence the "wisdom" of evolution [18].

And regarding the unformulated claim implicit in the use of the verb "counteract," response is given by James Watson, Nobel laureate for his discoveries concerning the structure of DNA:

> The two types of experiments under the original moratorium pose no real threat to the general public. Much too quickly, we concluded that it was dangerous to make bacteria which are resistant to many antibiotics or which *can* synthesize a deadly poison even though they normally don't. Instead, we should have focused upon the fact that most, if not all, bacterial species already exchange DNA with each other in nature—for example, through the infection process. Thus, if through recombinant DNA technology, we were to make an *E. coli* strain that, say, makes the cholera toxin, *we are very likely repeating what nature has done many times in the past.* There is every reason to believe that even if it did escape from the laboratory it would not pose any major public health threat. Even less convincing, especially in retrospect, were the arguments against putting the genes of tumor viruses into laboratory strains of bacteria. The argument here was that such cancer-gene-bearing bacteria might accidentally colonize parts of the human body, releasing cancer-causing DNA that might pass into our cells and initiate a cancer. By the time of Asilomar [February 1975], however, we already realized that many, if not all, of the so-called DNA tumor viruses were in fact ordinary animal viruses that routinely infect most of us early in life. By still unknown means, they remain latent in our bodies for the remainder of our lives, usually only expressing themselves as disease-causing agents under various physical conditions (such as Herpes virus-induced cold sores). So, the danger we face from our intestinal bacteria acquiring a little cancer virus DNA must be negligible compared to that we face every time we are infected with any of the innumerable DNA-containing viruses [19].

Such responses are by-passed, however, foreclosed by the suggestions implicit in the Chargaff question, whose ostensible thrust it is that terrible things are being done "in order to satisfy the ambition and the curiosity of a few scientists." Now there is no doubt that the results sought in this sphere would satisfy the ambition and curiosity of some scientists. Ambition and curiosity do play a role, sometimes an important role, in the motivation of scientists. Ambition can sometimes be overweening, but it is not in principle a wrongful motivator; curiosity on the part of the scientist is surely to be admired. All this simply obscures what is at issue here, however, since these motivations, even if their role could be ascertained, might only help to account for the personal efforts expended—while the matter before us is the justifiability of the research itself, or its restriction. That is a subject entirely separate from the motivation of those who may pursue it. Even if it were the case that this research were being pursued only by researchers having only those motives (another claim suggested but not asserted by the question), that would be almost entirely irrelevant to the justifiability of restricting it. In fact, although the frame of the question gives little opportunity to show it, many of those who pursue this work do so with the serious aim of advancing human knowledge, and the honorable motive of bettering the human condition.

It is at best awkward to sort all this out before getting to the question—which was about what? It was about rights! "Do we have the right to. . . ." Answering that question supposes a framework of moral theory the questioner has not even hinted at. Were such a framework put forward in response, it is doubtful that he would accept it. The point of the question, I submit, is not to inquire about rights, but in seeming to do so to insert a set of allegations about the imagined consequences of the research and the character of the researchers—so folded into one another that the answerer must drown in words while extricating himself. If the effort to extricate is made, it is almost certain to go unheard in the circles for which the impact of the question was intended. Anyway, the rejoinder is easy: repeat the question in different but similar form.

3. Some of these compound questions have the effect of attaching certain descriptive predicates to recombinant DNA research in the hearer's mind, without the questioner having to take responsibility for asserting their applicability. "Will it[genetic engineering] provide a new club in the hands of terrorists or dictatorial regimes?" (10, Grobstein) How to respond? No one can deny that any new, powerful instrument may someday be used by bad people. Since terrorists and dictators need instruments that pose immediate threat of pain or disaster, the vast array of clubs already at hand seem more suitable to their purpose than any uses—even far-fetched uses—of recombined DNA. But the association has been made; our imaginations have been set to work; no answer can erase the smirch.

Again, "Does anyone imagine that the roster of carcinogens or mutagens has been completed?" (17, Sinsheimer) "Could their ingredients [i.e., those of slow viruses] lurk in these random bits of genome we now juggle?" (18, Sinsheimer) Zap! Pow! No matter the responses calmly given, or the evaluation of these responses in the light of fullest knowledge; the damage has been done: "carcinogens," "mutagens," "lurk," "juggle." The very process of replying to such questions, or even of trying to reframe them so as to be able to reply fairly, obliges the researcher to crawl into the net thrown.

And again, when the Mayor of Cambridge brings to the legislative process a discussion of things "crawling out of the laboratories into the

sewers" [20], that is not different in tone from the query of a distinguished professor of biochemistry who asks, "If Dr. Frankenstein must
go on producing his little biological monsters . . . [h]ow can we be sure
what would happen once the little beasts escaped from the laboratory?"
(1, Chargaff) The only appropriate response to questions emotionally
loaded in this way is the exhibition of their character.

4. Some of the questions are purely rhetorical, of course, their
answers being either perfectly obvious or totally inaccessible. What
happens if the outcome of some experiment is disaster? "Where will we
be?" (21, King) In a disastrous situation, clearly. "Is there anything more
far-reaching than the creation of new forms of life?" (3, Chargaff) Who
can say? The destruction of all or much life might prove much more
far-reaching. But rational responses are not really sought by such questions, which is why giving such responses is pointless. Although in the
form of questions, these utterances actually function in discourse more
like expletives, or intensifiers. Because we respect their authors' honest
feelings of concern, and may share them, we are tempted to nod our
heads to the question—acquiescing inadvertently to the suggestions
embedded within them.

I turn to underlying issues of substance. Once having penetrated
the forms and rhetorical purposes of these heavy questions, we may
distinguish two kinds of such questions differing in thrust. The first asks
about some unknown mis*haps,* the second about some unknown mis-
uses [21].

Of questions anticipating mis*hap,* there are two subkinds. The first
asks about mis*fire* (achieving results different from and more dangerous
than those sought); the second asks about *accident* (resulting from inadvertent escape). Here are some examples from the list presented earlier:

Misfire: "Are we wise in getting ready to mix up . . . the genomes of
eukaryotic and prokaryotic cells?" (4, Chargaff) "Are there really no
evolutionary booby traps for unwary species?" (19, Sinsheimer)

Accident: "What scientist would claim that complete laboratory
containment is possible and that accident due to human fallibility and
technical failures will not inevitably occur?" (12, Simring) "How many
normal, healthy persons must be put at risk in order to advance the
possibility . . . of proceeding to cures?" (20, Wald)

Of questions anticipating mis*use,* three subkinds may be distinguished, though the lines between them are not sharp. The first asks
about *demonic* misuse, the second about *fortuitous* misuse, the third
about *wholly unknown* misuse. Here are some examples from the list
presented earlier:

Demonic misuse: "Will genetic engineering provide a new club in the hands of terrorists and dictators?" (10, Grobstein)

Fortuitous misuse: "Will genetic engineering widen or close the existing gap between knowledge-rich and knowledge-poor cultures and nations?" (9, Grobstein)

Wholly unknown misuse: "Shall we take into our hands our own future evolution?" (16, Sinsheimer) "Are there some kinds of knowledge, even though they offer health benefits, for which the price in other values is too high?" (6, Grobstein)

The fallacies of compounding and loading—needing no further exhibition—may and do infect questions in all these subtypes. Underlying all of them, however, lies a substantive issue of some complexity: where lies the burden of proof? Two parties are in deep disagreement. One, the researchers, would move ahead with inquiries using recombinant DNA; the other, their critics, raise heavy questions with which they hope to stop, delay, or otherwise restrict such inquiries. We need now to determine whether the burden of proof, with respect to the matters raised in these questions, rests upon the researchers or upon their critics. We need to decide whether there should be freedom to advance until restriction has been justified, or restraint in advancement until freedom has been justified.

Implicit in argumentative heavy questions is the strong suggestion that until the doubts they raise have been satisfactorily resolved, the research contemplated ought not go forward. That, I submit, is a serious mistake. The burden of showing that the answers to these questions (if they are determinable) would not seriously trouble us, or of showing that the questions cannot be answered, or of showing that the questions are in critical cases improperly framed, does not rest upon those who seek to advance the inquiry. To appreciate fully why the placement of the burden of proof is crucial in this matter, and why the burden does not rest upon the researchers, two features of the situation need extensive clarification.

Alpha. The first of these features is the nature and degree of our ignorance in this sphere, and the limits upon inferences to be drawn from it. Some of the heavy questions cited suggest grave but unspecified social consequences for which no grounds are given. Some raise questions about the course of evolution and our proper role in it. Some suggest catastrophic biological possibilities for whose realization there is no evidence. Some suggest disasters in terms so vague that evidence against them cannot even be described. This is the archetypical parade of horribles.

It is true, of course, that there is a world of knowledge about genetic molecules, their role in evolution, and the effects of their recombination, that we do not have. Many very awful things are logically *possible*. We do not know, and probably will not ever be certain, that these awful outcomes cannot transpire. When Sinsheimer asks whether the ingredients of slow viruses *could* "lurk in these random bits of genome we now juggle," the only rational answer is yes. When Chargaff asks, "How can we be sure what would happen once the little beasts escaped from the laboratory?" the only rational answer is that we cannot be sure—either that they are beasts, or that they will escape, or what will happen were those events to transpire. But nothing of consequence for the argument may properly be inferred from such ignorance. The future events we cannot be sure about are literally numberless. No argument against advancing on some line of scientific inquiry can be rationally grounded on the sheer possibility of misadventure. It is a truism that such possible misadventures "lurk" in every sphere, even in the business of everyday life, and certainly in the conduct of political affairs. If heavy questions such as these must be answered before the inquiry is allowed to proceed, advance on virtually every research frontier, in chemistry, physics, and biology, not to mention every bold political venture, is subject to the same blockage. Some inquiries, now critical in climatology, and in high-energy physics, and elsewhere, are far more vulnerable to such gloomy speculations than those in microbiology. Possible dangers are certainly worth thinking about—but the sheer possibility of them cannot be taken as a serious argument for restriction.

"But[the critic replies] the dangers upon which I speculate are more than sheer possibilities. There is some probability that they will be realized if this research goes forward. My heavy questions oblige you to consider the substantial weight of these factors before proceeding."

This objection is only superficially reasonable. It is in fact virtually impossible to weigh the "danger factors" suggested in these heavy questions. Any such weighing would presuppose both (a) some moderately specific description of the catastrophic eventuality envisaged, *and* (b) some reasonable estimate of the probability of that outcome. In any calculation of the advantages and disadvantages of following some line of conduct or not following it, the product of a given result multiplied by the likelihood of that result is required. Logically conceivable outcomes, for which no such product can be rationally estimated, cannot be serious considerations in appraising conduct. Most of the outcomes intimated in the heavy question of the critics cannot be described at all. Some are not even identified, but are left for the imagination to conjure up (e.g., "Is it safe, in the present state of our society, to provide means to intervene in the very essence of human individuality?" (7, Grobstein) Not one of the "danger factors" alluded to approaches the degree of specificity that would be required to estimate in the roughest way (much less calculate) its probability on present evidence. The thoroughly amorphous quality of the dangers suggested disqualifies them as serious "probabilities" to be weighed.

For those imagined outcomes having some very vague description, the task of estimating the likelihood of their occurrence remains. "Are there some kinds of knowledge, even though they offer health benefits, for which the price in other values is too high?" (6, Grobstein) How could one answer such a question? What are the prices paid in values? How are they estimated or compared? Is this a calculation we are seriously expected to undertake? Perhaps the questioner may respond as follows: "Well, it *is* a hard question. But that's the point. We don't know. We cannot even guess what the price of knowledge in this sphere will be. Oughtn't we try to find out?" No one will object to that exhortation. Any guidance we can get on the prices of knowledge, in whatever currency we can identify, will help us to decide whether we are prepared to pay for it. But having no idea at present what the price (in undescribed currency) we may (with indeterminate probability) be obliged to pay (in the unspecified future) cannot reasonably serve as grounds for restriction on the search for that knowledge.

Again, Professor Cavalieri, after asking how "all the possible growth conditions outside the laboratory can be simulated" for the development of new bacterial strains, and after speculating upon eventualities he himself calls "far-fetched," undoes his own argument with his own clincher: "There are no objective answers to these questions." [22]. Just so. That is why their being asked has no probative force in support of restriction.

"But [the critics may rejoin] you fail to appreciate the magnitude of the disasters we are asking about. It's true we can't describe them. We can just imagine them—great plagues, or blights, or the like. It's true also that we haven't any sound estimate of their likelihood. But suppose the likelihood is small—even very, very small. Won't you agree that the product of a catastrophe of enormous dimensions, multiplied by even a very small probability fraction, yields a significant danger factor? And doesn't that serve as a proper consideration in deciding whether to go ahead now?" [23].

Some argument of this kind, I submit, lies behind many of the almost inchoate questions that have been put before us. It embodies two serious mistakes.

First. The application of a probability calculus to such states of affairs is spurious. We may talk as though the event in question had a certain probability, although very small. In fact, we don't have any real event in mind, and we don't have any serious estimate of its likelihood, even that it is infinitesimal. The whole enterprise is one of wildest conjecture. It is—as mentioned above—a variety of conjecture that can go on, equally rationally, with respect to almost any form of human activity. The more vivid our imagination, and the more cautious our souls, the less likely we will be to undertake any inquiry, or any enterprise of any kind that could, with any degree of probability, yield the outcomes imagined. This game of frightening speculation can be played with respect to far-fetched outcomes of not acting, as well as of acting. As easily as we can dream up catastrophes flowing from the malevolent future use of recombinant DNA technology, we can imagine scenarios in which changes in our natural environment will require, for the very survival of the human species, a capacity to control genetic codes that is just now becoming possible. Extinction, or other disaster, *could* be the result of failing to develop that control[24]. Science fiction can be written in many ways.

Second. Arguments from ignorance really must, in this case, be turned back upon their proponents. The critic asks whether there may be some very small probability of some very great disaster flowing from recombinant DNA research. The researcher is fully as entitled to ask whether there may be some very small probability of some very great and wonderful boon for humankind flowing from that research. Might we soon eliminate all genetic diseases and malfunctions? Might we greatly increase, perhaps double, the length of productive life for all humans? Might we learn how wisely to stabilize the population of the globe, and through the mastery of genetic codes eliminate hunger forever? Might we learn how to open new realms of human consciousness? It would be irresponsible to argue that the hope of accomplishing these ends in the foreseeable future through the recombination of DNA and allied research *justifies* anything. It is fair to say, however, that since some of these outcomes may in fact become actual research objectives, the likelihood of their occurrence may be as great as or greater than that of imagined superdisasters. If the game of multiplying conjectured states of affairs by wholly conjectural quantities is taken as a serious enterprise on the one side, it can be taken as seriously on the other[25]. Against some unspecified but profound evolutionary disruption multiplied by its conjectured probability, let us put on the scales the elimination of all cancers and genetic diseases, multiplied by its conjectured probability. Both parties may find some satisfaction in the manufacture of descriptions and numbers. Neither, surely, will have proved anything [26]).

But if such speculations be taken seriously, one is probably better advised to stake his bets on the imaginative researchers than on the imaginative critics, not because the researchers' powers of clairvoyance are stronger, but because they have the advantage of being able to say, with some degree of concreteness, where they want to go, and to push, when pushing is possible, in that direction. They, at least, have good reasons to want to reach their envisaged destinations [27].

Both sides in these disputes are pained by their ignorance, of course, dismayed by it. It is precisely ignorance that we seek to overcome. Professor Sinsheimer expresses this dismay most poignantly:

> Can we predict the consequences? Except in the most general terms we are ignorant of the broad principles of evolution, of the factors that govern its rate and directions. We have no general theorems to account for the spectrum of organisms that we see and the gaps in between. . . . We simply do not know.
>
> We are ignorant of the relative importance of the various factors we currently perceive to participate in the evolutionary process. . . .
>
> We are ignorant of any absolute measure of adaptation. We are ignorant of the depth of security of our own environmental niche. . . . We do not know [28].

All profoundly true. But the inferences from such ignorance must be carefully drawn. If the resultant admonition is simply, "Be cautious" (as Sinsheimer himself appears in some contexts to conclude), no one will decry his wisdom. If sharp restriction upon inquiry is admonished, the argument is very different indeed. Premises that may support caution will not support anything close to prohibition [29].

Heavy questions seem persuasive because they play upon our ignorance. But ignorance of the kind they rely upon does not shift to any legitimate scientist the burden of proving his inquiries danger-free. Presenting the indeterminable possibility of some barely describable outcomes of some research cannot serve as an argument for restricting or prohibiting that research.

Beta. Does the breadth and depth of our ignorance leave the dispute a standoff? Since we are unable to do the multiplications of unknowns on both sides, are we left in neutral position, to decide upon advance or retreat only by resorting to our subjective feelings of probability, or by responding to whatever impulses we feel toward caution or venture? No, the case is not like that at all.

If consequentialist considerations left the decision to inquire or to restrict inquiry in perfect balance, there remains a second critical element affecting the placement of the burden of proof. It is the principle that freedom of inquiry must be protected. Before examining that principle and its applications here, however, it should be emphasized that the discussion of the argument from ignorance (under *Alpha,* above) does not lead to the conclusion that consequentialist considerations are in perfect balance. Quite the contrary, in fact. Best judgment after the most careful reviews, by outstanding individual scientists, by the most reputable groups of scientists, and by carefully charged mixed groups of scientists and laymen, comes down strongly, on utilitarian grounds, in support of the active continuation of research with recombinant DNA.

Conjectures from ignorance give weight to neither side, as I have argued above. But very reasonable estimates can be made of the progress to be anticipated with recombined DNA, both in applied medical science and in theoretical biology. There is substantial knowledge of the techniques of laboratory containment; there is substantial evidence regarding the weakness of certain laboratory strains of host bacteria and the odds against their survival outside the laboratory if containment failed. With such evidence in hand, and the evidence provided by progress already made, it is entirely reasonable to pursue certain moderate, intermediate objectives with recombinant DNA, and even to estimate the length of time that will be required to achieve them. We can now pass fully humane and rational judgment on this matter. If the decision to restrict or to advance research in this sphere were only a matter of weighing, as on a balance scale, the magnitude of the reasonably anticipated benefits multiplied by their reasonably anticipated probabilities and summed, against the magnitude of the reasonably anticipated risks multiplied by their reasonably estimated probabilities and summed, the scale falls heavily on the side of continuing this research vigorously [30].

That reasoned judgment is only part of the story, however. Suppose, *arguendo* (what is not the case), that the scales of estimated risks and benefits were exactly, or as nearly as could be determined, balanced. What role in the decision process should be played by our commitment to freedom of inquiry?

Freedom of inquiry—it will be agreed—is a value of the highest concern, lying at the heart of the scientific process itself. It is so profoundly important to all of us that it must not be abridged without the most compelling reasons. This proposition will be affirmed by virtually every scientist and scholar. In a society holding liberty as a paramount ideal, the centrality of this proposition is underscored. When, in such a society, the issue of restriction of inquiry arises within the precincts of universities and research institutions explicitly devoted to the quest for new knowledge, commitment to that freedom is further intensified. We are not faced with the need simply to choose between two sides of a balance scale. The decision *not* to restrict inquiry is one for which we have substantial grounds before the scales are put into action.

Those who argue, as I do, that the commitment to free inquiry is critical to the placement of the burden of proof, do not contend that never, under any circumstances, may restrictions upon inquiry be imposed. Other weighty moral principles, combined with overwhelming evidence on the other side, might in very special circumstances override [31]. But it is most important to see that we rightly do not proceed in these matters as though the decision to restrict or not to restrict were made by the crude weighing of outcomes and their utilities. The commitment to free inquiry (although itself defensible with a sophisticated utilitarianism) must enter with powerful effect. Given its entry, the burden of proof lies heavily upon those who would in some degree override it.

The advocates of restriction are not oblivious to this argument. They respond in two different ways, both mistaken.

1. The first response takes the form of a denial that restriction of this inquiry (into recombinant DNA) is a restriction of free inquiry. Professor Wald says: "I have heard the issue raised . . . of suppressing free scientific inquiry. . . . No one I know is trying to suppress the inquiry"[32]. But immediately preceding that disclaimer he expresses the hope that the "big question" would be tackled, not the question of how to do the research safely, "but whether to do it at all." Wald cannot have it both ways. If one seriously contemplates not permitting the research to be done at all, he considers the restriction of freedom. Conceivably, evidence believed sufficient to justify that restriction may be gathered. But to suggest that a certain line of inquiry may have to be formally closed, and in the same breath that no suppression is involved, is doublespeak.

Professor King defends the restrictive view similarly. He argues:

> This is not a question of freedom of inquiry. This is a question of freedom of manufacture, of modifying the environment, of modifying living organisms, not of asking questions about them, but of the route which you take in getting the answer [33].

But science is all routes. King's complaint suggests that here (and perhaps elsewhere in science) we face known sets of questions, and must choose among several paths to the answers sought, some of those paths being too risky. This implicit picture is a grave distortion of the actual state of affairs. In the first place, there is no set of questions—here or anywhere in science—that circumscribes the ventures of research. There are very many things we know that we do not know, and that we are deliberately seeking to find out. But we cannot begin to identify all the things we do not know. Science progresses as much in learning to ask the right questions as in getting answers to questions already well asked. It is particularly myopic to suppose that in a newly opening sphere of biological inquiry we can identify certain results to be sought and specify certain ways in which scientists are to be permitted to pursue them. In the second place, even where certain specifiable questions can be asked as a result of progress to date with recombined DNA, it is grossly misleading to suggest those same questions may be answered in some other way, and that what is involved in the restriction is merely the preference of one "route" to another. In fact, blocking this path of research—the use of recombined DNA—*amounts in practice to the suppression of inquiry into whole sets of questions for which there are no other known or foreseeable routes of investigation.*

The recombination of DNA molecules will certainly raise more questions, in the short and the long term, than it will answer. Every step in the advance will open new routes; many steps will suggest destinations now undreamed of. Hence our perennial anxiety about restriction. We must not foreclose—except rarely, perhaps, for reasons of demonstrably compelling force—lines of inquiry that may open doors to aspects of reality now so dimly understood that we cannot even frame the right questions about them.

The warning that certain "routes" in science are too dangerous to permit has a painful history. The case of Galileo is commonly mentioned in this connection and is indeed specially apt. Scientists and scholars who were not stupid and not pernicious argued seriously that he should not be allowed to pursue certain lines of inquiry, both because of the malevolent uses to which his results might be put and because, should he err in his researches, the damage might prove intolerable. The instruments of his inquiry were considered literally infernal. Speculation about the celestial bodies his critics did not forbid. But, said they, some outcomes are not to be tolerated, some routes not to be taken.

The critics of recombinant DNA rejoin: "To liken us to the Inquisition of the seventeenth century is absurd. What we would restrict is not inquiry, but manufacture. The asking is all right; it's the making that is too hazardous." That distinction, however, simply does not stand up under examination. Virtually every scientific inquiry (excepting the most purely theoretical) involves the manipulation of nature, not simply the raising of questions about it. If there is one feature that distinguishes modern science from its reflective ancestors, it is the drive to *experiment*, to lead nature as well as to be led by her. To the extent one does not permit nature to be led he abridges not only that experimental spirit, but also the very possibility of meaningful experiment in that sector. Imagine Louis Pasteur advised that inquiry into immunological systems is appropriate—but that manipulation of the microorganisms themselves is too dangerous to allow. Preventing disease raises questions very worth answering, his critics might have said, but not by that route. Indeed, the reasonable evidence that might have been registered then against permitting Pasteur to proceed was more substantial by far than any registered now in the case before us.

Part of the difficulty here arises from confusion about what "freedom of inquiry" is, and what it protects. Is it simply another version of the protection of free speech? Or is it something more than that, a branch or variant of academic freedom, offering special protections to persons professionally engaged in the pursuit of knowledge? It is the critic's assumption that freedom of inquiry is no more than the former that leads him to say, as King does, "This is not a question of freedom of inquiry. This is a question of freedom of manufacture." Underlying such remarks is the premise that the "inquiry" whose freedom we honor consists of speaking, expressing, asking—but not of doing. Freedom of speech, after all, is to be distinguished from freedom of action; it is one thing to advocate a revolution, another to begin one.

But the freedom of inquiry rightly protected in the context of scientific research is not simply free speech. It goes substantially beyond liberty with words, extending to *activities* of various kinds—so long as those activities are legitimately devised to advance the pursuit of truth and conducted by competent professionals guided by professional codes of conduct. A very great deal of the most valuable medical and biological research does and must involve "modifying the environment" and "modifying living organisms." Virtually all of modern molecular biology involves some form of manufacture or modification. These activities cannot be distinguished from scientific inquiry; they are parts of its essential nature.

Whether, from a jurisprudential point of view, freedom of inquiry thus understood is to be defended as flowing directly from the First Amendment of the U.S. Constitution, is a question I by-pass here. William Van Alstyne, Professor of Law at Duke University and former President of the American Association of University Professors, has argued persuasively that academic freedom, properly construed, should be so defended[34]. John Searle, Professor of Philosophy at the University of California at Berkeley, does not address the question of constitutional foundations but agrees with Van Alstyne that the freedom of academic inquiry provides protections going well beyond those afforded controversial speech. Searle distinguishes two (not necessarily inconsistent) theories of academic freedom, a "special theory" and a "general theory"[35]. The general theory ties some of the rights of scholars to the rights of expression enjoyed by all citizens. The special theory of academic freedom explains the protection of additional rights derived from our conception of the nature of the university, its objectives, and how those objectives can best be achieved. Put shortly, the account goes like this:

We begin with universal agreement that knowledge is of extraordinary value, and that the university has as one of its primary objectives the enlargement and dissemination of knowledge. We understand that knowledge is most likely to be advanced through inquiry in which knowledge claims are put to test, including experimental test. Professional experimentalists earn their respected position by diligent application and by their proved mastery of the techniques of investigation and validation in their discipline. We protect scientific investigators in their inquiries because they have special qualifications to advance the search for knowledge that amateurs have not [36].

With this "special" theory of academic freedom Van Alstyne is in accord. He finds the further development of a "general theory" to be misleading, in suggesting that academics are somehow more fully entitled to political freedoms than are ordinary folks. But he agrees that professional scientists and scholars are additionally protected by academic freedom, if the scope of that protection be carefully and properly understood. Van Alstyne writes:

> Insofar as it pertains to faculty members in institutions of higher learning, "academic freedom" is characterized by a personal liberty to pursue the investigation, research, teaching, and publication of any subject as a matter of professional interest without vocational jeopardy or threat of other sanction, save only upon adequate demonstration of an inexcusable breach of professional ethics in the exercise of that freedom. Specifically, that which sets academic freedom apart as a distinct freedom is its vocational claim of special and limited accountability in respect to all academically related pursuits of the teacher-scholar: an accountability not to any institutional or societal standard of economic benefit, acceptable interest, right thinking, or socially constructive theory, but solely to a fiduciary standard of professional integrity [37].

Two points respecting that freedom—the same freedom defended by Professor Searle and repeatedly by the AAUP—are properly underscored:

(a) It is a liberty, in the strict sense, marked by the absence of restraint or any threat against its exercise. It is not an enforceable claim against the assets of others. Freedom of inquiry is not a right in the strong sense that the scholar or scientist can demand the provision of means needed to pursue the particular line of research he thinks wise. Wisely or unwisely, institutions may decline to provide special support for their members without abridging their freedom of inquiry. Benefits or risks promised by a given line of research may reasonably enter into such decisions. Whether support be warranted must be determined through the review of the professional credentials of the investigator and of his research program by his scientific or scholarly peers.

(b) Assuming that there has been "no failure of professional integrity" in the pursuit of knowledge, academic freedom, in this restricted but critical sense, "protects the rights of faculty members to conduct whatever instruction and research they may be retained to provide" [38].

From this it follows that academic freedom—viewed as an identifiable subset of First Amendment freedom—"requires a significant modification in the standards of judicial review otherwise applicable to freedom of speech" [39]. Van Alstyne gives the illustration of the appropriateness of protecting the professional study of some literary material—say, the hardest-core pornography—even had the state generally criminalized its use. More to the present point would be our common conviction that scientific researchers may be permitted to handle, study, and experiment with drugs whose mere possession is generally a crime. There are rights of inquiry, in short, that flow from the nature of the knowledge-getting process and from our profound commitment to that process [40]. Van Alstyne is eloquent:

> The distinction of academic freedom from the general protection of free speech is precisely located in its immediate and indissoluble nexus with the cardinal social expectation laid upon the particular profession with which it is identified. . . . Professionally related efforts directed in good faith precisely to fulfill the social directive of the academic profession . . . will make the case appropriate for the constitutional protection of academic freedom when the absence of these elements might otherwise spell its failure [41].

The scientific researcher and scholar is *expected* to experiment, to test, to press the frontier of understanding about the natural world. It is, as Van Alstyne says, "simply contradictory to lay that expectation upon the profession and then to prevent its accomplishment by deterring its fulfillment through rules that punish its exercise" [42].

This does not mean, obviously, that the professional investigator is free from all restraints, that academic freedom is an absolute. Institutions may and do formulate rules for the guidance of research within their precincts. Medical experimentation with human subjects, an invaluable instrument for the achievement of humane and scholarly objectives, is yet properly subject to carefully formulated restrictions [43]. The freedom of academic inquiry does, however, go well beyond protecting speech. That key point is missed by those who suppose that the freedoms involved here are no more than those required by an open political forum. Freedom of inquiry going beyond that forum is involved here. It is very precious, for reasons already given. Only when this liberty is demonstrably outweighed can the restrictive judgment stand. It is not the burden of the scholar or scientist to show that such demonstration can*not* be given.

2. A second reply by some advocates of restriction is more forthright. They contend, not that the principle of free inquiry is not at stake, but that, although it clearly is at stake, it is simply overriden here by other considerations.

Shaw Livermore, Professor of History at The University of Michigan, concluding that recombinant DNA research should not now be pursued, writes as follows:

> The claims of free inquiry and individual initiative are among the most zealously guarded in a free society and they should remain so. Only when the continued vitality of that free society is threatened should one contemplate a moratorium, but wisdom and moral responsibility require that a choice be made when such a threat appears. Free inquiry must in the last instance be understood as a means by which we seek to achieve the most profound human ends [44].

But the extraordinary value of this means has been so long, so repeatedly, and so incontrovertibly established, that we cannot seriously doubt its status as valued *end* also. The continuum of ends and means is real. Free inquiry is not a means in the narrow sense that a specific tool is a means for some limited enterprise. It is the principle that grounds all other intellectual means; its power—and our willingness to abide for a while with what we may think mistaken—lies partly in the general confidence that free inquiry will remain operative and not be restricted save for reasons of the most urgent and compelling sort. Such reasons have not been given in this context.

Professor Liebe Cavalieri is more blunt. He writes:

> I am aware that these suggestions [given by him for assorted restrictions upon research with recombinant DNA] may be regarded in some circles as a threat to freedom of research. Freedom to search for truth has always been a precious academic right, and every scientist jealously guards it. But this venerated 19th-century idea can no longer be entertained in the light of this new technology [45].

Why not? Because new evidence has entered the picture? What new evidence? Cavalieri's minor premise takes just the form we will by now have come to expect. His reluctant abandonment of free inquiry is defended by what he clearly thinks his mightiest cannon:

> We must ask, with Professor Chargaff, "Have we the right to counteract, irreversibly, the evolutionary wisdom of millions of years, in order to satisfy the ambition and curiosity of a few scientists?" [46].

It would be error, both scientific and philosophical, to suppose that heavy questions such as this one and the others present *any* evidence. That they present evidence of such solidity and weight as to override the application of the principles of free scientific inquiry to research in the biochemistry of genetics is certainly and plainly false.

II. ARGUMENTS BY METAPHOR

An argument from analogy relies upon the likeness of two things, or two states of affairs, in some important respect(s), leading to the inference that the two cases are like in some further respect. But many analogies have no argumentative force, serving only to express feelings or to evoke them. It is unfortunate when, in the discussion of important matters, metaphorical figures of this second sort are taken as arguments. Bearing heavy emotional overtones, they elude reasoned response. Like the heavy questions discussed earlier, striking but fanciful metaphors create inappropriate conceptual frames. Likenesses that had never been straightforwardly claimed and could not be plausibly defended are implicitly supposed. Yet any effort to deny an implicit allegation serves to reinforce the association that the metaphor aimed to create. One could, of course, exhibit the enormous *dis*analogies between the two sets of affairs that had been conjoined. But such responses are largely ineffective, because the disanalogies are not denied. Identifying and discussing them serves indirectly to link more closely what is under attack with some figurative horror, while avoiding the use of plain declaratives.

While metaphors do not play the central role that heavy questions do in the argument over research with recombinant DNA, they are much leaned upon. Recall that minor premises are needed, in the restrictive arguments, establishing the grave defects of this research. If such premises cannot be asserted plainly, perhaps because the evidence for them is missing, they may be partially replaced by figures of speech (or questions) seeking to effect the linkage in the hearer's mind. Consider the following examples, taken from the contributions of the most respected scientific critics of research with recombinant DNA.

1. Professor George Wald opens his response to an advocate of recombinant DNA research by remarking that his own lifelong enterprise was that of trying to understand nature. Wald then continues with this analogy:

   > It [Wald's own work] did not involve the manipulation and deformation of nature. I keep, as I hear this kind of discussion [i.e., of the benefits of recombinant DNA research], thinking of that sad major in the American army in South Vietnam, who said of a city that had been bombed out by the Americans, "We had to destroy it in order to save it" [47].

2. If that seems a bit hard to get a handle on, consider this one. A distinguished scientist replies to Professor Wald that "while I admire Dr. Wald as being able to do experiments that don't perturb nature, most of us to some degree have to perturb nature." [48]: To this Professor Chargaff responds with this metaphorical riposte: "The trouble is, you see, the Virgin giving birth to a little baby, and saying that it is just a tiny baby, not very big."[49]: The relevance of this story is hard to see even when the mother giving that plaintive response is simply an unmarried girl. Is the point of telling it that every alteration of nature resulting from his experiments is as distressing to the scientist as the birth of an infant is to a guilt-ridden unwed mother? Not very plausible. That the mother is here supposed a Virgin (sic) convolutes the erstwhile argument to the point of impregnability.

3. In another place Professor Chargaff exhibits the rhetorical effectiveness of a strained metaphor that is wholly without probative force; he puts recombinant DNA research in a frame of downright evil. The object of attack here is the set of guidelines proposed by the National Institutes of Health for the conduct of research with recombinant DNA. Chargaff writes:

   > Although I do not think that a terrorist organization ever asked the Federal Bureau of Investigation to establish guidelines on the proper conduct of bombing experiments, I do not doubt what the answer would have been: namely, that they ought to refrain from doing anything unlawful [50].

An agency asked for guidelines in the conduct of work it properly oversees is here likened to a law-enforcement bureau charged with the apprehension of dangerous criminals. When the general principles the NIH has laid down are not specific enough, response to inquiries from scientists with careful guidelines (it is here implied) is no better than conniving with crooks. To suggest, by analogy with bombing experiments by terrorists, that this research activity is intrinsically bad and unlawful is grossly unfair. Discussion of the appropriate content of research guidelines is hardly possible in that spirit.

4. Professor Sinsheimer also finds metaphor useful. Seeking to express agreement (with Paul Berg) on research objectives, while disagreeing on the use of recombinant DNA techniques to reach them, Sinsheimer writes:

> In principle one can certainly have no objection to the project of mapping the eukaryotic chromosome any more than one might have to a project of mapping the Sierras. The dispute is perhaps about the technique to be used. One might object to mapping the Sierras if the technique were to move them mountain by mountain to Long Island so that each could be individually measured [51].

The stretch of fancy here we recognize with a smile. We do not need to be told how intolerable it would be to destroy beautiful mountains utterly, while at the same time clobbering populous suburbs. But whether recombinant DNA techniques for the mapping of chromosomes are injurious at all is precisely the point at issue! Moreover, no one supposes that by this technique irreplaceable resources will be irremediably raped. But no use arguing; to try to sort the matter out is only to reinforce the image.

5. If one believes that research with recombinant DNA is an unthinkable "attack on the biosphere," how make the allegation vividly, without asserting it? Here is one prescription. Use the classic story of Prometheus (who stole fire from the gods to bring it and its benefits to mankind) but suppose him infused with the spirit of Herostratus, the arrogant arsonist who deliberately set fire to the Temple of Diana at Ephesus in order to immortalize himself. Bemoaning the guilt of his generation, Chargaff writes: "The hybridization of Prometheus with Herostratus is bound to give evil results" [52].

What to say? Shall we point out that the act of Herostratus was one of absolutely callous self-aggrandizement? The differentiation may be granted—the picture of the recombinant DNA researcher as well-meaning incendiary is already injected and is enough for the purpose.

6. No one can surpass Chargaff at this business when he is fired up. What figure of speech will most effectively discredit one's opponents and their research? Picture them warmakers—and unjust,

aggressive warmakers at that. Chargaff's 1976 letter to *Science* ends
on a crescendo:

> My generation, and perhaps the one preceding mine, has been the
> first to engage, under the leadership of the exact sciences, in a de-
> structive colonial warfare against nature. The future will curse us for it
> [53].

Further analysis of these and like devices is not needed. The irrational character of these instruments, I have shown, shields them from
concise reasonable response. They have the practical effect in the public
forum of saying, or of seeming to say, to those who would advance
research with recombinant DNA:

> Show cause why you shall not be made to desist from this destruc-
> tive colonial warfare against nature. Prove that self-aggrandizement
> does not taint your motivation as you pursue this research. Dem-
> onstrate that nothing of serious consequence can happen in the
> course of the inquiry before you begin it. Do not ask for guidelines in
> doing the research, however, since we know beforehand that what
> you propose to do is satanic. You may experiment, surely (we honor
> freedom of inquiry), but do not perturb nature in the process. We
> realize that yours is a mission of arrogant and aggressive destruc-
> tiveness, of course—but you may proceed after you have proved that
> your hands are clean.

Had argument to this effect been incidental to the objections raised
against recombinant DNA research, it might have been dismissed with
short remark. In truth, such arguments—by heavy question and by
metaphor—have been the stock in trade of the leading scientific critics.
The technical credentials of these critics, combined with the inability of
nonscientists to grapple with often complex microbiological arguments in
response by scientists having equally good credentials, plus the general
thrill of sensational speculation, have led to the use of these very passages
ad nauseum in the public prints. Such arguments rely, I submit, upon
rhetorical excess, and more seriously upon a mistaken understanding of
where lies the burden of proof. To exhibit the character of these devices
makes it plain that the function they are introduced to perform has not
been fulfilled. They do not show, they cannot begin to show, that recombinant DNA research has the evil attributes that alone might (given some
true general principle) justify its prohibition or restriction. Neither arguments by metaphor, nor arguments by heavy question, can replace or
support the minor premises that would be required by a sound argument
defending the restriction of this research.

NOTES

1. An analysis of these principles appears in my essay, "When May Research Be Stopped?" reprinted in this volume.
2. Erwin Chargaff, "On the Dangers of Genetic Meddling," *Science*, **192**(4243) (June 4, 1976): 938.
3. See, for example, *Report of the Working Party on the Experimental Manipulation of the Genetic Composition of Micro-Organisms*, presented to Parliament by the Secretary of State for Education and Science, London, January 1975. This document is widely known as the Ashby Report, after the Chairman of the Working Party, Lord Eric Ashby.
4. Clifford Grobstein, "Recombinant DNA Research: Beyond the NIH Guidelines," *Science*, **194**(4270) (December 10, 1976): 1134.
5. Francis R. Simring, "On the Dangers of Genetic Meddling," *Science*, **192**(4243) (June 4, 1976): 940.
6. Liebe F. Cavalieri, "New Strains of Life—or Death," *The New York Times Magazine*, August 22, 1976: 67.
7. Research upon the aging process, upon techniques to determine the sex of children before birth, upon isotope separation, and upon efforts to contact extraterrestrial intelligences are also the targets of Professor Sinsheimer's heavy questions in "Inquiry Into Inquiry," *Hastings Center Report*, **6**(4) (1976): 18.
8. In the British journal, *New Scientist*, 1975, cited by Michael Rogers in *Biohazard* (New York: Alfred A. Knopf, 1977): 194.
9. Robert L. Sinsheimer, "Potential Risks," *Research with Recombinant DNA* (Washington, D.C.: National Academy of Sciences, 1977): 74–76.
10. George Wald, "Discussion of Potential Benefits," *Research with Recombinant DNA* (Washington, D.C.: National Academy of Sciences, 1977): 82–83.
11. J. King, "Discussion of the Involvement of the Public," *Research with Recombinant DNA* (Washington, D.C.: National Academy of Sciences, 1977): 40.
12. A representative of the People's Business Commission, Jeremy Rifkin, presented the following variants of the heavy question, arguing fervidly at the opening session of the National Academy of Sciences Forum, in March of 1977:

   (22) "How many scientists and corporate executives from the pharmaceutical companies in this room believe that they have a moral right and an authority to proceed on this experimental path before the American people, all 200 million, are fully informed about all, good and bad, of the long-range implications of this research?"

   (23) "What do you think the American public would say if they heard Abbot, Upjohn, Pfizer, Hoffmann-LaRoche, Lilly, say to the American public, 'We companies have the right to patent new forms of life?' "

   (24) "What does that [the patent on a new form of life] mean twenty years from now?"

   Jeremy Rifkin, "Priorities for Day I," *Research with Recombinant DNA* (Washington, D.C.: National Academy of Sciences, 1977): 20.

13. A professor in the School of Public Communication at Boston University, Caryl Rivers, in a national magazine, confuses recombinant DNA techniques with a wholly different process of cell fusion. She compounds new and heavier questions:

(25) "What if some 'genetic stew' escaped—could it turn into a 'Doomsday Bug'?"

The title of this piece gives its flavor.

(26) "Genetic Engineers—Now That They've Gone Too Far, Can They Stop?"

Caryl Rivers, "Genetic Engineers—Now That They've Gone Too Far, Can They Stop?" *Ms.* 4(12): 49: June 1976.

Headline writers for newspapers have the job of attracting the public eye. Guided in some measure by the scientists referred to above, they are not to be much faulted when, as did the *Washington Star* reporting the story of the dispute in Cambridge, Mass., they too indulge.

(27) "Is Harvard the Proper Place for Frankenstein Tinkering?"

Cited by Michael Rogers, *Biohazard* (New York: Alfred A. Knopf, 1977): 196.

14. The introduction of heavy questions into the legislative process was inevitable. It would be unfair to condemn the Mayor of Cambridge, Mass., Alfred E. Vellucci, for translating the words and spirit of his more sophisticated mentors into terms the members of his city council could react to viscerally:

(28) "I want to know about those things that may come crawling out of the laboratories into the sewers."

Cited by June Goodfield, *Playing God* (New York: Random House, 1977): 186.

15. The Democratic Party of Washtenaw County, Michigan (home of The University of Michigan), in plenary session, February, 1978, faced the following resolution:

"Whereas the debate in the scientific community concerning the hazards of recombinant DNA research continues,

Whereas some biologists of the highest standing in their profession believe that a public health or environemntal catastrophe might result from recombinant DNA research.

Whereas no one has given a rigorous demonstration that the risks are small,

Whereas the people of the world have not given their "informed consent" to recombinant DNA experiments which risk their health,

Be it resolved that recombinant DNA research be prohibited and that this prohibition be rescinded only if the people of the world freely and democratically give their consent to recombinant DNA research."
Ann Arbor (Mich.) *News,* 20 February, 1978, p. A-10.

16. See, for example, I. M. Copi, *Introduction to Logic,* 5th edition (New York: Macmillan, 1978:98 ff; and M. C. Beardsley, *Practical Logic* (Englewood Cliffs, N.J., Prentice Hall, 1950): 522ff.

17. "Now That They've Gone Too Far, Can They Stop?" (26, Rivers) and "Is Harvard the Proper Place for Frankenstein Tinkering?" (27, *Washington Star*)

are crude efforts, although possibly effective in the mass media for which they were designed. See note 13. The same compound structure is manifest in some of the more subtle efforts.

18. Stanley Cohen, "Potential Benefits of [DNA] Research," *Research with Recombinant DNA* (Washington, D.C.: National Academy of Sciences, 1977):56.
19. James D. Watson, "In Defense of DNA," *The New Republic* (June 25, 1977): 13. Second emphasis added.
20. See note 14.
21. For more on the categories of risk, see C. Cohen "When May Research Be Stopped?" *New England Journal of Medicine*, **296** (May 26, 1977):1203–1210.
22. Liebe F. Cavalieri, "New Strains of Life—or Death," *New York Times Magazine* (August 22, 1976):67.
23. For an elaboration of this argument, see Stephen P. Stich, "Forbidden Knowledge," unpublished paper delivered at an Indiana University Forum on Recombinant DNA, Bloomington, Indiana, November 12, 1977, to appear in the *Proceedings* of this Forum.
24. An illustration of the allure of argument from sheer conjecture is to be found even in the work of Daniel Callahan, of the Hastings Center, a generally careful analyst in such matters. With respect to basic scientific research and to applied scientific research, Callahan writes, "[T]he burden of proof will lie with those who propose that the research be carried forward" in cases "where serious potential harm to the general public can be hypothesized with some degree of probability greater than zero." D. Callahan, "Ethical Responsibility in Science in the Face of Uncertain Consequences," *Annals of the New York Academy of Science*, 1976, **265:** 10. He cannot seriously intend what he has written here. Since everything possible in the world can be hypothesized with a degree of probability greater than zero, it would be a consequence of this unfortunate principle that the burden of proof *invariably* lies on the researcher.
25. In the sphere of nuclear research similar problems arise. In some cases, Freeman Dyson writes, "The costs of saying yes can be calculated and demonstrated in a style that is familiar and congenial to lawyers, whereas the costs of saying no are a matter of conjecture and have no legal standing. . . . We must try to establish processes of decision-making that give the costs of yes and no an equal voice." F. J. Dyson, *Bulletin of ʌtomic Scientists*, June 1975: 23.
26. A thoughtful exploration of the uses of probability in this context is to be found in S. P. Stich, "The Recombinant DNA Debate," *Philosophy and Public Affairs*, **7**(3) (Spring 1978). Stich gives greater weight than I believe appropriate to the role of subjective probabilities in such matters. But he is sensitive to the distinction between objectively determined probabilities and subjective conjectures, and attends to the problem of whose conjectures ought to be weighed, if any. Respecting the sphere of recombinant DNA research, Stich concludes that appropriately selected subjective probabilities certainly do not justify restriction.
27. L. L. Larison Cudmore, after explaining at length the extraordinary difficul-

ties yet before us in making the genes we want, and doing the things we want with them, goes on to explain why our attainment of that end is so greatly to be prized:

> In spite of these barriers, impassable at the moment, there will be a day when we can get the gene of our choice into a human nucleus. I wish it were today. Our paradoxical genes bring us a lot of grief. Molecules are only human; they make mistakes. The structure of a gene can be changed spontaneously. These mistakes are supposed to happen, to feed the insatiable appetite of natural selection for new ideas, always more new ideas. But even a tiny molecular slip—the change of a single base in the DNA—can be fatal. The translating machinery is not programmed to detect mistakes, and we suffer for it. There are several thousand genetic diseases, some quite horrible. A gene has the wrong information; the protein comes out wrong, misshapen, and it cannot do what it is supposed to. Most of these genetic diseases cannot be cured; many are fatal. Many aren't fatal, but it would be kinder if they were. It hardly seems fair that our genes are an experimental playground for such a ruthless, impersonal force. Hardly any of the misspellings ever make a new kind of sense. Each organism alive has been honed into perfection by millions of years of evolution; the interrelationships of the systems are delicate and elaborate, and can be permanently disrupted by one single wrong base out of the three billion pairs we have. It seems a hard life this one, what with the competition for survival, millions of unsuccessful genetic experiments (one of which might be a child of ours), just so one experiment might work. But that's progress. And the experimentation continues. Evolution's and ours. *The aim of ours is to counteract the effects of evolution's.* To one day be able to cure genetic diseases, to repair genes gone wrong. That is a consummation devoutly to be wished.

The Center of Life (N.Y. Times Books, 1977): 100–101. Emphasis added.
28. Sinsheimer, "Potential Risks," *op. cit.:* 77.
29. Dr. Ruth Macklin, of the Hastings Center, writes:

> When such knowledge [of the probabilities of different outcomes and other relevant facts] is lacking there is no way of engaging in ethical decision-making of the sort where appeal to consequences [of this kind] is essential.

Ruth Macklin, "On the Ethics of *Not* Doing Research," *Hastings Center Report,* 7(6) (December 1977): 13.
30. See: *Report of the Working Party on the Experimental Manipulation of Micro-Organisms* (Ashby Report) presented to Parliament, London, 1975; *Report of the University Committee to Recommend Policy for the Molecular Genetics and Oncology Program* (Ann Arbor: The University of Michigan, March 1976); "Potential Benefits of the (DNA) Research," Dr. Daniel Nathans (Boury Pro-

fessor and Director, Department of Microbiology, The Johns Hopkins University School of Medicine); and "Mapping the Mammalian Genome," Dr. Paul Berg, Willson Professor of Biochemistry, Stanford University Medical Center, both appearing in *Research With Recombinant DNA* (Washington, D.C.: National Academy of Sciences, 1977), *inter alia.* The American Association for the Advancement of Science established a special committee on Scientific Freedom and Responsibility; it included distinguished scholars in both social and natural sciences, a former justice of the Supreme Court, and a former Secretary of the Interior. After reviewing the evidence for some years, the Committee's judgment on the continuation of research with recombinant DNA was unequivocal: "We hold that the dangers today are remote, and that they are decisively outweighed by the great benefits that such research can bring." "Scientific Freedom and Responsibility," Report of the AAAS Committee on Scientific Freedom and Responsibility, *Science,* **188:** 688 (1975).

31. See: Carl Cohen, *op. cit.:* 1208–1210.
32. Wald, *op. cit.:* 55.
33. Jonathan King, *op cit.:* 40.
34. William Van Alstyne, "The Specific Theory of Academic Freedom and the General Issue of Civil Liberty." Reprinted in *The Concept of Academic Freedom,* ed. E. Pincoffs (Austin: University of Texas Press, 1975): 59–85.
35. John Searle, "Academic Freedom," Chap. 6 of *The Campus War* (New York: World Publishing Company, 1971): 183–212.
36. *Ibid.:* 84–91.
37. Van Alstyne, *op. cit.:* 71.
38. *Ibid.:* 72.
39. *Ibid.:* 77.
40. Reinforcement of this view comes from the report of the Committee on Scientific Freedom and Responsibility of the American Association for the Advancement of Science. "Scientific freedom, like academic freedom, is an acquired right, generally accepted by society as necessary for the advancement of knowledge from which society may benefit. Scientists possess no rights beyond those of other citizens except those necessary to fulfill the responsibilities arising from their special knowledge, and from the insight arising from that knowledge." *Science,* **188**(16) (May 1975).
41. *Ibid.:* 77–78.
42. *Ibid.:* 77.
43. See: C. Cohen, "When May Research be Stopped?" *op. cit.;* see also: *Guidelines* for experimentation with human subjects, National Institutes of Health. Van Alstyne (*op. cit.,* p. 78) gives the example of a professor of psychology falsely shouting fire in a theater in order to observe crowd reaction firsthand and to achieve a general understanding of mass action under sudden stress. No one supposes that that researcher would be immunized by his intellectual objectives or professional status.
44. S. Livermore, "Statement of Dissent," *Report of Committee B* (Ann Arbor: The University of Michigan Committee to Recommend Policy for the Molecular Genetics and Oncology Program, March 1976): 47.
45. Cavalieri, *op. cit.:* p. 69.

46. *Ibid.*

47. Wald, *op. cit.:* 55.

48. Bernard D. Davis, Adele Lehman Professor of Bacterial Physiology, Harvard Medical School, in "Discussion of Potential Benefits" appearing in *Research with Recombinant DNA* (Washington, D.C.: National Academy of Sciences, 1977): 57.

49. Chargaff, *op. cit.,* in *Research with Recombinant DNA:* 57.

50. Chargaff, letter to *Science,* **192**(4243) (June 4, 1976): 938.

51. Sinsheimer, "Potential Risks," *op. cit.:* 74.

52. Chargaff, *Science, op. cit.:* 938.

53. *Ibid.:* 940.

Appendices

Appendix I
A Glossary
for Recombinant DNA Methodology

Allele A variant form of a gene. The form of a given gene that is found in an organism as isolated from nature is called the *wild-type allele;* variant forms arising by mutational changes in base-pair sequence are called *mutant alleles.*

Bacteria Unicellular microorganisms containing a single very large DNA molecule per cell (the *chromosome*). Many species of bacteria can be grown very easily, very inexpensively, in very large amounts, in solutions consisting of a few salts and a carbon source.

Bacteriophage A virus that grows in bacteria.

Bases, base pairs See *DNA.*

Biological containment Use of genetically altered bacteria, bacteriophage, and plasmids that are unable to perform essential functions such as growth, DNA replication, transfer of DNA to other cells, infection of cells, etc., except under rigidly specified laboratory conditions. An example of biological containment would be the use, as a host organism for recombinant DNA molecules, of an *E. coli* cell that can grow only at a temperature of less than $32°$ C and only if both streptomycin and diaminopimelic acid, neither of which is normally found in the environment, are provided in its growth medium.

Clone A group of genetically identical cells or organisms, all of which have arisen from the same cell by asexual reproduction.

Cohesive termini, sticky ends Single-stranded DNA sequences occurring at the ends of double-stranded DNA molecules, in which the sequence of bases is complementary to (and will therefore form a double-stranded structure with) another single-stranded region at the other end of that DNA molecule or on another DNA molecule.

Conjugation The transfer of DNA from one bacterial cell to another during bacterial mating. The DNA can be either chromosomal or plasmid DNA.

Conjugative plasmid, transmissible plasmid A plasmid that can spontaneously transfer its DNA to another cell.

Covalent bonds The type of relatively strong chemical bonds that hold together most of the atoms in a molecule. The bases within one strand of a DNA molecule are linked together by covalent bonds. The two strands of a double-stranded DNA molecule are held together by the hydrogen bonds in the specific base pairs.

Covalently closed DNA A DNA molecule that is circular and in which both strands are covalently continuous. *Plasmids* are examples of covalently closed DNA molecules.

Cut Breaks opposite one another in both strands of a DNA molecule.

DNA Deoxyribonucleic acid. The genetic material of all cells and many viruses. Most DNA molecules consist of two interwound chains or *strands* of four basic units called *nucleotides* or *bases*. These bases are *A, C, G, T*. Their structure is such that if *A* is found in one strand, *T* is always found opposite it in the other strand, and vice versa. Similarly, *C* and *G* always pair with each other. Thus the sequence of bases in one strand of a DNA molecule uniquely determines the sequence of bases in the other strand. A double-stranded DNA molecule thus consists of a sequence of *base pairs*. DNA molecules can be either linear or circular. In microorganisms (*bacteria* and *viruses*) DNA molecules are usually circular.

DNA polymerase An enzyme that can fill in single-stranded gaps in double-stranded DNA by inserting the proper complementary bases opposite the bases in the intact strand.

DNA replication The process by which the two complementary strands of a DNA molecule separate and a new complementary strand is synthesized by DNA polymerase on each of the separated strands. This process gives rise to two daughter DNA molecules, each of which has a nucleotide sequence identical to that of the parental molecule.

E. coli, Escherichia coli The most extensively studied species of bacteria. It can be isolated from the gastrointestinal tract of most mammals including man. Some varieties of it are pathogenic. Most laboratory strains of it are not. All humans and most common mammals are colonized by one or another or several strains of *E. coli*. The present degree of understanding of *E. coli* is probably an order of magnitude greater than that of any other bacterium, as is the capability for manipulating it in precise ways by genetic techniques.

EcoR$_I$ The most widely used restriction enzyme. It makes a staggered cut and leaves the sequence $\frac{G}{CTTAA}$ at both ends of every DNA fragment it generates. That is, *EcoR$_I$* catalyzes the following reaction:

$$
\begin{array}{ll}
5' \cdots \text{—G—A—A—T—T—C—}\cdots 3' & \textbf{Eco R}_1 \\
3' \cdots \text{—C—T—T—A—A—G—}\cdots 5' &
\end{array}
$$

nick ↓

$$
\begin{array}{l}
5' \cdots \text{—G} \quad \text{A—A—T—T—C—}\cdots 3' \\
3' \cdots \text{—C—T—T—A—A} \quad \text{G—}\cdots 5'
\end{array}
$$

↑ nick

$$
\begin{array}{l}
5' \cdots \text{— G} \\
3' \cdots \text{— C—T—T—A—A}
\end{array}
\quad + \quad
\begin{array}{l}
\text{A—A—T—T—C—}\cdots 3' \\
\text{G—}\cdots 5'
\end{array}
$$

Endonuclease A nuclease that puts nicks or cuts in DNA molecules. It does not require a free end to act.

Enzyme A protein molecule that catalyzes (increases the rate of) a chemical reaction.

Eukaryotes Organisms having cells containing a defined nucleus, multiple chromosomes, and a defined mitotic apparatus. Eukaryotes can be either unicellular (yeasts, protozoa) or multicellular (animals and plants).

Exonuclease A nuclease that removes bases sequentially from the ends of a linear DNA molecule.

Gap A break in one strand of a DNA molecule in which one or more bases is removed.

Gene A sequence of DNA base pairs that codes for a single species of protein, which is itself a sequence of amino acid units in a chain. Proteins (*enzymes*) are thus *gene products*.

Genome The total amount of nucleic acid (usually DNA, but RNA in the case of some viruses) used for genetic purposes in an organism.

Genotype The genetic constitution of an organism; its total array of genes.

Hydrogen bonds The type of relatively weak chemical bonds that hold *complementary base pairs* (A and T, C and G) together.

Ligase An enzyme that can reform normal covalent bonds at a nick—that is, join the two bases bounding a nick by a phosphodiester bond.

Macromolecule A large molecule, usually composed of a sequence of a limited number of different kinds of basic subunits. DNA, RNA, proteins, and polysaccharides are all examples of macromolecules.

mRNA, messenger RNA The kind of RNA that is synthesized from genes and that acts as the intermediate in the conversion of information stored as a base-pair sequence in DNA into an amino acid sequence in a protein.

Mutant   An organism with a mutation in it.

Mutation   A change in the genetic material of an organism. Mutations can be base-pair changes, deletions, additions, or inversions of a series of base pairs. Mutations can be deleterious, neutral, or advantageous, depending on their nature and on the environment in which the organism must survive.

Nick   A break in one strand of a DNA molecule in which no bases are removed; a phosphodiester bond break.

Nonconjugative plasmid, nontransmissible plasmid   A plasmid that cannot transfer its DNA to another cell. Nonconjugative plasmids can, however, be "mobilized" to transfer their DNA in the presence of a conjugative plasmid in the same cell.

Nuclease   An enzyme that acts to break down the structure of a DNA molecule.

Nucleotides   See DNA.

Phenotype   The total array of observable characteristics of an organism—its morphological and physiological properties. In a given environment, a given genotype will always determine the same phenotype. Any change in the phenotype in that given environment implies a change in the genotype—a mutation.

Phosphodiester bond   The type of covalent bonds that link nucleotides together to form the polynucleotide strands of a DNA molecule. Phosphodiester bonds can be cut by endonucleases and exonucleases, formed by DNA polymerase, and repaired (reformed) by DNA ligase.

Physical containment   Equipment or practices that put a physical barrier of some sort between the experimenter and part or all of his experiment. Examples of physical containment are the use of glove boxes and laminar air flow safety cabinets, the avoidance of mouth-pipetting, the autoclaving of contaminated material, the maintenance of a laboratory under negative air pressure with respect to surrounding laboratories, the wearing of lab coats and gloves.

Plasmid   A relatively small circular DNA molecule capable of autonomous self-replication within a bacterium.

Prokaryotes   Simple, unicellular organisms such as bacteria and blue-green algae. Prokaryotes have a single chromosome and lack a defined nucleus and nuclear membrane.

Protein   The major structural and catalytic macromolecules in cells. Proteins consist of linear chains of 20 different kinds of building blocks called amino acids. Each triplet of base pairs in a DNA molecule codes for a different amino acid, so that the sequence of base pairs is converted into a sequence of amino acids during protein synthesis. All enzymes are proteins, but not all proteins are enzymes. Some play structural roles in cell membranes and a variety of cellular organelles.

Recombinant DNA, chimeric DNA, hybrid DNA DNA molecules of different origin that have been joined together by biochemical techniques to make a single molecule, usually circular and usually capable of some specific biological function, especially *self-replication* in an appropriate cell.

Resistance transfer factor, RTF, R-factor A plasmid carrying a gene coding for a protein that makes the bacteria carrying the R-factor resistant to being killed by an antibiotic. Many different R-factors are known, carrying resistance to one or another or a group of several antibiotics. An R-factor is known for practically every known antibiotic. They can be transferred readily from one bacterium to another and are a serious public health problem.

Restriction enzyme, restriction endonuclease An endonuclease that makes cuts in DNA molecules only at specific base-pair sequences, and at every such sequence that occurs in a DNA molecule. The base-pair sequence at which cutting occurs differs from one restriction enzyme to another. Some restriction enzymes make *staggered cuts* and thus leave cohesive ends.

RNA Ribonucleic acid. All cells contain both DNA and RNA, and some viruses have RNA as their genetic material. RNA differs from DNA in that it has the sugar ribose instead of deoxyribose, it is usually single-stranded instead of double-stranded, and it has the base U (uracil) in place of the base T (thymine). RNA is used as the molecule that transmits information encoded in the DNA base sequence to the protein-synthesizing apparatus of the cell (see mRNA) and as major structural components of ribosomes, the cellular structures on which protein synthesis occurs.

Transcription The process of synthesizing an RNA chain using a DNA molecule as a template.

Transformation *When applied to bacteria,* this term means acquisition by a bacterium of new genes following infection of that bacterium by DNA carrying these genes. *When applied to animal cells,* this term means conversion of the cell from a normal, noncancerous cell to an abnormal, cancerous cell capable of causing a tumor when injected into an animal. Transformation in animal cells can be "spontaneous" or can be caused by certain *oncogenic* animal viruses or by *carcinogens.*

Translation The process of converting the information stored in the sequence of an mRNA molecule into a protein.

Vector, vehicle A plasmid or viral DNA molecule into which another DNA molecule can be inserted without disruption of the ability of the plasmid or viral DNA to replicate itself.

Virus A DNA or RNA molecule surrounded by a protein coat that protects it. The viral DNA or RNA is capable of replicating itself inside an appropriate type of cell. Most, but not all, viruses kill the cells in which they replicate. Some viral DNA's can be covalently inserted into the chromosomal DNA of the cells they infect, in which case the viral DNA is replicated as a part of the cellular DNA and does not kill the cell, although it may alter its properties. *Oncogenic* viruses can *transform* normal cells to *tumorigenic* or cancer-causing cells.

Wild type The genetic state of an organism isolated from nature. Often used to refer to an organism that has no obvious mutations in it.

Appendix II
NIH Guidelines for
Recombinant DNA Research

I. INTRODUCTION

The purpose of these guidelines is to recommend safeguards for research on recombinant DNA molecules to the National Institutes of Health and to other institutions that support such research. In this context we define recombinant DNAs as molecules that consist of different segments of DNA which have been joined together in cell-free systems, and which have the capacity to infect and replicate in some host cell, either autonomously or as an integrated part of the host's genome.

This is the first attempt to provide a detailed set of guidelines for use by study sections as well as practicing scientists for evaluating research on recombinant DNA molecules. We cannot hope to anticipate all possible lines of imaginative research that are possible with this powerful new methodology. Nevertheless, a considerable volume of written and

The portion of the NIH Guidelines for Recombinant DNA Research reproduced here is the operative section, in which the guidelines themselves are described. Not included is an introductory statement by Dr. Donald S. Frederickson, Director of the National Institutes of Health, nor a number of appendices dealing with such matters as postal regulations for the shipment of biological materials. The complete text of all this material can be found in the Federal Register, **41:** 27902–27943 (1976). The guidelines are currently in the process of being revised in light of new data. A draft version of the revised guidelines has been published in the Federal Register, **42:** 49596–49609 (1977).

verbal contributions from scientists in a variety of disciplines has been received. In many instances the views presented to us were contradictory. At present, the hazards may be guessed at, speculated about, or voted upon, but they cannot be known absolutely in the absence of firm experimental data—and, unfortunately, the needed data were, more often than not, unavailable. Our problem then has been to construct guidelines that allow the promise of the methodology to be realized while advocating the considerable caution that is demanded by what we and others view as potential hazards.

In designing these guidelines we have adopted the following principles, which are consistent with the general conclusions that were formulated at the International Conference on Recombinant DNA Molecules held at Asilomar Conference Center, Pacific Grove, California, in February 1975 (3): (i) There are certain experiments for which the assessed potential hazard is so serious that they are not to be attempted at the present time. (ii) The remainder can be undertaken at the present time provided that the experiment is justifiable on the basis that new knowledge or benefits to humankind will accrue that cannot readily be obtained by use of conventional methodology and that appropriate safeguards are incorporated into the design and execution of the experiment. In addition to an insistence on the practice of good microbiological techniques, these safeguards consist of providing both physical and biological barriers to the dissemination of the potentially hazardous agents. (iii) The level of containment provided by these barriers is to match the estimated potential hazard for each of the different classes of recombinants. For projects in a given class, this level is to be highest at initiation and modified subsequently only if there is a substantiated change in the assessed risk or in the applied methodology. (iv) The guidelines will be subjected to periodic review (at least annually) and modified to reflect improvements in our knowledge of the potential biohazards and of the available safeguards.

In constructing these guidelines it has been necessary to define boundary conditions for the different levels of physical and biological containment and for the classes of experiments to which they apply. We recognize that these definitions do not take into account existing and anticipated special procedures and information that will allow particular experiments to be carried out under different conditions than indicated here without sacrifice of safety. Indeed, we urge that individual investigators devise simple and more effective containment procedures and that study sections give consideration to such procedures which may allow change in the containment levels recommended here.

It is recommended that all publications dealing with recombinant

DNA work include a description of the physical and biological containment procedures practiced, to aid and forewarn others who might consider repeating the work.

II. CONTAINMENT

Effective biological safety programs have been operative in a variety of laboratories for many years. Considerable information therefore already exists for the design of physical containment facilities and the selection of laboratory procedures applicable to organisms carrying recombinant DNAs (4–17). The existing programs rely upon mechanisms that, for convenience, can be divided into two categories: (i) a set of standard practices that are generally used in microbiological laboratories, and (ii) special procedures, equipment, and laboratory installations that provide physical barriers which are applied in varying degrees according to the estimated biohazard.

Experiments on recombinant DNAs by their very nature lend themselves to a third containment mechanism—namely, the application of highly specific biological barriers. In fact, natural barriers do exist which either limit the infectivity of a vector or vehicle (plasmid, bacteriophage or virus) to specific hosts, or its dissemination and survival in the environment. The vectors that provide the means for replication of the recombinant DNAs and/or the host cells in which they replicate can be genetically designed to decrease by many orders of magnitude the probability of dissemination of recombinant DNAs outside the laboratory.

As these three means of containment are complementary, different levels of containment appropriate for experiments with different recombinants can be established by applying different combinations of the physical and biological barriers to a constant use of the standard practices. We consider these categories of containment separately here in order that such combinations can be conveniently expressed in the guidelines for research on the different kinds of recombinant DNA (Section III).

A. Standard Practices and Training

The first principle of containment is a strict adherence to good microbiological practices (4–13). Consequently, all personnel directly or indirectly involved in experiments on recombinant DNAs must receive adequate instruction. This should include at least training in aseptic techniques and instruction in the biology of the organisms used in the experiments so that the potential biohazards can be understood and appreciated.

Any research group working with agents with a known or potential biohazard should have an emergency plan which describes the procedures to be followed if an accident contaminates personnel or environment. The principal investigator must ensure that everyone in the laboratory is familiar with both the potential hazards of the work and the emergency plan. If a research group is working with a known pathogen for which an effective vaccine is available, all workers should be immunized. Serological monitoring, where appropriate, should be provided.

B. Physical Containment Levels

A variety of combinations (levels) of special practices, equipment, and laboratory installations that provide additional physical barriers can be formed. For example, 31 combinations are listed in "Laboratory Safety at the Center for Disease Control" (4); four levels are associated with the "Classification of Etiologic Agents on the Basis of Hazard" (5), four levels were recommended in the "Summary Statement of the Asilomar Conference on Recombinant DNA Molecules" (3); and the National Cancer Institute uses three levels for research on oncogenic viruses (6). We emphasize that these are an aid to, and not a substitute for, good technique. Personnel must be competent in the effective use of all equipment needed for the required containment level as described below. We define only four levels of physical containment here, both because the accuracy with which one can presently assess the biohazards that may result from recombinant DNAs does not warrant a more detailed classification, and because additional flexibility can be obtained by combination of the physical with the biological barriers. Though different in detail, these four levels (P1< P2< P3< P4) approximate those given for human etiologic agents by the Center for Disease Control (i.e., classes 1 through 4; ref. 5), in the Asilomar summary statement (i.e., minimal, low, moderate, and high; ref. 3), and by the National Cancer Institute for oncogenic viruses (i.e., low, moderate and high; ref. 6), as is indicated by the P-number or adjective in the following headings. It should be emphasized that the descriptions and assignments of physical containment detailed below are based on existing approaches to containment of hazardous organisms.

We anticipate, and indeed already know of, procedures (14) which enhance physical containment capability in novel ways. For example, miniaturization of screening, handling, and analytical procedures provides substantial containment of a given host-vector system. Thus, such procedures should reduce the need for the standard types of physical containment, and such innovations will be considered by the Recombinant DNA Molecule Program Advisory Committee.

The special practices, equipment and facility installations indicated for each level of physical containment are required for the safety of laboratory workers, other persons, and for the protection of the environment. Optional items have been excluded; only those items deemed absolutely necessary for safety are presented. Thus, the listed requirements present basic safety criteria for each level of physical containment. Other microbiological practices and laboratory techniques which promote safety are to be encouraged. Additional information giving further guidance on physical containment is provided in a supplement to the guidelines (Appendix D).

P1 Level (Minimal). A laboratory suitable for experiments involving recombinant DNA molecules requiring physical containment at the P1 level is a laboratory that possesses no special engineering design features. It is a laboratory commonly used for microorganisms of no or minimal biohazard under ordinary conditions of handling. Work in this laboratory is generally conducted on open bench tops. Special containment equipment is neither required nor generally available in this laboratory. The laboratory is not separated from the general traffic patterns of the building. Public access is permitted.

The control of biohazards at the P1 level is provided by standard microbiological practices of which the following are examples: (i) Laboratory doors should be kept closed while experiments are in progress. (ii) Work surfaces should be decontaminated daily and following spills of recombinant DNA materials. (iii) Liquid wastes containing recombinant DNA materials should be decontaminated before disposal. (iv) Solid wastes contaminated with recombinant DNA materials should be decontaminated or packaged in a durable leakproof container before removal from the laboratory. (v) Although pipetting by mouth is permitted, it is preferable that mechanical pipetting devices be used. When pipetting by mouth, cotton-plugged pipettes shall be employed. (vi) Eating, drinking, smoking and storage of food in the working area should be discouraged. (vii) Facilities to wash hands should be available. (viii) An insect and rodent control program should be provided. (ix) The use of laboratory gowns, coats, or uniforms is discretionary with the laboratory supervisor.

P2 level (Low). A laboratory suitable for experiments involving recombinant DNA molecules requiring physical containment at the P2 level is similar in construction and design to the P1 laboratory. The P2 laboratory must have access to an autoclave within the building; it may have a Biological Safety Cabinet.[1*] Work which does not produce a considerable aerosol is conducted on the open bench. Although this laboratory is not separated from the general traffic patterns of the building, access to the laboratory is limited when experiments requiring P2 level physical containment are being conducted. Experiments of lesser biohazard potential can be carried out concurrently in carefully demarcated areas of the same laboratory.

The P2 laboratory is commonly used for experiments involving microorganisms of low biohazard such as those which have been classified by the Center for Disease Control as Class 2 agents (5).

The following practices shall apply to all experiments requiring P2 level physical containment: (i) Laboratory doors shall be kept closed while experiments are in progress. (ii) Only persons who have been advised of the potential biohazard shall enter the laboratory. (iii) Children under 12 years of age shall not enter the laboratory. (iv) Work surfaces shall be decontaminated daily and immediately following spills of recombinant DNA materials. (v) Liquid wastes of recombinant DNA materials shall be decontaminated before disposal. (vi) Solid wastes contaminated with recombinant DNA materials shall be decontaminated or packaged in a durable leak-proof container before removal from the laboratory. Packaged materials shall be disposed of by incineration or sterilized before disposal by other methods. Contaminated materials that are to be processed and reused (i.e., glassware) shall be decontaminated before removal from the laboratory. (vii) Pipetting by mouth is prohibited; mechanical pipetting devices shall be used. (viii) Eating, drinking, smoking, and storage of food are not permitted in the working area. (ix) Facilities to wash hands shall be available within the laboratory. Persons handling recombinant DNA materials should be encouraged to wash their hands frequently and when they leave the laboratory. (x) An insect and rodent control program shall be provided. (xi) The use of laboratory gowns, coats, or uniforms is required. Such clothing shall not be worn to the lunch room or outside the building. (xii) Animals not related to the experiment shall not be permitted in the laboratory. (xiii) Biological Safety Cabinets[1*] and/or other physical containment equipment shall be used to minimize the hazard of aerosolization of recombinant DNA materials from operations or devices that produce a considerable aerosol (e.g., blender, lyophilizer, sonicator, shaking machine, etc.). (xiv) Use of the

*Numbered footnotes appear at the end of the Appendix.

hypodermic needle and syringe shall be avoided when alternate methods are available.

P3 Level (Moderate). A laboratory suitable for experiments involving recombinant DNA molecules requiring physical containment at the P3 level has special engineering design features and physical containment equipment. The laboratory is separated from areas which are open to the general public. Separation is generally achieved by controlled access corridors, air locks, locker rooms or other double-doored facilities which are not available for use by the general public. Access to the laboratory is controlled. Biological Safety Cabinets[1] are available within the controlled laboratory area. An autoclave shall be available within the building and preferably within the controlled laboratory area. The surfaces of walls, floors, bench tops, and ceilings are easily cleanable to facilitate housekeeping and space decontamination.

Directional air flow is provided within the controlled laboratory area. The ventilation system is balanced to provide for an inflow of supply air from the access corridor into the laboratory. The general exhaust air from the laboratory is discharged outdoors and so dispersed to the atmosphere as to prevent reentry into the building. No recirculation of the exhaust air shall be permitted without appropriate treatment.

No work in open vessels involving hosts or vectors containing recombinant DNA molecules requiring P3 physical containment is conducted on the open bench. All such procedures are confined to Biological Safety Cabinets.[1]

The following practices shall apply to all experiments requiring P3 level physical containment: (i) The universal biohazard sign is required on all laboratory access doors. Only persons whose entry into the laboratory is required on the basis of program or support needs shall be authorized to enter. Such persons shall be advised of the potential biohazards before entry and they shall comply with posted entry and exit procedures. Children under 12 years of age shall not enter the laboratory. (ii) Laboratory doors shall be kept closed while experiments are in progress. (iii) Biological Safety Cabinets[1] and other physical containment equipment shall be used for all procedures that produce aerosols of recombinant DNA materials (e.g., pipetting, plating, flaming, transfer operations, grinding, blending, drying, sonicating, shaking, etc.). (iv) The work surfaces of Biological Safety Cabinets[1] and other equipment shall be decontaminated following the completion of the experimental activity contained within them. (v) Liquid wastes containing recombinant DNA materials shall be decontaminated before disposal. Solid wastes contaminated with recombinant DNA materials shall be decontaminated or packaged in a durable leak-proof container before removal from the laboratory. Packaged material shall be sterilized before disposal. Contaminated materials that are to be processed and reused (i.e., glassware) shall be sterilized in the controlled laboratory area or placed in a durable leak-proof container before removal from the controlled laboratory area. This container shall be sterilized before the materials are processed. (vi) Pipetting by mouth is prohibited; mechanical pipetting devices shall be used. (vii) Eating, drinking, smoking, and storage of food are not permitted in the laboratory. (ix) Facilities to wash hands shall be available within the laboratory. Persons shall wash hands after experiments involving recombinant DNA materials and before leaving the laboratory. (x) An insect and rodent control program shall be provided. (xi) Laboratory clothing that protects street clothing (i.e., long sleeve solid-front or wrap-around gowns, no-button or slipover jackets, etc.) shall be worn in the laboratory. FRONT-BUTTON LABORATORY COATS ARE UNSUITABLE. Gloves shall be worn when handling recombinant DNA materials. Provision for laboratory shoes is recommended. Laboratory clothing shall not be worn outside the laboratory and shall be decontaminated before it is sent to the laundry. (xii) Raincoats, overcoats, topcoats, coats, hats, caps, and such street outerwear shall not be kept in the laboratory.

(xiii) Animals and plants not related to the experiment shall not be permitted in the laboratory. (xiv) Vacuum lines shall be protected by filters and liquid traps. (xv) Use of the hypodermic needle and syringe shall be avoided when alternate methods are available. (xvi) If experiments of lesser biohazard potential are to be conducted in the same laboratory concurrently with experiments requiring P3 level physical containment they shall be conducted only in accordance with all P3 level requirements. (xvii) Experiments requiring P3 level physical containment can be conducted in laboratories where the directional air flow and general exhaust air conditions described above cannot be achieved, provided that this work is conducted in accordance with all other requirements listed and is contained in a Biological Safety Cabinet[1] with attached glove ports and gloves. All materials before removal from the Biological Safety Cabinet[1] shall be sterilized or transferred to a non-breakable, sealed container, which is then removed from the cabinet through a chemical decontamination tank, autoclave, ultraviolet air lock, or after the entire cabinet has been decontaminated.

P4 Level (High). Experiments involving recombinant DNA molecules requiring physical containment at the P4 level shall be confined to work areas in a facility of the type designed to contain microorganims that are extremely hazardous to man or may cause serious epidemic disease. The facility is either a separate building or it is a controlled area, within a building, which is completely isolated from all other areas of the building. Access to the facility is under strict control. A specific facility operations manual is available. Class III Biological Safety Cabinets[1] are available within work areas of the facility.

A P4 facility has engineering features which are designed to prevent the escape of microorganisms to the environment (14,15,16,17). These features include: (i) Monolithic walls, floors, and ceilings in which all penetrations such as for air ducts, electrical conduits, and utility pipes are sealed to assure the physical isolation of the work area and to facilitate housekeeping and space decontamination; (ii) air locks through which supplies and materials can be brought safely into the facility; (iii) continuous clothing change and shower rooms through which personnel enter into and exit from the facility; (iv) double-door autoclaves to sterilize and safely remove wastes and other materials from the facility; (v) a biowaste treatment system to sterilize liquid effluents if facility drains are installed; (vi) a separate ventilation system which maintains negative air pressures and directional air flow within the facility; and (vii) a treatment system to decontaminate exhaust air before it is dispersed to the atmosphere. A central vacuum utility system is not encouraged; if one is installed, each branch line leading to a laboratory shall be protected by a high efficiency particulate air filter.

The following practices shall apply to all experiments requiring P4 level physical containment: (i) The universal biohazard sign is required on all facility access doors and all interior doors to individual laboratory rooms where experiments are conducted. Only persons whose entry into the facility or individual laboratory rooms is required on the basis of program or support needs shall be authorized to enter. Such persons shall be advised of the potential biohazards and instructed as to the appropriate safeguards to ensure their safety before entry. Such persons shall comply with the instructions and all other posted entry and exit procedures. Under no condition shall children under 15 years of age be allowed entry. (ii) Personnel shall enter into and exit from the facility only through the clothing change and shower rooms. Personnel shall shower at each exit from the facility. The air locks shall not be used for personnel entry or exit except for emergencies. (iii) Street clothing shall be removed in the outer facility side of the clothing change area and kept there. Complete laboratory clothing including undergarments, pants and shirts or jumpsuits, shoes, head cover, and gloves shall be provided and used by all persons who enter into the facility. Upon exit, this clothing shall be stored in lockers provided for this purpose or discarded into collection hampers before personnel enter into the shower area. (iv)

Supplies and materials to be taken into the facility shall be placed in an entry air lock. After the outer door (opening to the corridor outside of facility) has been secured, personnel occupying the facility shall retrieve the supplies and materials by opening the interior air lock door. This door shall be secured after supplies and materials are brought into the facility. (v) Doors to laboratory rooms within the facility shall be kept closed while experiments are in progress. (vi) Experimental procedures requiring P4 level physical containment shall be confined to Class III Biological Safety Cabinets.[1] All materials, before removal from these cabinets, shall be sterilized or transferred to a nonbreakable sealed container, which is then removed from the system through a chemical decontamination tank, autoclave, or after the entire system has been decontaminated.

(vii) No materials shall be removed from the facility unless they have been sterilized or decontaminated in a manner to prevent the release of agents requiring P4 physical containment. All wastes and other materials and equipment not damaged by high temperature or steam shall be sterilized in the double-door autoclave. Biological materials to be removed from the facility shall be transferred to a nonbreakable sealed container which is then removed from the facility through a chemical decontamination tank or a chamber designed for gas sterilization. Other materials which may be damaged by temperature or steam shall be sterilized by gaseous or vapor methods in an air lock or chamber designed for this purpose. (viii) Eating, drinking, smoking, and storage of food are not permitted in the facility. Foot-operated water fountains located in the facility corridors are permitted. Separate potable water piping shall be provided for these water fountains. (ix) Facilities to wash hands shall be available within the facility. Persons shall wash hands after experiments. (x) An insect and rodent control program shall be provided. (xi) Animals and plants not related to the experiment shall not be permitted in the facility. (xii) If a central vacuum system is provided, each vacuum outlet shall be protected by a filter and liquid trap in addition to the branch line HEPA filter mentioned above. (xiii) Use of the hypodermic needle and syringe shall be avoided when alternate methods are available. (xiv) If experiments of lesser biohazard potential are to be conducted in the facility concurrently with experiments requiring P4 level containment, they shall be confined in Class I or Class II Biological Safety Cabinets[1] or isolated by other physical containment equipment. Work surfaces of Biological Safety Cabinets[1] and other equipment shall be decontaminated following the completion of the experimental activity contained within them. Mechanical pipetting devices shall be used. All other practices listed above with the exception of (vi) shall apply.

C. Shipment

To protect product, personnel, and the environment, all recombinant DNA material will be shipped in containers that meet the requirements issued by the U.S. Public Health Service (Section 72.25 of Part 72, Title 42, Code of Federal Regulations), Department of Transportation (Section 173.387(b) of Part 173, Title 49, Code of Federal Regulations) and the Civil Aeronautics Board (C.A.B. No. 82, Official Air Transport Restricted Articles Tariff No. 6-D) for shipment of etiologic agents. Labeling requirements specified in these Federal regulations and tariffs will apply to all viable recombinant DNA materials in which any portion of the material is derived from an etiologic agent listed in paragraph (c) of 42 CFR 72.25. Additional information on packing and shipping is given in a supplement to the guidelines (Appendix D, part X).

D. Biological Containment Levels

Biological barriers are specific to each host-vector system. Hence the criteria for this mechanism of containment cannot be generalized to the same extent as for physical containment. This is particularly true at the present time when our experience with existing host-vector systems and our predictive knowledge about projected systems are sparse. The classification of experiments with recombinant DNAs that is necessary for the construction of the experimental guidelines (Section III) can be accomplished with least confusion if we use the host-vector system as the primary element and the source of the inserted DNA as the secondary element in the classification. It is therefore convenient to specify the nature of the biological containment under host-vector headings such as those given below for *Escherichia coli* K-12.

III. EXPERIMENTAL GUIDELINES

A general rule that, though obvious, deserves statement is that the level of containment required for any experiment on DNA recombinants shall never be less than that required for the most hazardous component used to construct and clone the recombinant DNA (i.e., vector, host, and inserted DNA). In most cases the level of containment will be greater, particularly when the recombinant DNA is formed from species that ordinarily do not exchange genetic information. Handling the purified DNA will generally require less stringent precautions than will propagating the DNA. However, the DNA itself should be handled at least as carefully as one would handle the most dangerous of the DNAs used to make it.

The above rule by itself effectively precludes certain experiments—namely, those in which one of the components is in Class 5 of the "Classification of Etiologic Agents on the Basis of Hazard" (5), as these are excluded from the United States by law and USDA administrative policy. There are additional experiments which may engender such serious biohazards that they are not to be performed at this time. These are considered prior to presentation of the containment guidelines for permissible experiments.

A. Experiments That are Not to be Performed

We recognize that it can be argued that certain of the recombinants placed in this category could be adequately contained at this time. Nonetheless, our estimates of the possible dangers that may ensue if that containment fails are of such a magnitude that we consider it the wisest policy to at least defer experiments on these recombinant DNAs until there is more information to accurately assess that danger and to allow the construction of more effective biological barriers. In this respect, these guidelines are more stringent than those initially recommended (1).

The following experiments are not to be initiated at the present time: (i) Cloning of recombinant DNAs derived from the pathogenic organisms in Classes 3, 4, and 5 of "Classification of Etiologic Agents on the Basis of Hazard" (5), or oncogenic viruses classified by NCI as moderate risk (6), or cells known to be infected with such agents, regardless of the host-vector system used. (ii) Deliberate formation of recombinant DNAs containing genes for the biosynetheis of potent toxins (e.g., botulinum or diphtheria toxins; venoms from insects, snakes, etc.). (iii) Deliberate creation from plant pathogens of recombinant DNAs that are likely to increase virulence and host range. (iv) Deliberate release into the environment of any organism containing a recombinant DNA molecule. (v) Transfer of a drug resistance trait to microorganisms that are not known to acquire it naturally if such acquisition could compromise the use of a drug to control disease agents in human or veterinary medicine or agriculture.

In addition, at this time large-scale experiments (e.g., more than 10 liters of culture) with recombinant DNAs known to make harmful products are not to be carried out. We differentiate between small- and large-scale experiments with such DNAs because the probability of escape from containment barriers normally increases with increasing scale. However, specific experiments in this category that are of direct societal benefit may be excepted from this rule if special biological containment precautions and equipment designed for large-scale operations are used, and provided that these experiments are expressly approved by the Recombinant DNA Molecule Program Advisory Committee of NIH.

B. Containment Guidelines for Permissible Experiments

It is anticipated that most recombinant DNA experiments initiated before these guidelines are next reviewed (i.e., within the year) will employ *E. coli* K-12 host-vector systems. These are also the systems for which we have the most experience and knowledge regarding the effectiveness of the containment provided by existing hosts and vectors necessary for the construction of more effective biological barriers.

For these reasons, *E. coli* K-12 appears to be the system of choice at this time, although we have carefully considered arguments that many of the potential dangers are compounded by using an organism as intimately connected with man as is *E. coli*. Thus, while proceeding cautiously with *E. coli*, serious efforts should be made toward developing alternate host-vector systems; this subject is discussed in considerable detail in Appendix A.

We therefore consider DNA recombinants in *E. coli* K-12 before proceeding to other host-vector systems.

1. Biological Containment Criteria Using E. coli K-12 Host-Vectors—EK1 Host-Vectors.

These are host-vector systems that can be estimated to already provide a moderate level of containment, and include most of the presently available systems. The host is always *E. coli* K-12, and the vectors include nonconjugative plasmids [e.g., pSC101, ColE1 or derivatives thereof (19–26)] and variants of bacteriophage λ (27–29).

The *E. coli* K-12 nonconjugative plasmid system is taken as an example to illustrate the approximate level of containment referred to here. The available data from experiments involving the feeding of bacteria to humans and calves (30–32) indicate that *E. coli* K-12 did not usually colonize the *normal* bowel, and exhibited little, if any, multiplication while passing through the alimentary tract even after feeding high doses (i.e., 10^9 to 10^{10} bacteria per human or calf). However, general extrapolation of these results may not be warranted because the implantation of bacteria into the intestinal tract depends on a number of parameters, such as the nature of the intestinal flora present in a given individual and the physiological state of the inoculum. Moreover, since viable *E. coli* K-12 can be found in the feces after humans are fed 10^7 bacteria in broth (30) or 3×10^4 bacteria protected by suspension in milk (31), transductional and conjugational transfer of the plasmid vectors from *E. coli* K-12 to resident bacteria in the fecal matter before and after excretion must also be considered.

The nonconjugative plasmid vectors cannot promote their own transfers, but require the presence of a conjugative plasmid for mobilization and transfer to other bacteria. When present in the same cell with derepressed conjugative plasmids such as F or R1*drd19*, the nonconjugative ColE1, colE1-*trp* and pSC101 plasmids are transferred to suitable recipient strains under ideal laboratory conditions at frequencies of about 0.5, 10^{-4} to 10^{-5}, and 10^{-6} per donor cell, respectively. These frequencies are reduced by another factor of 10^2 to 10^4 if the conjugative plasmid employed is repressed with respect to expression of donor fertility.

The experimental transfer system which most closely resembles nonconjugative plasmid transfer in nature is a triparental mating. In such matings, the bacterial cell possessing the nonconjugative plasmid must first acquire a conjugative plasmid from another cell before it can transfer the nonconjugative plasmid to a secondary recipient. With ColE1, the frequencies of transfer are 10^{-2} and 10^{-4} to 10^{-5} when using conjugative plasmid donors possessing derepressed and repressed plasmids, respectively. Mobilization of ColE1-*trp* and pSC101 under similar laboratory conditions is so low as to be usually undetectable (33). Since most conjugative plasmids in nature are repressed for expression of donor fertility, the frequency at which nonconjugative plasmids are mobilized and transferred by this sequence of events *in vivo* is difficult to estimate. However, in calves fed on an antibiotic-supplemented diet, it has been estimated that such triparental nonconjugative R plasmid transfer occurs at frequencies of no more than 10^{-10} to 10^{-12} per 24 hours per calf (32). In terms of considering other means for plasmid transmission in nature, it should be noted that transduction does operate *in vivo* for *Staphylococcus aureus* (34) and probably for *E. coli* as well. However, no data are available to indicate the frequencies of plasmid transfer *in vivo* by either transduction or transformation.

These observations indicate the low probabilities for possible dissemination of such plasmid vectors by accidental ingestion, which would probably involve only a few hundred or thousand bacteria provided that at least the standard practices (Section II—A above) are followed, particularly the avoidance of mouth pipetting. The possibility of colonization and hence of transfer are increased, however, if the normal flora in the bowel is disrupted by, for example, antibiotic therapy (35).

For this reason, persons receiving such therapy must not work with DNA recombinants formed with any *E. coli* K-12 host-vector system during the therapy period and for seven days thereafter; similarly, persons who have achlorhydria or who have had surgical removal of part of the stomach or bowel should avoid such work, as should those who require large doses of antacids.

The observations on the fate of *E. coli* K-12 in the human alimentary tract are also relevant to the containment of recombinant DNA formed with bacteriophage λ variants. Bacteriophage can escape from the laboratory either as mature infectious phage particles or in bacterial host cells in which the phage genome is carried as a plasmid or prophage. The fate of *E. coli* K-12 host cells carrying the phage genome as a plasmid or prophage is similar to that for plasmid-containing host cells as discussed above. The survival of the λ phage genome when released as infectious particles depends on their stability in nature, their infectivity and on the probability of subsequent encounters with naturally occurring λ-sensitive *E. coli* strains. Although the probability of survival of λ and its infection of resident intestinal *E. coli* in animals and humans has not been measured, it is estimated to be small given the high sensitivity of λ to the low pH of the stomach, the insusceptibility to λ infection of smooth *E. coli* cells (the type that normally resides in the gut), the infrequency of naturally occurring λ-sensitive *E. coli* (36) and the failure to detect infective λ particles in human feces after ingestion of up to 10^{11} λ particles (37). Moreover, λ particles are very sensitive to desiccation.

Establishment of λ as a stable lysogen is a frequent event (10^0 to 10^{-1}) for the att^+ int^+ cI^+ phage so that this mode of escape would be the preponderant laboratory hazard; however, most EK1 λ vectors currently in use lack the *att* and *int* functions (27–29), thus reducing the probability of lysogenization to about 10^{-8} to 10^{-6} (38–40). The frequency for the conversion of λ to a plasmid state for persistence and replication is also only about 10^{-6} (41). Moreover, the routine treatment of phage lysates with chloroform (42) should eliminate all surviving bacteria including lysogens and λ plasmid carriers. Lysogenization could also occur when an infectious λ containing cloned DNA infects a λ-sensitive cell in nature, and recombines with a resident lambdoid prophage. Although λ-sensitive *E. coli* strains seem to be rare, a significant fraction do carry lambdoid prophages (43–44) and thus this route of escape should be considered.

While not exact, the estimates for containment afforded by using these host-vectors are at least as accurate as those for physical containment, and are sufficient to indicate that currently employed plasmid and λ vector systems provide a moderate level of biological containment. Other nonconjugative plasmids and bacteriophages that, in association with *E. coli* K-12, can be estimated to provide the same approximate level of moderate containment are included in the EK1 class.

EK2 Host-Vectors. These are host-vector systems that have been genetically constructed and shown to provide a high level of biological containment as demonstrated by data from suitable tests performed in the laboratory. The genetic modifications of the *E. coli* K12 host and/or the plasmid or phage vector should not permit survival of a genetic marker carried on the vector, preferably a marker within an inserted DNA fragment, in other than specially designed and carefully regulated laboratory environments at a frequency greater than 10^{-8}. This measure of biological containment has been selected because it is a measurable entity. Indeed, by testing the contributions of preexisting and newly introduced genetic properties of vectors and hosts, individually or in various combinations, it should be possible to estimate with considerable precision that the specially designed host-vector system can provide a margin of biological containment in excess of that required. For the time being, no host-vector system will be considered to be a bona fide EK2 host-vector system until it is so certified by the NIH Recombinant DNA Molecule Program Advisory Committee.

For EK2 host-vector systems in which the vector is a plasmid, no more than one in 10^{8} host cells should be able to perpetuate the vector and/or a cloned DNA fragment under nonpermissive conditions designed to represent the natural environment either by survival of the original host or as a consequence of transmission of the vector and/or a cloned DNA fragment by transformation, transduction or conjugation to a host with properties common to those in the natural environment.

In terms of potential EK2 plasmid-host systems, the following types of genetic modifications should reduce survival of cloned DNA. *The examples given are for illustrative purposes and should not be construed to encompass all possibilities.* The presence of the nonconjugative plasmids ColE1-*trp* and pSC101 in an *E. coli* K-12 strain possessing a mutation eliminating host-controlled restriction and modification (*hsdS*) results in about 10^{3}-fold reduction in mobilization to restriction-proficient recipients. The combination of the *dapD8*, Δ*bioH-asd*, Δ*gal-chl*^{r} and *rfb* mutations in *E. coli* K-12 results in no detectable survivors in feces of rats following feeding by stomach tube of 10^{10} cells in milk and similarly leads to complete lysis of cells suspended in broth medium lacking diaminopimelic acid. *E. coli* K-12 strains with Δ*thyA* and *deoC (dra)* mutations undergo thymineless death in growth medium lacking thymine and give a 10^{5}-fold reduced survival during passage through the rat intestine compared to wild-type *thy*$^{+}$ *E. coli* K-12. (However, the Δ*thyA* mutation alone or in combination with a *deoB(drm)* mutation only reduces *in vivo* survival by a factor of 10^2.) Other host mutations, as yet untested, that might further reduce survival of the plasmid-host system or reduce plasmid transmission are: the combination *polA*(TS)*recA*(TS) Δ*thyA* which might interfere with ColE1 replication and lead to DNA degradation at body temperatures; Con⁻ mutations that reduce the ability of conjugative plasmids to enter the plasmid-host complex and thus should reduce mobilization of the cloned DNA to other strains; and mutations that confer resistance to known transducing phages. Mutations can also be introduced into the plasmid to cause it to be dependent on a specific host, to make its replication thermosensitive and/or to endow it with a killer capability such that all cells (other than its host) into which it might be transferred will not survive.

In the construction of EK2 plasmid-host systems it is important to use the most stable mutations available, preferably deletions. Obviously, the presence of all mutations contributing to higher degrees of biological containment must be verified periodically by appropriate tests. In testing the level of biological containment afforded by a proposed EK2 plasmid-host system, it is important to design relevant tests to evaluate the survival of the vector and/or a cloned DNA fragment under conditions that are possible in nature and that are also most advantageous for its perpetuation. For example, one might conduct a triparental mating with a primary donor possessing a derepressed F-type or I-type conjugative plasmid, the safer host with Δ*bioH-asd*, *dapD8*, Δ*gal-chl^r*, *rfb*, Δ*thyA*, *deoC*, *trp* and *hsdS* mutations and a plasmid vector carrying an easily detectable marker such as for ampicillin resistance or an inserted gene such as *trp*⁺, and a secondary recipient that is Su⁺ *hsdS trp* (i.e., permissive for the recombinant plasmid). Such matings would be conducted in a medium lacking diaminopimelic acid and thymine, and survival of the Ap^r or *trp*⁺ marker in any of the three strains followed as a function of time. Survival of the vector and/or a cloned marker by transduction could also be evaluated by introducing a known generalized transducing phage into the system. Similar experiments should also be done using a secondary recipient that is restrictive for the plasmid vector as well as with primary donors possessing repressed conjugative plasmids with incompatibility group properties like those commonly found in enteric microorganisms. Since a common route of escape of plasmid-host systems in the laboratory might be by accidental ingestion, it is suggested that the same types of experiments be conducted in suitable animal-model systems. In addition to these tests on survival of the vector and/or a cloned DNA fragment, it would be useful to determine the survival of the host strain under nongrowth conditions such as in water and as a function of drying time after a culture has been spilled on a lab bench.

For EK2 host-vector systems in which the vector is a phage, no more than one in 10^8 phage particles should be able to perpetuate itself and/or a cloned DNA fragment under nonpermissive conditions designed to represent the natural environment either (a) as a prophage or plasmid in the laboratory host used for phage propagation or (b) by surviving in natural environments and transferring itself and/or a cloned DNA fragment to a host (or its resident lambdoid prophage) with properties common to those in the natural environment.

In terms of potential EK2 λ-host systems, the following types of genetic modification should reduce survival of cloned DNA. *The examples given are for illustrative purposes and should not be construed to encompass all possibilities.* The probability of establishing λ lysogeny in the normal laboratory host should be reduced by removal of the phage *att* site, the Int function, the repressor gene(s) and adding virulence-enhancing mutations. The frequency of plasmid formation, although normally already less than 10^{-6}, could be further reduced by defects in the $p_R Q$ region, including mutations such as *vir-s*, *cro*(TS), c^{17}, *ri*^{c}, *O*(TS), *P*(TS), and *nin*. Moreover, chloroform treatment used routinely following cell lysis would reduce the number of surviving cells, including possible lysogens or plasmid carriers, by more than 10^8. The host may also be modified by deletion of the host λ *att* site and inclusion of one or more of the mutations described above for plasmid-host systems to further reduce the chance of formation and survival of any lysogen or plasmid carrier cell.

The survival of escaping phage and the chance of encountering a sensitive host in nature are very low, as discussed for EK1 systems. The infectivity of the phage particles could be further reduced by introducing mutations (e.g., suppressed ambers) which would make the phage particles extremely unstable except under special laboratory conditions (e.g., high concentrations of salts or putrescine). Another means would be to make the phage itself a two-component system, by eliminating the tail genes and reproducing the phage as heads packed with DNA; when necessary and under specially controlled conditions, these heads could be made infective by adding tail preparations. An additional safety factor in this regimen is the extreme instability of the heads, unless they are stored in 10mM putrescine, a condition easy to obtain in the laboratory but not in nature. The propagation of the escaping phage in nature could further be blocked by adding various conditional mutations which would permit growth only under special laboratory conditions or in a special permissive laboratory host with suppressor or *gro*-type (*mop*, *dnaB*, *rpoB*) mutations. An additional safety feature would be the use of an r⁻m⁻ (*hsdS*) laboratory host, which produces phage with unmodified DNA which should be restricted in r⁺m⁺ bacteria that are probably prevalent in nature. The likelihood of recombination between the λ vector and lambdoid prophages which are present in some *E. coli* strains might be reduced by elimination of the Red function and the presence of the recombination-reducing Gam function together with mutations contributing to the high lethality of the λ phage. However, these second-order precautions might not be relevant if the stability and infectivity of the escaping λ particles are reduced by special mutations or by propagating the highly unstable heads.

Despite multiple mutations in the phage vectors and laboratory hosts, the yield of phage particles under suitable laboratory conditions should be high (10^{10}–10^{11} particles/ml). This permits phage propagation in relatively small volumes and constitutes an additional safety feature.

The phenotypes and genetic stabilities of the mutations and chromosome alterations included in these λ-host systems indicate that containment well in excess of the required 10^{-8} or lower survival frequency for the λ vector with or without a cloned DNA fragment should be attained. Obviously the presence of all mutations contributing to this high degree of biological containment must be verified periodically by appropriate tests. Laboratory tests should be performed with the bacterial host to measure all possible routes of escape such as the frequency of lysogen formation, the frequency of plasmid formation and the survival of the lysogen or carrier bacterium. Similarly, the potential for perpetuation of a cloned DNA fragment carried by infectious phage particles can be tested by challenging typical wild-type *E. coli* strains or a λ-sensitive nonpermissive laboratory K-12 strain, especially one lysogenic for a lambdoid phage.

In view of the fact that accurate assessment of the probabilities for escape of infectious λ grown on r^-m^- Su^+ hosts is dependent upon the frequencies of r^-, Su^+, and λ-sensitive strains in nature, investigators need to screen *E. coli* strains for these properties. These data will also be useful in predicting frequencies of successful escape of plasmid cloning vectors harbored in r^-m^- Su^+ strains.

When any investigator has obtained data on the level of containment provided by a proposed EK2 system, these should be reported as rapidly as possible to permit general awareness and evaluation of the safety features of the new system. Investigators are also encouraged to make such new safer cloning systems generally available to other scientists. NIH will take appropriate steps to aid in the distribution of these safer vectors and hosts.

EK3 Host-Vectors. These are EK2 systems for which the specified containment shown by laboratory tests has been independently confirmed by appropriate tests in animals, including humans or primates, and in other relevant environments in order to provide additional data to validate the levels of containment afforded by the EK2 host-vector

systems. Evaluation of the effects of individual or combinations of mutations contributing to the biological containment should be performed as a
means to confirm the degree of safety provided and to further advance
the technology of developing even safer vectors and hosts. For the time
being, no host-vector system will be considered to be a bona fide EK3
host-vector system until it is so certified by the NIH Recombinant DNA
Molecule Program Advisory Committee.

**2. Classification of Experiments Using the E. Coli K-12 Contain-
ment Systems.**   In the following classification of containment criteria for
different kinds of recombinant DNAs, the stated levels of physical and
biological containment are minimums. Higher levels of biological containment (EK3 > EK2 > EK1) are to be used if they are available and are
equally appropriate for the purposes of the experiment.

(*a*) *Shotgun Experiments.*   These experiments involve the production
of recombinant DNAs between the vector and the total DNA or (preferably) any partially purified fraction thereof from the specified cellular
source.

(*i*) *Eukaryotic DNA Recombinants—Primates.*   P3 physical containment
+ an EK3 host-vector, or P4 physical containment + an EK2 host-vector,
except for DNA from uncontaminated embryonic tissue or primary tissue cultures therefrom, and germ-line cells for which P3 physical containment + an EK2 host-vector can be used. The basis for the lower
estimated hazard in the case of DNA from the latter tissues (if freed of
adult tissue) is their relative freedom from horizontally acquired adven-
titious viruses.

Other mammals. P3 physical containment + an EK2 host-vector.

Birds. P3 physical containment + an EK2 host-vector.

Cold-blooded vertebrates. P2 physical containment + an EK2 host-
vector except for embryonic or germ-line DNA which require P2 physical containment + an EK1 host-vector. If the eukaryote is known to
produce a potent toxin, the containment shall be increased to P3 + EK2.

Other cold-blooded animals and lower eukaryotes. This large class of
eukaryotes is divided into the following two groups:

(*1*) Species that are known to produce a potent toxin or are known
pathogens (i.e., an agent listed in Class 2 of ref. 5 or a plant pathogen) or
are known to carry such pathogenic agents must use P3 physical containment + an EK2 host-vector. Any species that has a demonstrated
capacity for carrying particular pathogenic agents is included in this
group unless it has been shown that those organisms used as the source

of DNA do not contain these agents; in this case they may be placed in the second group.

(2) The remainder of the species in the class can use P2 + EK1. However, any insect in this group should have been grown under laboratory conditions for at least 10 generations prior to its use as a source of DNA.

Plants. P2 physical containment + an EK1 host-vector. If the plant carries a known pathogenic agent or makes a product known to be dangerous to any species, the containment must be raised to P3 physical containment + an EK2 host-vector.

(ii) Prokaryotic DNA Recombinants—Prokaryotes that Exchange Genetic Information with E. coli.[2] The level of physical containment is directly determined by the rule of the most dangerous component (see introduction to Section III). Thus P1 conditions can be used for DNAs from those bacteria in class 1 of ref. 5 ("Agents of no or minimal hazard . . .") which naturally exchange genes with *E. coli*; and P2 conditions should be used for such bacteria if they fall in Class 2 of ref. 5 ("Agents of ordinary potential hazard . . ."), or are plant pathogens or symbionts. EK1 host-vectors can be used for all experiments requiring only P1 physical containment; in fact, experiments in this category can be performed with *E. coli* K-12 vectors exhibiting a lesser containment (e.g., conjugative plasmids) than EK1 vectors. Experiments with DNA from species requiring P2 physical containment which are of low pathogenicity (for example, enteropathogenic *Escherichia coli*, *Salmonella typhimurium*, and *Klebsiella pneumoniae*) can use EK1 host-vectors, but those of moderate pathogenicity (for example, *Salmonella typhi*, *Shigella dysenteriae* type I, and *Vibrio cholerae*) must use EK2 host-vectors.[3] A specific pathogen requiring P2 physical containment + an EK2 host-vector would be cloning the tumor gene of *Agrobacterium tumefaciens*.

Prokaryotes that do not exchange genetic information with E. coli. The minimum containment conditions for this class consist of P2 physical containment + an EK2 host-vector or P3 physical containment + an EK1 host-vector, and apply when the risk that the recombinant DNAs will increase the pathogenicity or ecological potential of the host is judged to be minimal. Experiments with DNAs from pathogenic species (Class 2 ref. 5 plus plant pathogens) must use P3 + EK2.

(iii) Characterized Clones of DNA Recombinants Derived from Shotgun Experiments. When a cloned DNA recombinant has been rigorously characterized[4] and there is sufficient evidence that it is free of harmful genes,[4] then experiments involving this recombinant DNA can be car-

ried out under P1 + EK1 conditions if the inserted DNA is from a species that exchanges genes with *E. coli*, and under P2 + EK1 conditions if not.

(b) Purified Cellular DNAs Other than Plasmids, Bacteriophages, and Other Viruses. The formation of DNA recombinants from cellular DNAs that have been enriched[5] by physical and chemical techniques (i.e., not by cloning) and which are free of harmful genes can be carried out under lower containment conditions than used for the corresponding shotgun experiment. In general, the containment can be decreased one step in physical containment (P4 → P3 → P2 → P1) while maintaining the biological containment specified for the shotgun experiment, or one step in biological containment (EK3 → EK2 → EK1) while maintaining the specified physical containment—provided that the new condition is not less than that specified above for characterized clones from shotgun experiments (Section (a)—iii).

(c) Plasmids, Bacteriophages, and Other Viruses. Recombinants formed between EK-type vectors and other plasmid or virus DNAs have in common the potential for acting as double vectors because of the replication functions in these DNAs. The containment conditions given below apply only to propagation of the DNA recombinants in *E. coli* K-12 hosts. They do not apply to other hosts where they may be able to replicate as a result of functions provided by the DNA inserted into the EK vectors. These are considered under other host-vector systems.

(i) Animal Viruses. P4 + EK2 or P3 + EK3 shall be used to isolate DNA recombinants that include *all or part* of the genome of an animal virus. This recommendation applies not only to experiments of the "shotgun" type but also to those involving partially characterized subgenomic segments of viral DNAs (for example, the genome of defective viruses, DNA fragments isolated after treatment of viral genomes with restriction enzymes, etc). When cloned recombinants have been shown by suitable biochemical and biological tests to be free of harmful regions, they can be handled in P3 + EK2 conditions. In the case of DNA viruses, harmless regions include the late region of the genome; in the case of DNA copies of RNA viruses, they might include the genes coding for capsid proteins or envelope proteins.

(ii) Plant Viruses. P3 + EK1 or P2 + EK2 conditions shall be used to form DNA recombinants that include *all or part* of the genome of a plant virus.

(iii) Eukaryotic Organelle DNAs. The containment conditions given below apply only when the organelle DNA has been purified[6] from isolated organelles. Mitochondrial DNA from primates: P3+ EK1 or P2+ EK2. Mitochondrial or chloroplast DNA from other eukaryotes: P2+ EK1. Otherwise, the conditions given under shotgun experiments apply.

(iv) Prokaryotic Plasmid and Phage DNAs—Plasmids and Phage from Hosts that Exchange Genetic Information with E. coli. Experiments with DNA recombinants formed from plasmids or phage genomes that have not been characterized with regard to presence of harmful genes or are known to contribute significantly to the pathogenicity of their normal hosts must use the containment conditions specified for shotgun experiments with DNAs from the respective host. If the DNA recombinants are formed from plasmids or phage that are known not to contain harmful genes, or from purified[6] and characterized plasmid or phage DNA segments known not to contain harmful genes, the experiments can be performed with P1 physical containment + an EK1 host-vector.

Plasmids and phage from hosts that do not exchange genetic information with E. coli. The rules for shotgun experiments with DNA from the host apply to their plasmids or phages. The minimum containment conditions for this category (P2+ EK2, or P3+ EK1) can be used for plasmid and phage, or for purified[6] and characterized segments of plasmid and phage DNAs, when the risk that the recombinant DNAs will increase the pathogenicity or ecological potential of the host is judged to be minimal.

NOTE.—Where applicable, cDNAs (i.e., complementary DNAs) synthesized *in vitro* from cellular or viral RNAs are included within each of the above classifications. For example, cDNAs formed from cellular RNAs that are not purified and characterized are included under *(a)* shotgun experiments; cDNAs formed from purified and characterized RNAs are included under *(b)*; cDNAs formed from viral RNAs are included under *(c)*; etc.

3. Experiments with Other Prokaryotic Host-Vectors. Other prokaryotic host-vector systems are at the speculative, planning, or development stage, and consequently do not warrant detailed treatment here at this time. However, the containment criteria for different types of DNA recombinants formed with *E. coli* K-12 host-vectors can, with the aid of some general principles given here, serve as a guide for containment conditions with other host-vectors when appropriate adjustment is made for their different habitats and characteristics. The newly developed host-vector systems should offer some distinct advantage over the *E. coli* K-12 host-vectors—for instance, thermophilic organisms or other host-vectors whose major habitats do not include humans and/or economically important animals and plants. In general, the strain of any prokaryotic species used as the host is to conform to the definition of class 1 etiologic agents given in ref. 5 (i.e., "Agents of no or minimal hazard, . . ."), and the plasmid or phage vector should not make the host more hazardous. Appendix A gives a detailed discussion of the *B. subtilis* system, the most promising alternative to date.

At the initial state, the host-vector must exhibit at least a moderate level of biological containment comparable to EK1 systems, and should be capable of modification to obtain high levels of containment comparable to EK2 and EK3. The type of confirmation test(s) required to move a host-vector from an EK2-type classification to an EK3-type will clearly depend upon the preponderant habitat of the host-vector. For example, if the unmodified host-vector propagates mostly in, on, or around higher plants, but not appreciably in warm-blooded animals, modification should be designed to reduce the probability that the host-vector can escape to and propagate in, on, or around such plants, or transmit recombinant DNA to other bacterial hosts that are able to occupy these ecological niches, and it is these lower probabilities which must be confirmed. The following principles are to be followed in using the containment criteria given for experiments with *E. coli* K-12 host-vectors as a guide for other prokaryotic systems. Experiments with DNA from prokaryotes (and their plasmids or viruses) are classified according to whether the prokaryote in question exchanges genetic information with the host-vector or not, and the containment conditions given for these two classes with *E. coli* K-12 host-vectors applied. Experiments with recombinants between plasmid or phage vectors and DNA that extends the range of resistance of the recipient species to therapeutically useful drugs must use P3 physical containment + a host-vector comparable to EK1 or P2 physical containment + a host-vector comparable to EK2. Transfer of recombinant DNA to plant pathogens can be made safer by using nonreverting, doubly auxotrophic, nonpathogenic variants. Experiments using a plant pathogen that affects an element of the local flora will require more stringent containment than if carried out in areas where the host plant is not common.

Experiments with DNAs from eukaryotes (and their plasmids or viruses) can also follow the criteria for the corresponding experiments with *E. coli* K-12 vectors if the major habitats of the given host-vector overlap those of *E. coli*. If the host-vector has a major habitat that does not overlap those of *E. coli* (e.g., root nodules in plants), then the containment conditions for some eukaryotic recombinant DNAs need to be increased (for instance, higher plants and their viruses in the preceding example), while others can be reduced.

4. Experiments with Eukaryotic Host-Vectors

(a) Animal Host-Vector Systems. Because host cell lines generally have little if any capacity for propagation outside the laboratory, the primary focus for containment is the vector, although cells should also be derived from cultures expected to be of minimal hazard. Given good microbiological practices, the most likely mode escape of recombinant DNAs from a physically contained laboratory is carriage by humans; thus vectors should be chosen that have little or no ability to replicate in human cells. To be used as a vector in a eukaryotic host, a DNA molecule needs to display all of the following properties:

1. It shall not consist of the whole genome of any agent that is infectious for humans or that replicates to a significant extent in human cells in tissue culture.
2. Its functional anatomy should be known—that is, there should be a clear idea of the location within the molecule of:
   (a) The sites at which DNA synthesis originates and terminates.
   (b) The sites that are cleaved by restriction endonucleases.
   (c) The template regions for the major gene products.
3. It should be well studied genetically. It is desirable that mutants be available in adequate number and variety, and that quantitative studies of recombination have been performed.
4. The recombinant must be defective, that is, its propagation as a virus is dependent upon the presence of a complementing helper genome. This helper should either (a) be integrated into the genome of a stable line of host cells (a situation that would effectively limit the growth of the vector to that particular cell line) or (b) consist of a defective genome or an appropriate conditional lethal mutant virus (in which case the experiments would be done under nonpermissive conditions), making vector and helper dependent upon each other for propagation. However, if none of these is available, the use of a nondefective genome as helper would be acceptable.

Currently only two viral DNAs can be considered as meeting these requirements: these are the genomes of polyoma virus and SV40.

Of these, polyoma virus is highly to be preferred. SV40 is known to propagate in human cells, both *in vivo* and *in vitro*, and to infect laboratory personnel, as evidenced by the frequency of their conversion to producing SV40 antibodies. Also, SV40 and related viruses have been found in association with certain human neurological and malignant diseases. SV40 shares many properties, and gives complementation, with the common human papova viruses. By contrast, there is no evidence that polyoma infects humans, nor does it replicate to any significant extent in human cells *in vitro*. However, this system still needs to be studied more extensively. Appendix B gives further details and documentation.

Taking account of all these factors:

(1) Polyoma Virus. *(a)* Recombinant DNA molecules consisting of defective polyoma virus genomes plus DNA sequences of any non-pathogenic organism, including Class 1 viruses (5), can be propagated in or used to transform cultured cells. P3 conditions are required. Appropriate helper virus can be used if needed. Whenever there is a choice, it is urged that mouse cells, derived preferably from embryos, be used as the source of eukaryotic DNA. Polyoma virus is a mouse virus and recombinant DNA molecules containing both viral and cellular sequences are already known to be present in virus stocks grown at a high multiplicity. Thus, recombinants formed *in vitro* between polyoma virus DNA and mouse DNA are presumably not novel from an evolutionary point of view.

(b) Such experiments are to be done under P4 conditions if the recombinant DNA contains segments of the genomes of Class 2 animal viruses (5). Once it has been shown by suitable biochemical and biological tests that the cloned recombinant contains only harmless regions of the viral genome (see Section IIIB—2—c—i) and that the host range of the polyoma virus vector has not been altered, experiments can be continued under P3 conditions.

(2) SV40 Virus.

(a) Defective SV40 genomes, with appropriate helper, can be used as a vector for recombinant DNA molecules containing sequences of any nonpathogenic organism or Class I virus (5) (i.e., a shotgun type experiment). P4 conditions are required. Established lines of cultured cells should be used.

(b) Such experiments are to be carried out in P3 (or P4) conditions if the non-SV40 DNA segment is (a) a purified segment of prokaryotic DNA lacking toxigenic genes, or (b) a segment of eukaryotic DNA whose function has been established, which does not code for a toxic product, and which has been previously cloned in a prokaryotic host-vector system. It shall be confirmed that the defective virus-helper virus system does not replicate significantly more efficiently in human cells in

tissue culture than does SV40, following infection at a multiplicity of infection of one or more helper SV40 viruses per cell.

(*c*) A recombinant DNA molecule consisting of defective SV40 DNA lacking substantial segments of the late region, plus DNA from non-pathogenic organisms or Class I viruses (5) can be propagated as an autonomous cellular element in established lines of cells under P3 conditions provided that there is no exogenous or endogenous helper, and that it is demonstrated that *no* infectious virus particles are being produced. Until this has been demonstrated, the appropriate containment conditions specified in 2. *a.* and 2. *b.* shall be used.

(*d*) Recombinant DNA molecules consisting of defective SV40 DNA and sequences from nonpathogenic prokaryotic or eukaryotic organisms or Class I viruses (5) can be used to transform established lines of non-permissive cells under P3 conditions. It must be demonstrated that no infectious virus particles are being produced; rescue of SV40 from such transformed cells by co-cultivation or transfection techniques must be carried out in P4 conditions.

(3) Efforts are to be made to ensure that all cell lines are free of virus particles and mycoplasma.

Since SV40 and polyoma are limited in their scope to act as vectors, chiefly because the amount of foreign DNA that the normal virions can carry probably cannot exceed 2×10^6 daltons, the development of systems in which recombinants can be cloned and propagated purely in the form of DNA, rather than in the coats of infectious agents is necessary. Plasmid forms of viral genomes or organelle DNA need to be explored as possible cloning vehicles in eukaryotic cells.

(*b*) *Plant Host-Vector Systems.* For cells in tissue cultures, seedlings, or plant parts (e.g., tubers, stems, fruits, and detached leaves) or whole mature plants of small species (e.g., *Arabidopsis*) the P1–P4 containment conditions that we have specified previously are relevant concepts. However, work with most plants poses additional problems. The greenhouse facilities accompanying P2 laboratory physical containment conditions can be provided by: (i) Insect-proof greenhouses, (ii) appropriate sterilization of contaminated plants, pots, soil, and runoff water, and (iii) adoption of the other standard practices for microbiological work. P3 physical containment can be sufficiently approximated by confining the operations with whole plants to growth chambers like those used for work with radioactive isotopes, provided that (i) such chambers are modified to produce a negative pressure environment with the exhaust air appropriately filtered, (ii) that other operations with infectious materials are carried out under the specified P3 conditions, and (iii) to guard against inadvertent insect transmission of recombinant DNA, growth chambers are to be routinely fumigated and only used in insect proof rooms. The P2 and P3 conditions specified earlier are therefore extended to include these cases for work on higher plants.

The host cells for experiments on recombinant DNAs may be cells in culture, in seedling or plant parts. Whole plants or plant parts that cannot be adequately contained shall not be used as hosts for shotgun experiments at this time, and attempts to infect whole plants with recombinant DNA shall not be initiated until the effects on host cells in culture, seedlings or plant parts have been thoroughly studied.

Organelle or plasmid DNAs or DNAs of viruses of restricted host range may be used as vectors. In general, similar criteria for selecting host-vectors to those given in the preceding section on animal systems are to apply to plant systems.

DNA recombinants formed between the initial moderately contained vectors and DNA from cells of species in which the vector DNA can replicate require P2 physical containment. However, if the source of the DNA is itself pathogenic or known to carry pathogenic agents, or to produce products dangerous to plants, or if the vector is an unmodified virus of unrestricted host range, the experiments shall be carried out under P3 conditions.

Experiments on recombinant DNAs formed between the above vectors and DNAs from other species can also be carried out under P2 if that DNA has been purified[6] and determined not to contain harmful genes. Otherwise, the experiments shall be carried out under P3 conditions if the source of the inserted DNA is not itself a pathogen, or known to carry such pathogenic agents, or to produce harmful products—and under P4 conditions if these conditions are not met.

The development and use of host-vector systems that exhibit a high level of biological containment permit a decrease of one step in the physical containment specified above (P4 → P3 → P2 → P1).

(c) Fungal or Similar Lower Eukaryotic Host-Vector Systems. The containment criteria for experiments on recombinant DNAs using these host-vectors most closely resemble those for prokaryotes, rather than those for the preceding eukaryotes, in that the host cells usually exhibit a capacity for dissemination outside the laboratory that is similar to that for bacteria. We therefore consider that the containment guidelines given for experiments with *E. coli* K-12 and other prokaryotic host-vectors (Sections IIIB-1 and -2, respectively) provide adequate direction for experiments with these lower eukaryotic host-vectors. This is particularly true at this time since the development of these host-vectors is presently in the speculative stage.

IV. ROLES AND RESPONSIBILITIES

Safety in research involving recombinant DNA molecules depends upon how the research team applies these guidelines. Motivation and critical judgment are necessary, in addition to specific safety knowledge, to ensure protection of personnel, the public, and the environment.

The guidelines given here are to help the principal investigator determine the nature of the safeguards that should be implemented. These guidelines will be incomplete in some respects because all conceivable experiments with recombinant DNAs cannot now be anticipated. Therefore, they cannot substitute for the investigator's own knowledgeable and discriminating evaluation. Whenever this evaluation calls for an increase in containment over that indicated in the guidelines, the investigator has a responsibility to institute such an increase. In contrast, the containment conditions called for in the guidelines should not be decreased without review and approval at the institutional and NIH levels.

The following roles and responsibilities define an administrative framework in which safety is an essential and integrated function of research involving recombinant DNA molecules.

A. Principal Investigator

The principal investigator has the primary responsibility for: (i) Determining the real and potential biohazards of the proposed research, (ii) determining the appropriate level of biological and physical containment, (iii) selecting the microbiological practices and laboratory techniques for handling recombinant DNA materials, (iv) preparing procedures for dealing with accidental spills and overt personnel contamination, (v) determining the applicability of various precautionary medical practices, serological monitoring, and immunization, where available, (vi) securing approval of the proposed research prior to initiation of work, (vii) submitting information on purported EK2 and EK3 systems to the NIH Advisory Committee and making the strains available to others, (viii) reporting to the institutional biohazards committee and the NIH Office of Recombinant DNA Activities new information bearing on the guidelines, such as technical information relating to hazards and new safety procedures or innovations, (ix) applying for approval from the NIH Recombinant DNA Molecule Program Advisory Committee for large scale experiments with recombinant DNAs known to make harmful products (i.e., more than 10 liters of culture), and (x) applying to NIH for approval to lower containment levels when a cloned DNA recombinant derived from a shotgun experiment has been rigorously characterized and there is sufficient evidence that it is free of harmful genes.

Before work is begun, the principal investigator is responsible for: (i) Making available to program and support staff copies of those protions of the approved grant application that describe the biohazards and the precautions to be taken, (ii) advising the program and support staff of the nature and assessment of the real and potential biohazards, (iii) instructing and training this staff in the practices and techniques required to ensure safety, and in the procedures for dealing with accidentally created biohazards, and (iv) informing the staff of the reasons and provisions for any advised or requested precautionary medical practices, vaccinations, or serum collection.

During the conduct of the research, the principal investigator is responsible for: (i) Supervising the safety performance of the staff to ensure that the required safety practice and techniques are employed, (ii) investigating and reporting in writing to the NIH Office of Recombinant DNA Activities and the institutional biohazards committee any serious or extended illness of a worker or any accident that results in (a) inoculation of recombinant DNA materials through cutaneous penetration, (b) ingestion of recombinant DNA materials, (c) probable inhalation of recombinant DNA materials following gross aerosolization, or (d) any incident causing serious exposure to personnel or danger of environmental contamination, (iii) investigating and reporting in writing to the NIH Office of Recombinant DNA Activities and the institutional biohazards committee any problems pertaining to operation and implementation of biological and physical containment safety practices and procedures, or equipment or facility failure, (iv) correcting work errors and conditions that may result in the release of recombinant DNA materials, and (v) ensuring the integrity of the physical containment (e.g., biological safety cabinets) and the biological containment (e.g., genotypic and phenotypic characteristics, purity, etc.).

B. Institution

Since in almost all cases, NIH grants are made to institutions rather than to individuals, all the responsibilities of the principal investigator listed above are the responsibilities of the institution under the grant, fulfilled on its behalf by the principal investigator. In addition, the institution is responsible for establishing an institutional biohazards committee[7] to: (i) Advise the institution on policies, (ii) create and maintain a central reference file and library of catalogs, books, articles, newsletters, and other communications as a source of advice and reference regarding, for example, the availability and quality of the safety equipment, the availability and level of biological containment for various host-vector systems, suitable training of personnel and data on the potential biohazards associated with certain recombinant DNAs, (iii) develop a safety and operations manual for any P4 facility maintained by the institution and used in support of recombinant DNA research, (iv) certify to the NIH on applications for research support and annually thereafter, that facilities, procedures, and practices and the training and expertise of the personnel involved have been reviewed and approved by the institutional biohazards committee.

The biohazards committee must be sufficiently qualified through the experience and expertise of its membership and the diversity of its membership to ensure respect for its advice and counsel. Its membership should include individuals from the institution or consultants, selected so as to provide a diversity of disciplines relevant to recombinant DNA technology, biological safety, and engineering. In addition to possessing the professional competence necessary to assess and review specific activities and facilities, the committee should possess or have available to it the competence to determine the acceptability of its findings in terms of applicable laws, regulations, standards of practices, community attitudes, and health and environmental considerations. Minutes of the meetings should be kept and made available for public inspection. The institution is responsible for reporting names of and relevant background information on the members of its biohazards committee to the NIH.

C. NIH Initial Review Groups (Study Sections)

The NIH Study Sections, in addition to reviewing the scientific merit of each grant application involving recombinant DNA molecules, are responsible for: (i) Making an independent evaluation of the real and potential biohazards of the proposed research on the basis of these guidelines, (ii) determining whether the proposed physical containment safeguards certified by the institutional biohazards committee are appropriate for control of these biohazards, (iii) determining whether the proposed biological containment safeguards are appropriate, (iv) referring to the NIH Recombinant DNA Molecule Program Advisory Committee or the NIH Office of Recombinant DNA activities those problems pertaining to assessment of biohazards or safeguard determination that cannot be resolved by the Study Sections.

The membership of the Study Sections will be selected in the usual manner. Biological safety expertise, however, will be available to the Study Sections for consultation and guidance.

D. NIH Recombinant DNA Molecule Program Advisory Committee

The Recombinant DNA Molecule Program Advisory Committee advises the Secretary, Department of Health, Education, and Welfare, the Assistant Secretary for Health, Department of Health, Education, and Welfare, and the Director, National Institutes of Health, on a program for the evaluation of potential biological and ecological hazards of recombinant DNAs (molecules resulting from different segments of DNA that have been joined together in cell-free systems, and which have the capacity to infect and replicate in some host cell, either autonomously or as an integrated part of their host's genome), on the development of procedures which are designed to prevent the spread of such molecules within human and other populations, and on guidelines to be followed by investigators working with potentially hazardous recombinants.

The NIH Recombinant DNA Molecule Program Advisory Committee has responsibility for: (i) Revising and updating guidelines to be followed by investigators working with DNA recombinants, (ii) for the time being, receiving information on purported EK2 and EK3 systems and evaluating and certifying that host-vector systems meet EK2 or EK3 criteria, (iii) resolving questions concerning potential biohazard and adequacy of containment capability if NIH staff or NIH Initial Review Group so request, and (iv) reviewing and approving large scale experiments with recombinant DNAs known to make harmful products (e.g., more than 10 liters of culture).

E. NIH Staff

NIH Staff has responsibility for: (i) Assuring that no NIH grants or contracts are awarded for DNA recombinant research unless they (a) conform to these guidelines, (b) have been properly reviewed and recommended for approval, and (c) include a properly executed Memorandum of Understanding and Agreement, (ii) reviewing and responding to questions or problems or reports submitted by institutional biohazards committees or principal investigators, and disseminating findings, as appropriate, (iii) receiving and reviewing applications for approval to lower containment levels when a cloned DNA recombinant derived from a shotgun experiment has been rigorously characterized and there is sufficient evidence that it is free of harmful genes, (iv) referring items covered under (ii) and (iii) above to the NIH Recombinant DNA Molecule Program Advisory Committee, as deemed necessary, and (v) performing site inspections of all P4 physical containment facilities, engaged in DNA recombinant research, and of other facilities as deemed necessary.

V. FOOTNOTES

1. Biological Safety Cabinets referred to in this section are classified as *Class I, Class II* or *Class III* cabinets. A *Class I* cabinet is a ventilated cabinet for personnel protection having an inward flow of air away from the operator. The exhaust air from this cabinet is filtered through a high effiency or high efficiency particulate air (HEPA) filter before being discharged to the outside atmosphere. This cabinet is used in three operational modes; (1) with an 8 inch high full width open front, (2) with an installed front closure panel (having four eight inch diameter openings) without gloves, and (3) with an installed front closure panel equipped with arm length rubber gloves. The face velocity of the inward flow of air through the full width open front is 75 feet per minute or greater. A *Class II* cabinet is a ventilated cabinet for personnel and product protection having an open front with inward air flow for personnel protection, and HEPA filtered mass recirculated air flow for product protection. The cabinet exhaust air is filtered through a HEPA filter. The face velocity of the inward flow of air through the full width open front is 75 feet per minute or greater. Design and performance specifications for *Class II* cabinets have been adopted by the National Sanitation Foundation, Ann Arbor, Michigan. A *Class III* cabinet is a closed front ventilated cabinet of gas tight construction which provides the highest level of personnel protection of all Biohazard Safety Cabinets. The interior of the cabinet is protected from contaminants exterior to the cabinet. The cabinet is fitted with arm length rubber gloves and is operated under a negative pressure of at least 0.5 inches water gauge. All supply air is filtered through HEPA filters. Exhaust air is filtered through HEPA filters or incinerated before being discharged to the outside environment.
2. Defined as observable under optimal laboratory conditions by transformation, transduction, phage infection and/or conjugation with transfer of phage, plasmid and/or chromosomal genetic information.
3. The bacteria which constitute class 2 of ref. 5 ("Agents of ordinary potential hazard. . . .") represent a broad spectrum of etiologic agents which possess different levels of virulence and degrees of communicability. We think it appropriate for our specific purpose to further subdivide the agents of Class 2 into those which we believe to be of relatively low pathogenicity and those which are moderately pathogenic. The several specific examples given may suffice to illustrate the principle.
4. The terms "characterized" and "free of harmful genes" are unavoidably vague. But, in this instance, before containment conditions lower than the ones used to clone the DNA can be adopted, the investigator must obtain approval from the National Institutes of Health. Such approval would be contingent upon data concerning: (a) the absence of potentially harmful genes (e.g., sequences contained in indigenous tumor viruses or which code for toxic substances), (b) the relation between the recovered and desired segment (e.g., hybridization and restriction endonuclease fragmentation analysis where applicable), and (c) maintenance of the biological properties of the vector.

5. A DNA preparation is defined as enriched if the desired DNA represents at least 99% (w/w) of the total DNA in the preparation. The reason for lowering the containment level when this degree of enrichment has been obtained is based on the fact that the total number of clones that must be examined to obtain the desired clone is markedly reduced. Thus, the probability of cloning a harmful gene could, for example, be reduced by more than 10^5-fold when a nonrepetitive gene from mammals was being sought. Furthermore, the level of purity specified here makes it easier to establish that the desired DNA does not contain harmful genes.
6. The DNA preparation is defined as purified if the desired DNA represents at least 99% (w/w) of the total DNA in the preparation, provided that it was verified by more than one procedure.
7. In special circumstances, in consultation with the NIH Office of Recombinant DNA Activities, an area biohazards committee may be formed, composed of members from the institution and/or other organizations beyond its own staff, as an alternative when additional expertise outside the institution is needed for the indicated reviews.

VI. REFERENCES

1. Berg, P., D. Baltimore, H. W. Boyer, S. N. Cohen, R. W. Davis, D. S. Hogness, D. Nathans, R. O. Roblin, J. D. Watson, S. Weissman and N. D. Zinder (1974). *Potential Biohazards of Recombinant DNA Molecules.* Science *185,303.*
2. Advisory Board for the Research Councils. *Report of a Working Party on the Experimental Manipulation of the Genetic Composition of Micro-Organisms.* Presented to Parliament by the Secretary of State for Education and Science by Command of Her Majesty. January, 1975. London: Her Majesty's Stationery Office, 1975.
3. Berg, P., D. Baltimore, S. Brenner, R. O. Roblin and M. F. Singer (1975). *Summary Statement of the Asilomar Conference on Recombinant DNA Molecules.* Science *188,* 991; Nature *225,* 442; *Proc. Nat. Acad. Sci. 72,* 1981.
4. *Laboratory Safety at the Center for Diseased Control* (Sept., 1974). U.S. Department of Health, Education, and Welfare Publication No. CDC 75–8118.
5. *Classification of Etiologic Agents on the Basis of Hazard.* (4th Edition, July, 1974). U.S. Department of Health, Education, and Welfare. Public Health Service. Center for Disease Control, Office of Biosafety, Atlanta, Georgia 30333.
6. *National Cancer Institute Safety Standards for Research Involving Oncogenic Viruses* (Oct., 1974). U.S. Department of Health, Education, and Welfare Publication No. (NIH) 75–790.
7. *National Institutes of Health Biohazards Safety Guide* (1974). U.S. Department of Health, Education, and Welfare. Public Health Service, National Institutes of Health. U.S. Government Printing Office Stock No. 1740–00383.
8. *Biohazards in Biological Research* (1973). A. Hellman, M. N. Oxman and R. Pollack (ed.). Cold Spring Harbor Laboratory.

9. *Handbook of Laboratory Safety* (1971; 2nd Edition). N. V. Steere (ed.). The Chemical Rubber Co., Cleveland.

10. Bodily, H. L. (1970). *General Administration of the Laboratory.* H. L. Bodily, E. L. Updyke and J. O. Masons (eds.), Diagnostic Procedures for Bacterial, Mycotic and Parasitic Infections. American Public Health Association, New York. pp. 11–28.

11. Darlow, H. M. (1969). *Safety in the Microbiological Laboratory.* In J. R. Norris and D. W. Robbins (ed.), Methods in Microbiology. Academic Press, Inc. New York. pp. 169–204.

12. *The Prevention of Laboratory Acquired Infection (1974).* C. H. Collins, E. G. Hartley, and R. Pilsworth, Public Health Laboratory Service, Monograph Series No. 6.

13. Chatigny, M. A. (1961). *Protection Against Infection in the Microbiological Laboratory: Devices and Procedures.* In W. W. Umbreit (ed.). Advances in Applied Microbiology. Academic Press, New York, N.Y. 3: 131–192.

14. *Design Criteria for Viral Oncology Research Facilities,* U. S. Department of Health, Education, and Welfare, Public Health Service, National Institutes of Health, DHEW Publication No. (NIH) 75–891, 1975.

15. Kuehne, R. W. (1973). *Biological Containment Facility for Studying Infectious Disease.* Appl. Microbiol. 26 : 239–243.

16. Runkle, R. S. and G. B. Phillips, (1969). *Microbial Containment Control Facilities.* Van Nostrand Reinhold, New York.

17. Chatigny, M. A. and D. I. Clinger (1969). *Contamination Control in Aerobiology.* In R. L. Dimmick and A. B. Akers (eds.). An Introduction to Experimental Aerobiology. John Wiley & Sons, New York, pp. 194–263.

18. Grustein, M. and D. S. Hogness (1975). *Colony Hybridization: A Method for the Isolation of Cloned DNAs That Contain a Specific Gene.* Proc. Nat. Acad. Sci. U.S.A. 72, 3961–3965.

19. Morrow, J. F., S. N. Cohen, A. C. Y. Chang, H. W. Boyer, H. M. Goodman and R. B. Helling (1974). *Replication and Transcription of Eukaryotic DNA in Escherichia coli.* Proc. Nat. Acad. Sci. USA 71, 1743–1747.

20. Hershfield, V., H. W. Boyer, C. Yanofsky, M. A. Lovett and D. R. Helinski (1974). *Plasmid ColE1 as a Molecular Vehicle for Cloning and Amplification of DNA.* Proc. Nat. Acad. Sci. USA 71, 3455–3459.

21. Wensink, P. C., D. J. Finnegan, J. E. Donelson, and D. S. Hogness (1974). *A System for Mapping DNA Sequences in the Chromosomes of Drosophila melanogaster.* Cell 3, 315–325.

22. Timmis, K., F. Cabello and S. N. Cohen (1974). *Utilization of Two District Modes of Replication by a Hybrid Plasmid Constructed In Vitro from Separate Replicons.* Proc. Nat. Acad. Sci. USA 71, 4556–4560.

23. Glover, D. M., R. L. White, D. J. Finnegan and D. S. Hogness (1975). *Characterization of six Cloned DNAs from Drosophila melanogaster, Including One that Contains the Genes for rRNA.* Cell 5, 1949–155.

24. Kedes, L. H., A. C. Y. Chang, D. Houseman and S. N. Cohen (1975). *Isolation of Histone Genes from Unfractionated Sea Urchin DNA by Subculture Cloning in E. coli.* Nature 255, 533.

25. Tanaka, T. and B. Weisblum (1975). *Construction of a Colicin E1-R Factor Composite*

Plasmid In Vitro: Means for Amplification of Deoxyribonucleic Acid. J. Bacteriol. 121, 354–362.

26. Tanaka, T., B. Weisblum, M. Schnoss and R. Inman (1975). *Construction and Characterization of a Chimeric Plasmid Composed of DNA from Escherichia coli and Drosophila melanogaster.* Biochemistry 14, 2064–2072.

27. Thomas, M., J. R. Cameron and R.W. Davis (1974). *Viable Molecular Hybrids of Bacteriophage Lambda and Eukaryotic DNA.* Proc. Nat. Acad. Sci. USA 71, 4579–4583.

28. Murray, N. E. and K. Murray (1974). *Manipulation of Restriction Targets in Phage* λ *to Form Receptor Chromosomes for DNA Fragments.* Nature 251, 476–481.

29. Rambach, A. and P. Tiollais (1974). *Bacteriophage* λ *Having EcoR1 Endonuclease Sites only in the Non-essential Region of the Genome.* Proc. Nat. Acad. Sci. USA 71, 3927–3930.

30. Smith, H. W. (1975). *Survival of Orally-Administered Escherichia coli K12 in the Alimentary Tract of Man.* Nature 255, 500–502.

31. Anderson, E. S. (1975). *Viability of and Transfer of a Plasmid from Escherichia coli K12 in the Human Intestine.* Nature 255, 502–504.

32. Falkow, S. (1975). Unpublished experiments quoted in Appendix D of the *Report of the Organizing Committee of the Asilomar Conference on Recombinant DNA Molecules.* (P. Berg., D. Baltimore, S. Brenner, R. O. Roblin and M. Singer, eds.) submitted to the National Academy of Sciences.

33. R. Curtiss III, personal communication.

34. Novick, R. P. and S. I. Morse (1967). *In Vivo Transmission of Drug Resistance Factors between Strains of Staphylococcus aureus.* J. Exp. Med. 125, 45–59.

35. Anderson, J. D., W. A. Gillespie and M. H. Richmond. (1974). *Chemotherapy and Antibiotic Resistance Transfer between Enterobacteria in the Human Gastrointestinal Tract.* J. Med. Microbiol. 8, 461–473.

36. Ronald Davis, personal communication.

37. K. Murray, personal communication; W. Szybalski, personal communication.

38. Manly, K. R., E. R. Signer and C. M. Radding (1969). *Nonessential Functions of Bacteriophage* λ Virology 37, 177.

39. Gottesman, M.E. and R. A. Wiesberg (1971). *Prophage Insertion and Excision.* In The Bacteriophage Lambda (A. D. Hershey, ed.). Cold Spring Harbor Laboratory, pp. 113–138.

40. Shimada, K., R. A. Weisberg and M. E. Gottesman (1972). *Prophage Lambda at Unusual Chromosomal Locations: I. Location of the Secondary Attachment Sites and the Properties of the Lysogens.* J. Mol. Biol. 63, 483–503.

41. Signer, E. (1969). *Plasmid Formation: A New Mode of Lysogeny by Phage* λ Nature 223, 158––160.

42. Adams, M. H. (1959). *Bacteriophages.* Intersciences Publishers, Inc., New York.

43. Jacob, F. and E. L. Wollman. (1956). *Sur les Processes de Conjugaison et de Recombinaison chez Escherichia coli. I. L'induction par Conjugaison ou Induction Zygotique.* Ann. Inst. Pasteur 91, 486–510.

44. J. S. Parkinson as cited (p. 8) by Hershey, A. D. and W. Dove (1971). *Introduction to Lambda. In:* The Bacteriophage λ. A. D. Hershey, ed. Cold Spring Harbor Laboratory, New York.

Index